Selected Titles in This Series

154 **V. V. Prasolov and A. B. Sossinsky,** Knots, links, braids and 3-manifolds. An introduction to the new invariants in low-dimensional topology, **1997**

153 **S. Kh. Aranson, G. R. Belitsky, and E. V. Zhuzhoma,** Introduction to the qualitative theory of dynamical systems on surfaces, 1996

152 **R. S. Ismagilov,** Representations of infinite-dimensional groups, 1996

151 **S. Yu. Slavyanov,** Asymptotic solutions of the one-dimensional Schrödinger equation, 1996

150 **B. Ya. Levin,** Lectures on entire functions, 1996

149 **Takashi Sakai,** Riemannian geometry, 1996

148 **Vladimir I. Piterbarg,** Asymptotic methods in the theory of Gaussian processes and fields, 1996

147 **S. G. Gindikin and L. R. Volevich,** Mixed problem for partial differential equations with quasihomogeneous principal part, 1996

146 **L. Ya. Adrianova,** Introduction to linear systems of differential equations, 1995

145 **A. N. Andrianov and V. G. Zhuravlev,** Modular forms and Hecke operators, 1995

144 **O. V. Troshkin,** Nontraditional methods in mathematical hydrodynamics, 1995

143 **V. A. Malyshev and R. A. Minlos,** Linear infinite-particle operators, 1995

142 **N. V. Krylov,** Introduction to the theory of diffusion processes, 1995

141 **A. A. Davydov,** Qualitative theory of control systems, 1994

140 **Aizik I. Volpert, Vitaly A. Volpert, and Vladimir A. Volpert,** Traveling wave solutions of parabolic systems, 1994

139 **I. V. Skrypnik,** Methods for analysis of nonlinear elliptic boundary value problems, 1994

138 **Yu. P. Razmyslov,** Identities of algebras and their representations, 1994

137 **F. I. Karpelevich and A. Ya. Kreinin,** Heavy traffic limits for multiphase queues, 1994

136 **Masayoshi Miyanishi,** Algebraic geometry, 1994

135 **Masaru Takeuchi,** Modern spherical functions, 1994

134 **V. V. Prasolov,** Problems and theorems in linear algebra, 1994

133 **P. I. Naumkin and I. A. Shishmarev,** Nonlinear nonlocal equations in the theory of waves, 1994

132 **Hajime Urakawa,** Calculus of variations and harmonic maps, 1993

131 **V. V. Sharko,** Functions on manifolds: Algebraic and topological aspects, 1993

130 **V. V. Vershinin,** Cobordisms and spectral sequences, 1993

129 **Mitsuo Morimoto,** An introduction to Sato's hyperfunctions, 1993

128 **V. P. Orevkov,** Complexity of proofs and their transformations in axiomatic theories, 1993

127 **F. L. Zak,** Tangents and secants of algebraic varieties, 1993

126 **M. L. Agranovskiĭ,** Invariant function spaces on homogeneous manifolds of Lie groups and applications, 1993

125 **Masayoshi Nagata,** Theory of commutative fields, 1993

124 **Masahisa Adachi,** Embeddings and immersions, 1993

123 **M. A. Akivis and B. A. Rosenfeld,** Élie Cartan (1869–1951), 1993

122 **Zhang Guan-Hou,** Theory of entire and meromorphic functions: Deficient and asymptotic values and singular directions, 1993

121 **I. B. Fesenko and S. V. Vostokov,** Local fields and their extensions: A constructive approach, 1993

120 **Takeyuki Hida and Masuyuki Hitsuda,** Gaussian processes, 1993

119 **M. V. Karasev and V. P. Maslov,** Nonlinear Poisson brackets. Geometry and quantization, 1993

(Continued in the back of this publication)

Knots, Links, Braids and 3-Manifolds

An Introduction to the New Invariants in Low-Dimensional Topology

TRANSLATIONS OF
MATHEMATICAL
MONOGRAPHS
VOLUME 154

V. V. Prasolov
A. B. Sossinsky

Knots, Links, Braids and 3-Manifolds

An Introduction to the New Invariants in Low-Dimensional Topology

American Mathematical Society
Providence, Rhode Island

Translated by A. B. SOSSINSKY
from an original Russian manuscript

1991 *Mathematics Subject Classification.* Primary 53Mxx.

ABSTRACT. This book is an introduction to the remarkable work of Vaughan Jones and Victor Vassiliev on knot and link invariants and its recent modifications and generalizations, including a mathematical treatment of Jones-Witten invariants. It emphasizes the geometric aspects of the theory, and treats such topics as braids, homeomorphisms of surfaces, surgery of 3-manifolds (Kirby calculus), and branched coverings. This attractive geometric material, interesting in itself but not previously gathered in book form, constitutes the basis of the last two chapters, where the Jones-Witten invariants are constructed via the rigorous skein algebra approach (mainly due to the Saint Petersburg school).

Unlike several recent monographs, where all these invariants are introduced by using the sophisticated abstract algebra of quantum groups and representation theory, here the mathematical prerequisites are minimal, so that this book is accessible to most undergraduate math majors, as well as to physicists, for whom the subject matter is particularly fashionable. Numerous figures and problems make the book suitable both as a course text and for self-study.

Library of Congress Cataloging-in-Publication Data

Prasolov, V. V. (Viktor Vasil′evich)

Knots, links, braids and 3-manifolds : an introduction to the new invariants in low-dimensional topology / V. V. Prasolov, A. B. Sossinsky : [translated by A. B. Sossinsky from an original Russian manuscript].

p. cm. — (Translations of mathematical monographs; v. 154)

Includes bibliographical references (p. –) and index.

ISBN 0-8218-0588-6 (alk. paper)

1. Low-dimensional topology. I. Sossinsky, A. B. II. Title. III. Series.

QA612.14.P73 1996

514′.2—dc20 96-22164

CIP

---

10 9 8 7 6 5 4 3 2 02 01 00 99 98 97

Contents

Foreword

This book is primarily an elementary introduction to the remarkable work of Vaughan Jones and Victor Vassiliev on knot and link invariants and its more recent modifications and generalizations, including a mathematical treatment of the Jones–Witten invariants. It may also be viewed as an introduction to some of the most attractive geometric aspects of three-dimensional topology, including braid theory, surgery of 3-manifolds, and branched coverings.

The original work of Jones and Vassiliev (see [**Jon1**], [**Vas1**]) involved some rather sophisticated mathematics, such as representation theory, von Neumann algebras, Ocneanu traces and Markov's braid theorem for Jones, and spectral cohomology sequences and singularity theory for Vassiliev. Subsequent work by other authors has led to crucial simplifications, so that now both the Jones one-variable link polynomial and the Vassiliev knot invariants can be introduced and studied in an elementary way (the only exception being the existence proof for Vassiliev invariants). Thus both topics are treated at the very beginning of the book, in Chapter II (after a geometric introduction to knot theory presented in Chapter I), and are accessible to graduate students, or even to undergraduate math majors with a minimal background in basic topology.

In contrast to this, the present situation with the Jones–Witten invariants for links in manifolds is not as simple. Witten's beautiful definition (see [**Wit2**]) is not only quite complicated (it involves the ideology of gauge theory, Chern–Simons Lagrangians, and integrals over all affine connections), but is not rigorous enough for the mathematician. At the end of the book (Chapter IX), we give a mathematically correct version of this definition (developed in the work of A. Reshetikhin, V. Turaev, O. Viro, and others), following (not too closely) W. B. R. Lickorish's article [**Lic4**]. This approach is also quite complicated and requires a lot of preliminary material (e.g. surgery presentation of 3-manifolds, the Kirby calculus, and Temperley–Lieb algebras), which is studied in the intermediate chapters (III–VIII) of the book.

The material of these chapters should not, by any means, be regarded merely as the preparation to Chapter IX. On the contrary, this is classical and often beautiful three-dimensional topology. Chapter III is an introduction to braid theory, including some recent computer-implementable results on the word problem and Alexander's braiding theorem, and some results on pure braids. The latter provide a natural transition to the description of the homeomorphisms of surfaces (the Dehn–Lickorish theorem) in Chapter IV, which in turn lead to Heegaard decompositions (Chapter V) and surgery presentations (Chapter VI) of 3-manifolds.

Let us describe the contents of Chapters IV–VI more explicitely. A Heegaard decomposition is a presentation of a 3-manifold as the union of two identical handlebodies attached by a homeomorphism of their boundaries. Thus homeomorphisms of surfaces encode 3-manifolds. This encoding is quite effective, because homeomorphisms of surfaces are compositions of certain elementary homeomorphisms, called Dehn twists (§12). The latter easily lead to another way of presenting orientable 3-manifolds: by removing solid tori from the 3-sphere S^3 and pasting them back along different homeomorphisms of the boundary tori. Thus any orientable 3-manifold is encoded by a framed link, i.e., a link in S^3 with rational numbers (describing the corresponding torus homeomorphism) on its components; this encoding is called a surgery presentation (Chapter VI). The natural question that arises then is when do two different framed links present one and the same manifold? The answer is that they do if and only if one framed link can be taken to the other by a finite sequence of elementary operations called Kirby moves; this is known as the Kirby calculus and is described in §19.

In contrast to Chapters III–VI, Chapter VII, which deals with branched coverings, is not needed in the sequel. It includes an exposition of the theory of branched coverings of surfaces (a lovely topic, which we feel any student of topology should learn at an early stage) and a systematic treatment of the deep results on branched coverings of 3-manifolds due to W. Hilden and J. M. Montesinos and their collaborators.

The concluding chapter, besides the Jones–Witten invariants, deals with Yu. Drobotukhina's polynomial of links in projective 3-space (a very simple generalization of the Jones polynomial), and concludes with a brief survey of the relationship between knot invariants and physics.

* * *

It should be pointed out that the interconnections between physics and the theory of knot invariants have recently led to a surge in the popularity of knot theory among nontopologists and mathematical physicists. In particular, this relationship appears in V. Drinfeld's quantum group approach to knot invariants (via Lie algebra representation theory), in the Kontsevich integral defining Vassiliev invariants (via the Knizhnik–Zamolodchikov connection and Gauss diagram bialgebras), and in M. Atiyah's category theory approach to topological quantum field theories (via the Kauffman bracket). These exciting topics are not elementary and are already the subject matter of several books (see [**Tur3**], [**Ati1**], [**BM**], [**Fu**], [**Kau1**], [**CP**], [**Lus**]), and for these reasons are not treated in ours.

Many other important topics in three-dimensional topology also do not appear in our exposition. In choosing the subject matter of this book, besides the material needed for the Jones–Witten invariants, we gave preference to those topics that are not technically too difficult and that we find aesthetically pleasing. But there is much more to three-dimensional topology than than one can find here. In particular, we do not even mention the fundamental work of S. Novikov, S. Matveev, W. Thurston, M. Friedman, D. Gabai, A. Casson, W. Jaco, W. Haken, F. Waldhausen, J. Rubinstein, A. Thompson, and others in this field.

* * *

We have tried to make the book reasonably self-contained and, whenever possible, to avoid too many internal references to preceding chapters. The Appendix aims at achieving the first objective; it briefly reviews basic topology (giving specific references to accessible books). Regretfully, in two important cases we did not succeed in finding sufficiently elementary proofs and simply omitted them; these are the existence theorem for Vassiliev invariants and the "only if" part of Kirby's theorem.

Concerning the second objective, since it was of course impossible to make each chapter independent, we have taken pains to refer only to explicit parts of previous chapters, in such a way that the reader would understand the contents of the reference without having to read too much of the previous material. We have not always succeeded in doing this; in particular, it is improbable that the reader will understand Chapter VI without first reading most of Chapter V, or follow §31 in Chapter IX without working through most of Chapter VIII.

Nevertheless, there are several ways of reading our book other than continuously from beginning to end. For an elementary introduction to the theory of knot invariants, the material to read is Chapters I–III and §30 of Chapter IX. For a brief introduction to some geometric topics in low-dimensional topology, we suggest Chapters I, III–V, and §§20–21 in Chapter VII; for a more substantial first acquaintance to these topics, several sections from Chapters VI and VII may be added. A reader primarily interested in learning about the new invariants can start with Chapter II (refering back to Chapter I when necessary) and then try reading Chapters VIII and IX (where back references to Chapters V and VI will possibly incite him/her to a more systematic study of these intermediate geometric chapters).

* * *

The reader will undoubtedly have noticed the large number of figures appearing in this book. They are not only meant to be visual illustrations to the text, but in many cases (e.g. in definitions and proofs) they play the main role, allowing us to avoid a lot of verbose explanations. The reader should be warned that she/he will not be able to understand many of the key places of the book without scrutinizing the pictures, sometimes even working on them with a pencil. After some hesitation, we have chosen to draw the figures by hand, leaving to computer graphics the subordinate role of producing the little standard diagrams incorporated in the text.

Another important feature of the book, also necessitating personal work by the reader, is the presence of numerous problems. The solutions (sometimes in the form of answers or hints) appear at the end of the book, but we hope that the reader will not deprive him/herself of the pleasure of solving the problems without consulting the solution section.

Finally, let us note that each chapter (except the last one) ends with a section called Comments. Each such section contains brief and informal remarks about the history of the questions treated in that chapter and an indication of our main sources. We should stress that our historical remarks make no pretense to completeness. They are not the result of systematic research in the literature, and are of course rather subjective. But neither are they just the reproduction of hearsay or published opinions, always being based on concrete references.

* * *

The starting point of this book was a seminar that the authors conducted jointly with S. Chmutov at the Independent University of Moscow in 1993-94; its proceedings were partially published as lecture notes of the IUM (see [**PS**]) and constitute a first very preliminary version of this text. The authors are grateful to all the participants of this seminar, in particular to S. Chmutov, whose talk was the source of much of the material of §4. It is with great pleasure that we acknowledge our gratitude to V. A. Vassiliev: presence at his talks, discussions with him, as well his assistance in the organizational aspects of writing this book, were essential in its being completed. We are also grateful to M. M. Vinogradov, who created the knot.tex fonts used to produce the in-text diagrams appearing throughout this book, to E. N. Efimova for her assistance with the keyboarding and other work on the computer, and to O. Sipacheva for her help with the figures. Finally, we would like to note the amiable attention and competence that the American Mathematical Society has shown while working with us on the publication of this book. While the book was being written, both authors benefitted from AMS fSU grants, International Science Foundation grants (MQO-000 in 1994, MQO-300 in 1995), and grants from the Russian Fund for Basic Research (95-01-00846).

CHAPTER I

Knots, Links, and Ribbons

This chapter introduces one of the classical topics of three-dimensional topology – the theory of knots and links. In §1 we define and study these objects from the geometric point of view, while §2 deals with some real life applications. The latter may be called physical (if one wishes to take them seriously) or may be described as tricks with strings and ribbons for the amateur magician (if one prefers to stress their unexpected and amusing aspects).

§1. The topology of knots and links

1.1. You can imagine a *knot* as a thin tangled rope in space whose ends are glued together. (We postpone rigorous definitions to 1.2 below.) The simplest (untangled) knot is shown in Fig.1.1,a; it is known as the *unknot* or *trivial knot.* Fig.1.1,b,c shows two nontrivial knots, called the *trefoil* and *figure eight* knots respectively.

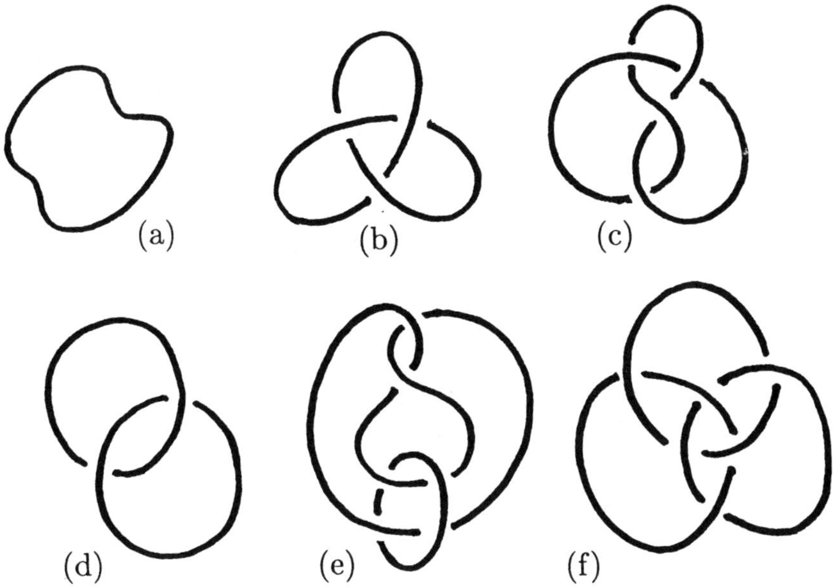

FIGURE 1.1. Examples of knots and links

A *link* can be imagined as several tangled ropes (that may also tangle with each other) each of which has its ends glued together. A link consisting of several unknots not linked together will be called a *trivial* link. Fig.1.1,d,e,f shows three nontrivial links, called the *Hopf link*, the *Whitehead link*, and the *Borromeo rings*. The first two, known long ago (and used in particular by Gauss), are named after two topologists, Heinz Hopf and J. W. H. Whitehead, who employed the corresponding links in very beautiful topological constructions, while the third appeared back in the Middle Ages on the coat of arms of the Borromeos, a famous family from the Italian nobility.

To represent knots (and links), it is convenient to use (as we in fact already have) their perpendicular projections on the plane. In so doing, we must take care to choose the projection plane so that:

(1) the tangent lines to the link at all points are projected onto lines on the plane (i.e., the projections of the tangents never degenerate into points);

(2) no more than two distinct points of the link are projected on one and the same point of the plane;

(3) the set of *crossing points* (those on which two points project) is finite and at each crossing point the projections of the two tangents do not coincide.

In particular, the two following situations are forbidden: when points of three distinct branches of the link project on one and the same point (as in Fig.1.2,a), or when the projections of two branches are tangent (Fig.1.2,b).

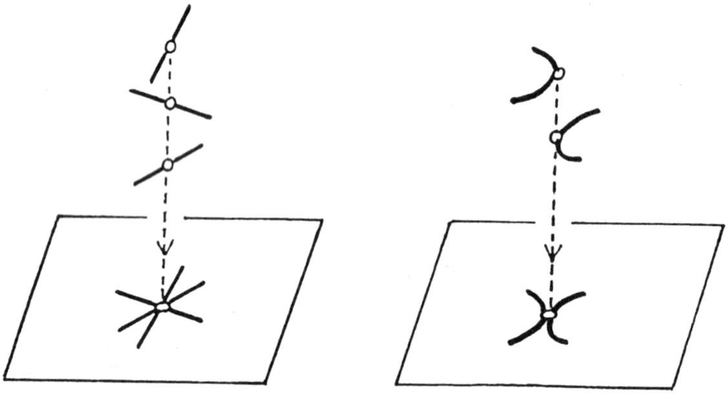

(a) (b)

FIGURE 1.2. Two forbidden projections

It is intuitively obvious that by small local moves of our link we can always ensure that conditions (1)–(3) hold. Therefore we can assume that the projection of any link consists of smooth curves with transversal intersections and self-intersections such that the projections of only two branches meet at each crossing point, *overpasses* and *underpasses* at crossing points being shown in the obvious way (as in Fig.1.1).

1.2. Definitions. Now we are ready to give the formal definitions of knots and links, as well as of their equivalence. The simplest approach is the following. Let us call any closed nonselfintersecting polygonal line in Euclidean space $\mathbb{R}^3$ a *polygonal knot*. Examples of polygonal knots are shown in Fig.1.3. A *polygonal link* is defined in a similar way; the reader will have no trouble transforming the links pictured in Fig.1.1 into polygonal ones by using a ruler and a pencil (or just imagination).

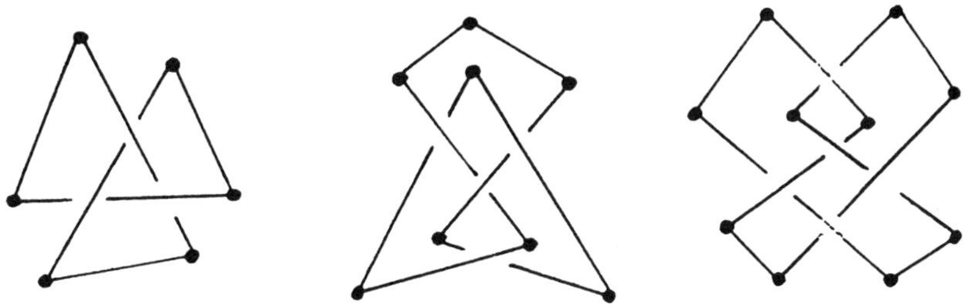

FIGURE 1.3. Examples of polygonal knots

In order to define the equivalence of knots (links), we shall use the following *elementary move*. Suppose that the sides AC and CB of the triangle ABC are edges of a polygonal link L that does not intersect the triangle ABC at any other points. Replace the two edges AC and CB by the edge AB, obtaining the link L_1 (Fig.1.4). Two links (in particular two knots) K and K' are called *equivalent* if they can be joined by a sequence of links $K_0 = K, K_1, ..., K_n = K'$ in which each subsequent link is obtained from the previous one by an elementary move (of the type described above) or its inverse.

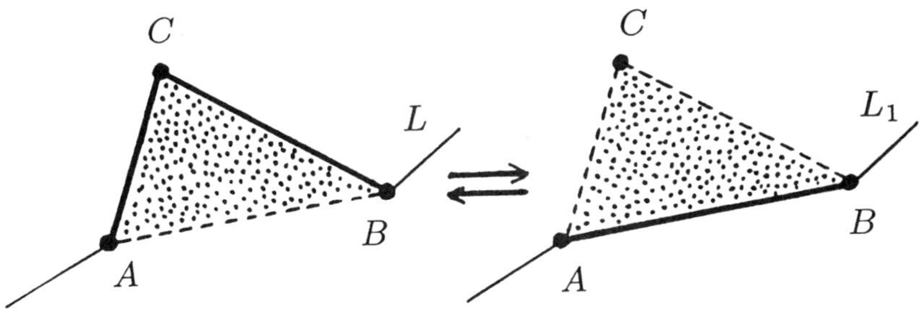

FIGURE 1.4. Elementary move

Now that we have a precise definition of knots and their equivalence, we can rigorously prove that conditions (1)–(3) from subsection 1.1 above hold for the knot (or link) projections that we consider. Indeed, for polygonal links, condition (3) is fulfilled automatically; if condition (1) breaks down for some edge, it can be satisfied by a small move of one of the endpoints of this edge; finally, the projection of any edge on the intersection point of two other edge projections can be pushed off that point by a tiny move of one of its endpoints, so that condition (2) can also be assumed to hold.

The notion of polygonal knot does not really reflect our intuitive perception of a tangled rope. A better mathematical model is the notion of smooth knot. Define an *embedding of the circle* S^1 in $\mathbb{R}^3$ as any continuous map $f : S^1 \to \mathbb{R}^3$ such that no pair of distinct points of the circle is mapped into the same point in space. A *smooth knot* is defined as the image of the circle in $\mathbb{R}^3$ under an infinitely differentiable embedding with nonvanishing differential:

$$f(t) = (x(t), y(t), z(t)), \quad \left(\frac{dx}{dt}, \frac{dy}{dt}, \frac{dz}{dt}\right) \neq (0, 0, 0).$$

Two smooth knots K_0 and K_1 are called *equivalent* if there exists a one-parameter family $f_t : \mathbb{R}^3 \to \mathbb{R}^3$, $t \in [0,1]$, of diffeomorphisms smoothly depending on the parameter t, taking the knot K_0 to the knot K_1, i.e., such that f_0 is the identity and $f_1(K_0) = K_1$. (Here "smoothly" means that the map $F : \mathbb{R}^3 \times [0,1] \to \mathbb{R}^3$ given by $(x,t) \mapsto f_t(x)$ is differentiable.) The family of diffeomorphisms f_t is said to be an *isotopy* joining the knots K_0 and K_1; for this reason equivalent knots are also called *isotopic* or *ambient isotopic.*

Remark. It can be shown that two isotopic knots with a common segment can be isotoped to each other by an isotopy that does not move points in some neighborhood of this segment. We shall not prove this fact (see [**BZ**]).

The definitions of smooth link and of the isotopy of links are similar to those for knots, and are left to the reader.

It follows from the definition of isotopy that if two knots (or links) K_0 and K_1 are isotopic, then there is a homeomorphism (namely f_1) of $\mathbb{R}^3$ taking K_0 to K_1. In particular, this implies that the complements to the knots, i.e., the sets $\mathbb{R}^3 - K_0$ and $\mathbb{R}^3 - K_1$, are homeomorphic. The same is true for links, i.e., isotopic links have homeomorphic complements. The converse statement for links, however, is false: there exist nonisotopic links with homeomorphic complements. Such an example appears in §12 below (see Problem 12.4). On the other hand, knots are isotopic if and only if their complements are homeomorphic. This difficult theorem was only proved in 1989 (see [**GL**]).

1.3. A natural question that arises at this point is whether the two approaches to the theory of knots and links, namely the polygonal approach and the smooth one, give rise to the same theory. They do. To be more precise, there is a natural one-to-one correspondence (called "smoothing") taking equivalence classes of polygonal links to isotopy classes of smooth links. We shall not prove this fact (see, for example, [**BZ, Prop. 1.10**]), but will use the two approaches indiscriminately.

1.4. Instead of the polygonal and smooth approaches, one might try the purely topological approach to knots and links, defining them by means of continuous embeddings (without smoothness or polygonality conditions). But this will not yield the same theory. Indeed, both smooth and polygonal knots can always be presented by diagrams (projections) with a finite number of crossings, while there exist topological embeddings of S^1 in $\mathbb{R}^3$ whose projections always have infinitely many crossings and are not isotopic to any smooth knot. A topological embedding of S^1 in $\mathbb{R}^3$ is said to be *tame* if it is isotopic to a smooth knot (or, which is the same, to a polygonal one). Otherwise it is called *wild.* Examples of wild knots (due to E. Artin and R. Fox, [**AF**]) are shown in Fig.1.5.

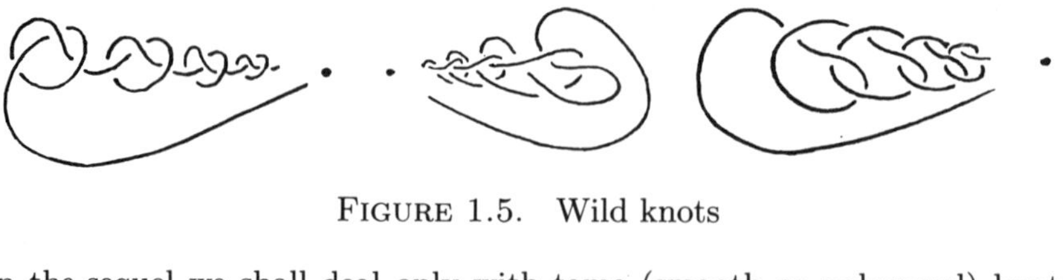

FIGURE 1.5. Wild knots

In the sequel we shall deal only with tame (smooth or polygonal) knots and links, usually omitting the adjectives. But one example of the application of wild knots to prove a theorem about tame knots is worth mentioning.

The *composition* or *connected sum* of two knots K_1 and K_2 is the rope on which we have successively tied the knot K_1 and then tied K_2 (this description can be transformed into a mathematical definition by the reader after glancing at Fig.1.6).

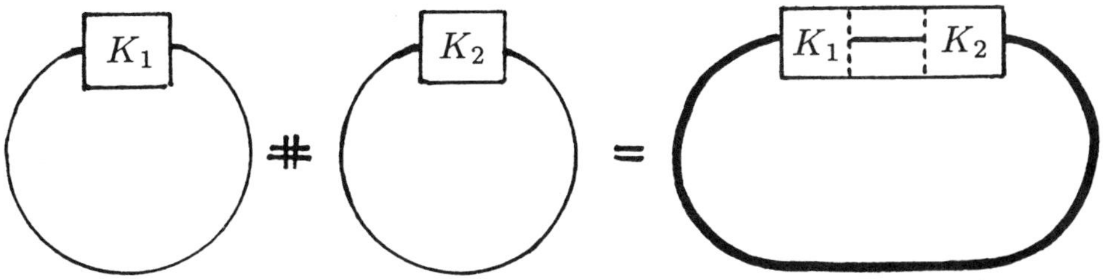

FIGURE 1.6. Connected sum of two knots

The composition of the knots K_1 and K_2 is denoted by $K_1 \# K_2$. It is commutative in the following strong sense.

Problem 1.1. Prove that the knots $K_1 \# K_2$ and $K_2 \# K_1$ are isotopic by an isotopy that is the identity on the "outside part" of the composite knot (the thickened part in Fig.1.6).

1.5. THEOREM. *If the knot K_1 is nontrivial, then so is its connected sum $K_1 \# K_2$ with any knot K_2.*

Proof. Assume the converse: suppose $K_1 \# K_2$ is trivial. Then there exists an isotopy f_t such that f_1 takes the knot $K_1 \# K_2$ to the unknot K_0. By the Remark in 1.2, we can (and will) assume that this isotopy does not move any points outside the cube C in Fig.1.7,a. By the result of Problem 1.1, there is an isotopy g_t of $K_2 \# K_1$ to K_0 that does not move any points outside the cube C' (Fig.1.7,b).

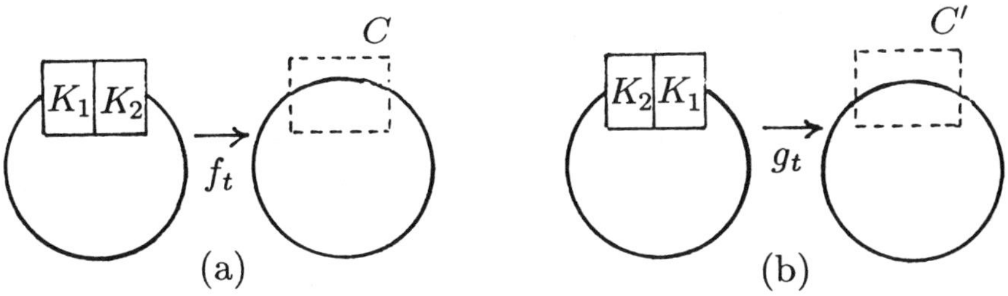

FIGURE 1.7. Unknotting inside a cube

Now consider the (possibly wild) knot W shown in Fig.1.8,a. It can be equivalently represented as shown in Fig.1.8,b. Consider the isotopy F_t that moves no points outside the union of the cubes C_i and acts inside each of these cubes just as f_t does inside C. Clearly this isotopy takes W to the unknot. On the other hand, the knot W can be represented as shown in Fig.1.8,c. Consider the isotopy G_t that moves no point outside the union of cubes C_j' and acts inside each of these cubes just as g_t does inside C'. Obviously this isotopy takes the knot W to the knot K_1. Thus we have $K_1 \sim W \sim K_0$, i.e., the knot K_1 is trivial, contrary to assumption. This contradiction shows that the knot $K_1 \# K_2$ is nontrivial. □

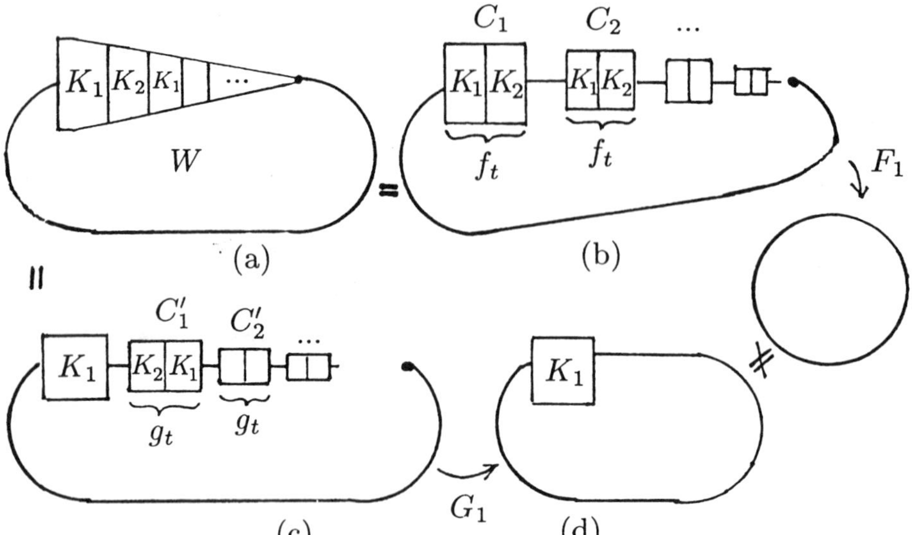

FIGURE 1.8. Connected summing does not unknot

1.6. Two diagrams of polygonal links are called *equivalent* or *plane isotopic* if they can be obtained from each other by a finite sequence of moves of the type shown in Fig.1.9.

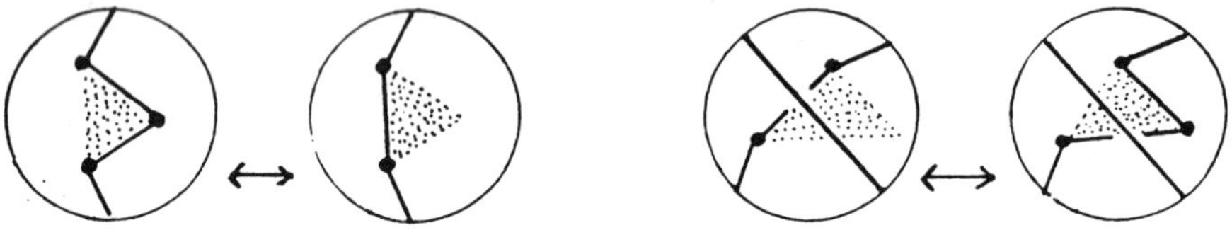

FIGURE 1.9. Moves used in plane isotopy

Note that these moves do not change the number of crossing points, so that, say, the transformation appearing in Fig.1.10 is not a plane isotopy. However, this transformation produces equivalent links in space, because the second link is obtained from the first by the elementary move replacing two edges of the shaded triangle by the third.

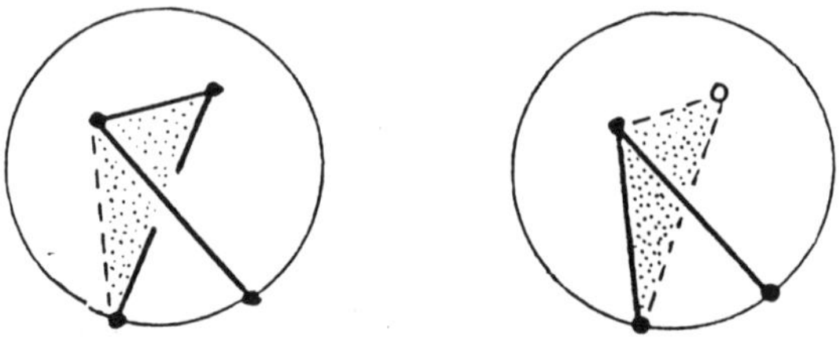

FIGURE 1.10. Nonplanar space isotopies

Plane isotopy can also be defined for diagrams of smooth links (by similar smooth moves). During a plane isotopy the following situations can never arise:

1) a new crossing point appears or an old one disappears;
2) the projections of two branches of the link become tangent;
3) the points of three branches of the link are projected on the same point.

These forbidden (by the definition of plane isotopy) situations occur in the space isotopies represented in Fig.1.11. The corresponding transformations are called the *first, second, third Reidemeister moves* and are denoted by Ω_1, Ω_2, Ω_3. They play an important role in the theory because, in a certain sense, they reduce space isotopy to plane isotopy, as the following theorem shows.

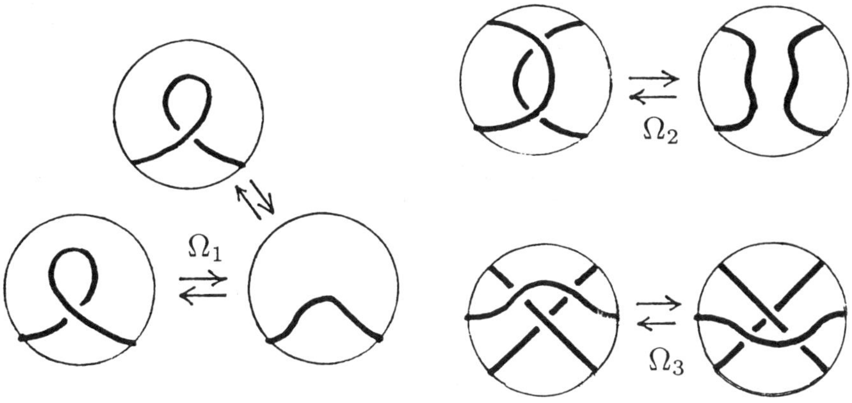

FIGURE 1.11. The Reidemeister moves

1.7. REIDEMEISTER THEOREM. *Two link diagrams correspond to isotopic links if and only if one can be obtained from the other by a finite sequence of Reidemeister moves and plane isotopies.*

Proof. The "if part" of the theorem is obvious. Indeed, the reader will have no trouble checking that each of the Reidemeister moves (see Fig.1.11) and the moves defining plane isotopy (see Fig.1.9) can be obtained by one or several elementary moves (see Fig.1.4). And so they generate a space isotopy.

To prove the "only if part", assume that the (polygonal) links with plane diagrams are isotopic (in space). By the definition of space isotopy for polygonal links, it suffices to consider the case when the two links are obtained from each other by one elementary move $[AB] \mapsto [AC] \cup [CB]$.

First, we can assume that the edges $[DA]$ and $[BE]$ ending at A and issuing from B do not intersect the interior of triangle $[ABC]$. Indeed, if, say, the edge $[DA]$ enters $[ABC]$, then we can choose a point A' near A on $[AC]$ and perform an Ω_1 move as shown on Fig 1.12.b, obtaining a new triangle $[A'BC]$ without an entering edge at A'.

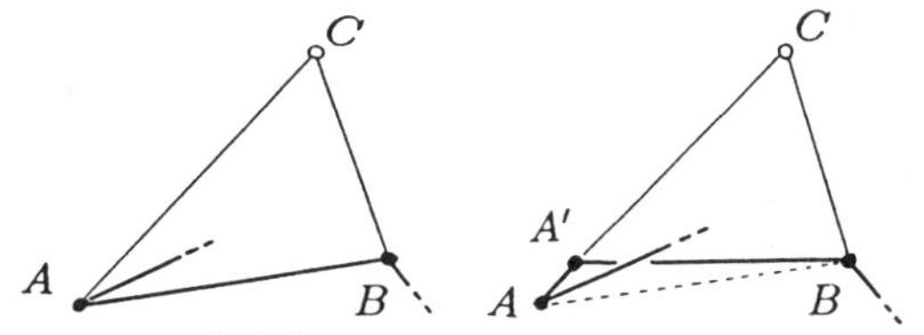

FIGURE 1.12. Eliminating an entering edge

Now note that the components of $\mathrm{int}[ABC] \cap L_0$, where L_0 denotes the projection of our link L on the plane ABC, are of two kinds: the *upper* and *lower* ones, corresponding to branches of L located respectively above and below the plane ABC; we can imagine that they are painted in red and green, respectively. (Recall that by the definition of elementary moves, the interior of triangle $[ABC]$ contains no points of the link L, but of course can intersect its projection.)

Let us subdivide triangle $[ABC]$ into smaller triangles (whose sides contain no vertices of $L_0 \cap \mathrm{int}[ABC]$) of four types (see Fig.1.13). Type I are tiny triangles Δ whose intersection with L_0 is the neighborhood of a crossing point with branches intersecting only two sides of Δ. Type II triangles contain only one vertex of L_0 and parts of the two edges issuing from it. Type III triangles contain only part of one edge of L_0 and none of its vertices. Type IV triangles contain nothing, i.e., the intersection of L_0 with their interiors is empty. The construction of such a triangulation is straightforward: first construct pairwise nonintersecting triangles of types I and II for all crossing points and all vertices of L_0 inside $[ABC]$, then triangulate the rest into triangles of types III and IV.

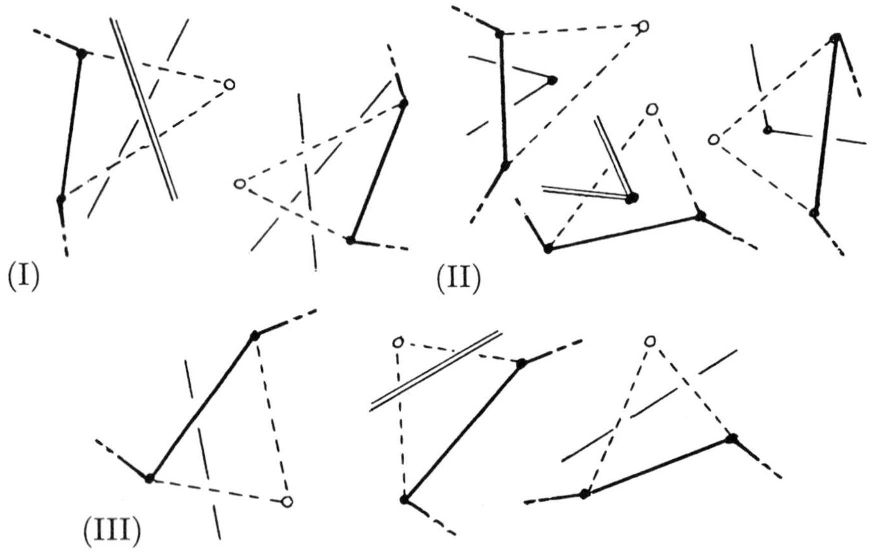

FIGURE 1.13. Examples of small triangles of types I, II, and III

Now instead of the original elementary move through $[ABC]$, let us perform a sequence of moves through the small triangles, starting from $[AB]$ and working up to $[AC] \cup [CB]$. In the process, we must remember that all the triangles are located below the red and above the green parts of $L_0 \cap \mathrm{int}[ABC]$, and specify the new over(under)crossings accordingly.

It is clear (look at Fig.1.13 again), that a move through a triangle of type I corresponds to the Reidemeister move Ω_3; of type II, to Ω_2 or to a plane isotopy; of type III, to Ω_2 or to a plane isotopy; finally, a move through a type IV triangle is, of course, a plane isotopy. This concludes the proof. □

Problem 1.2. (a) Prove that all the knots shown in Fig.1.14,a are isotopic to each other.

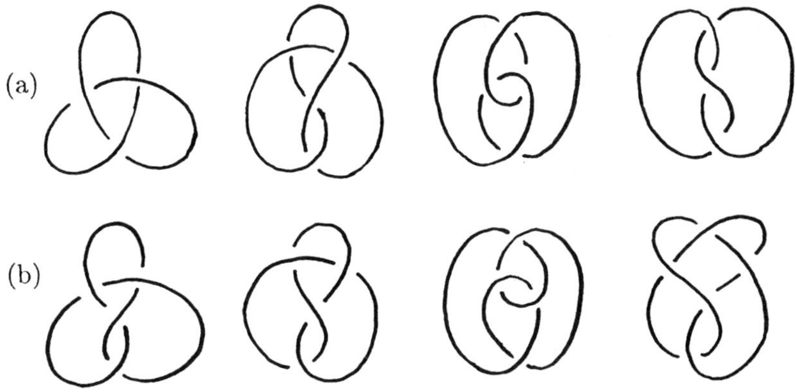

FIGURE 1.14. Various diagrams of the trefoil (a) and the eight (b)

(b) Prove that all the knots shown in Fig.1.14,b are isotopic to each other.

Problem 1.3. Prove that the two-component links shown in Fig.1.15 are all isotopic (and hence are all diagrams of the Whitehead link).

FIGURE 1.15. Various diagrams of the Whitehead link

Problem 1.4. A knot is called *amphicheiral* or *mirror symmetric* if it is isotopic to its mirror image. Show that the figure eight knot is amphicheiral. (The trefoil is not, as we shall see in §3, but at this point we don't have the means of proving this fact, as well as most other "negative" facts.)

Problem 1.5. An oriented knot (i.e., a knot supplied with a prefered direction, usually shown by an arrow) is called *chiral* or *invertible* if it is isotopic to itself with the orientation reversed. Prove that the trefoil is chiral, i.e., that the two oriented knots shown in Fig.1.16,a can be transformed to each other by Reidemeister moves and plane isotopies.

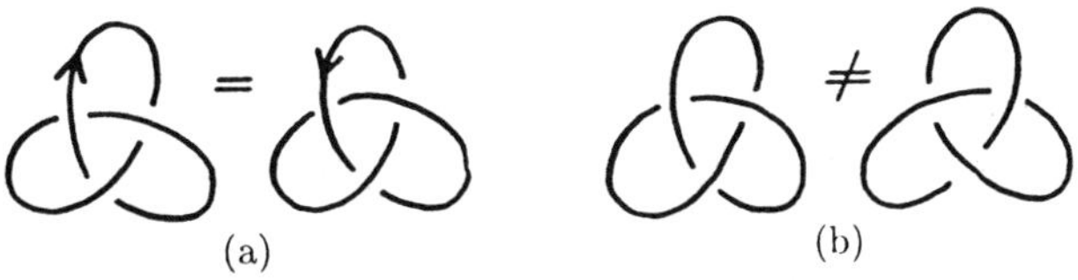

FIGURE 1.16. The trefoil is chiral but not amphicheiral

1.8. The arithmetic of knots. A knot is called *prime* if it is nontrivial and is not the connected sum of two nontrivial knots. The trefoils are examples of prime knots, while Figure 1.17 shows two *composite* (i.e., nonprime) knots, the oft used *square knot* (well known to boy scouts) and the *granny knot* (often produced by incorrectly tieing the more secure square knot). Although (isotopy classes of) knots do not constitute a group with respect to connected summing (no inverse elements, see Theorem 1.5 above), they have a nice arithmetic. It includes a prime factor decomposition theorem, proved by H. Schubert, saying that *every nontrivial knot has a unique (up to order) decomposition into the connected sum of prime knots.* Thus knots form a semigroup isomorphic to the multiplicative semigroup of nonnegative integers (but there is no canonical isomorphism, because there is no known canonical linear order in the set of knots).

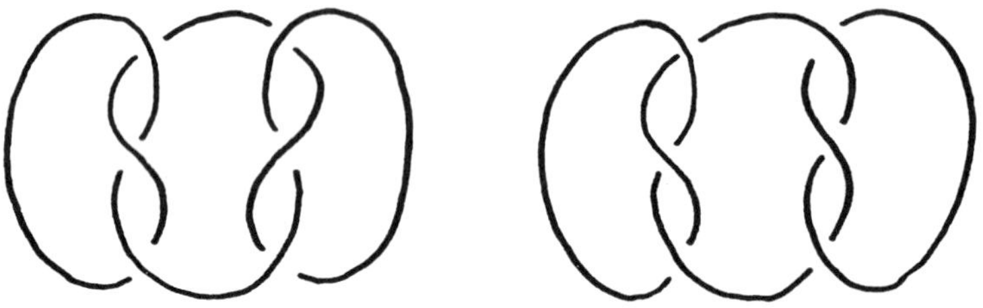

FIGURE 1.17. Composite knots

We do not present the proof of the prime factor decomposition theorem here. The proof of the existence (i.e., finiteness) of the decomposition uses an infinite summing argument (similar to the one used in the proof of Theorem 1.5) followed by a straightforward computation of the fundamental group (see the Appendix) using the van Kampen theorem. Uniqueness is more difficult (see [**Sch**]).

Remark. Prime knots (embeddings $S^n \subset \mathbb{R}^{n+2}$) exist in higher dimensions as well, but the uniqueness of prime factor decomposition is no longer true [**Kea**]. The reader who would like to get an idea of what multidimensional knots are like (a topic beyond the scope of this book) is refered to [**Zee**], [**FM**], and [**Bri**] for a start.

§2. Tricks with strings and ribbons

This section is written in a style perhaps more appropriate for a magician's handbook than for a serious mathematical monograph. This does not mean, however, that the facts discussed here are trite or trivial. On the contrary, some of the deepest and most mysterious properties of three-dimensional space appear in this section in the guise of experimental facts about strings, ropes, belts, wires, ribbons, and the like.

Although this section is not a formal prerequisite for any of the subsequent chapters of our book, it does provide a sort of intuitive background for some parts of the mathematical theory that follows. Moreover, these parts of the theory will often allow us to unravel (literally or figuratively) the mysteries of twisting and tangling that are described here.

2.1. The rope trick. A string a couple of feet long (a shoelace will do) is placed, unknotted, on a table; the magician challenges anyone present to pick up the string by its ends and then tie a knot on it without letting go of the extremities (see Fig.2.1,a). When all the spectators reach the conclusion that this can't be done, he proudly crosses his arms on his chest, leans down over the table (without uncrossing his arms), grabs the ends of the string, and dramatically pulls his arms apart (Fig.2.1,b-d). Presto: a little trefoil knot has appeared in the middle of the string!

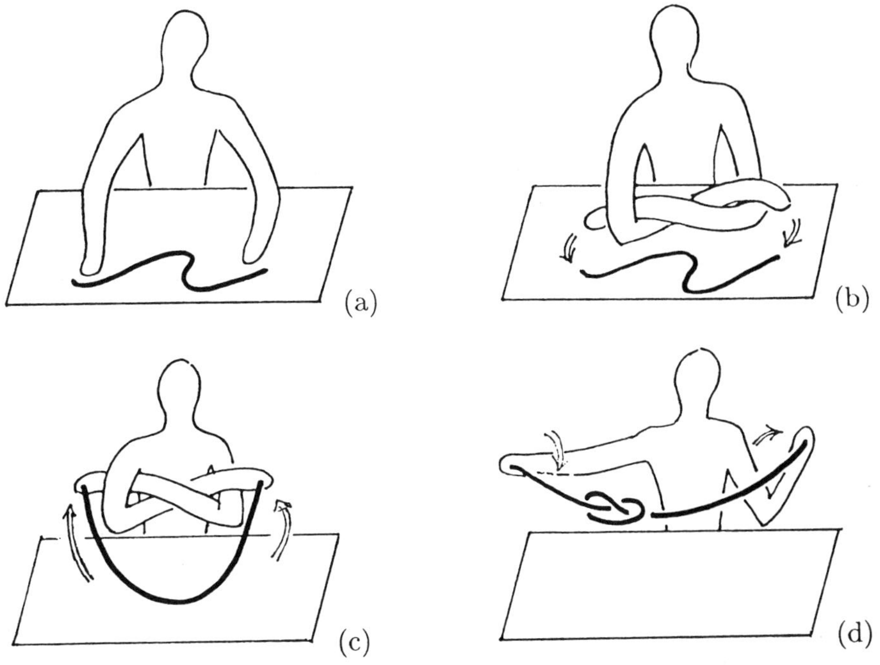

FIGURE 2.1. The rope trick

There is nothing particularly mysterious about this old trick for the topologist. It illustrates the commutativity of the connected sum operation for knots, saying, more specifically, that $K\#\bigcirc = \bigcirc\#K$, where K is the trefoil knot (represented first by the magician with his arms crossed, then by the knotted string) and $\bigcirc$ is the unknot (represented first by the unknotted string, then by the magician with his arms thrown apart). Schematically this is shown in Fig.2.2,a-d.

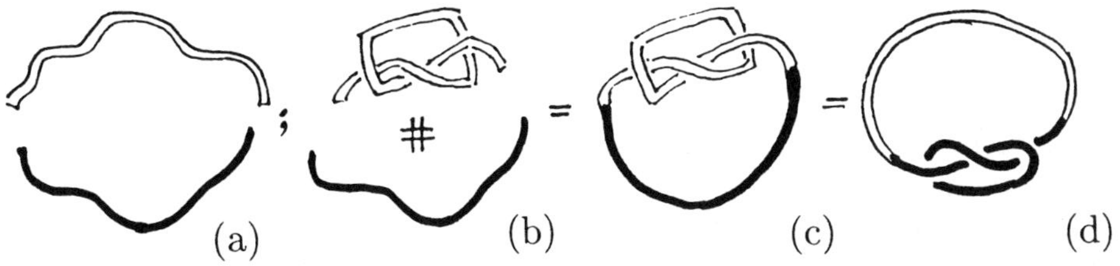

FIGURE 2.2. Mathematical version of the rope trick

2.2. Unknotting vs unbraiding. Once the rope trick has been demonstrated, the magician can go on to the next little experiment. He places the string (with the little trefoil still tied on it) on the table and challenges anyone to pick up the string by its ends and then untie the knot without letting go of the string's extremities. This usually produces some laughs from the spectators, when volunteers twist and cross their limbs in different ways in fruitless efforts to produce the desired unravelling knot.

Of course the reader of this book realizes that these efforts are doomed to failure because of Theorem 1.5 of the previous section, which says precisely that a nontrivial knot cannot be unknotted by taking its connected sum with another knot.

Another version of this same trick can then be demonstrated by the magician as follows. The string (with the stubborn trefoil still on it) is placed on the table, one extremity secured by a tack or by some scotch tape, so that the knotted part of the string lies in the half of the table nearest to the fixed end. Then the magician challenges anyone to tie a knot on the other end so that it will "cancel out" the first one (which or course he is not allowed to touch in the process of tieing the new knot, but which must disappear, together with the new knot, when the free end of the string is pulled).

A reasonably intelligent audience will usually protest that this is impossible (and will be right!). This reaction should not disappoint the magician, however, because this version of the you-can't-unknot-by-adding principle is actually a foil for the next trick, which is about unbraiding. It goes like this.

Three strings are secured to the edge of the table by tacks or scotch tape and braided as shown in Fig.2.3,a or in a more complicated way. The challenge is the same as in the previous demo: to construct a braid on the other half of the table (without touching the first half) that will cancel out with the first. A normal audience will usually protest that this is impossible. Then the magician (who has practiced in advance) rapidly constructs the required "cancelling braid", pulls at the ends of the three strings, and – zap! – both braids disappear (Fig.2.3,b-c).

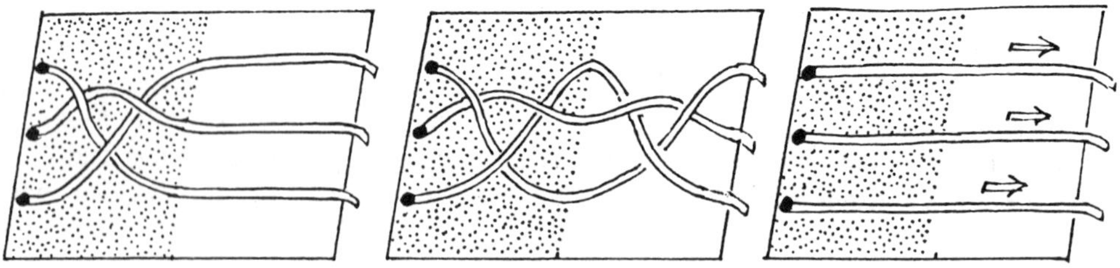

(a) (b) (c)

FIGURE 2.3. The braid trick

Mathematically, the results of the two last experiments express the fact that under composition braids form a group (in particular, any braid has an inverse braid), while knots do not (in fact, no knot cancels under composition with a nontrivial knot), only constituting a semigroup. We postpone all explanations of the unbraiding effect until the chapter on braids (Chapter III), where the mathematical definition of this important concept appears.

2.3. The belt trick. This is one of the most famous topological tricks. It may be demonstrated as follows: take a belt (which shouldn't be too stiff or too short), place it flat on a table so that it forms a loop (Fig.2.4,a), pick it up by the two ends, and pull it taut. What will happen? Unpredictably (unless of course you've done this before), a full twist (360°) will appear in the middle of the belt (Fig.2.4,b). Conversely, a belt with a full twist can placed flat on a table (without rotating the extremities) only by making a loop with it.

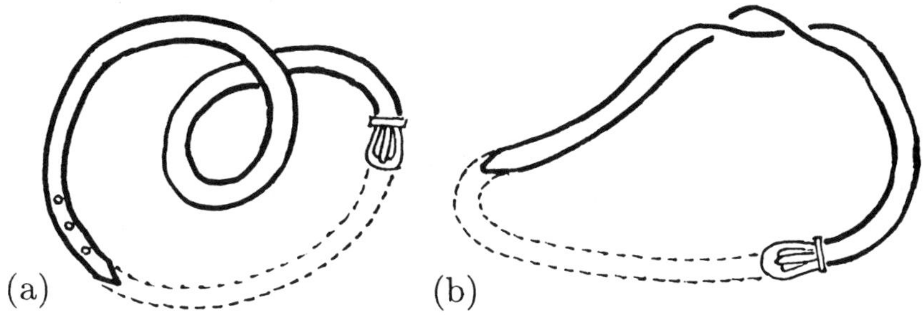

FIGURE 2.4. The belt trick

Note that the extremities of a belt are not equivalent (because of the buckle), and so we can distinguish "right-hand" and "left-hand" twists, as well as the corresponding "right-hand" and "left-hand" loops.

Mathematically, the belt trick expresses the fact that the two positions of the belt (with the addition of the two strips shown by dotted lines) are isotopic embeddings of $S^1 \times [0,1]$ in $\mathbb{R}^3$, and nothing more. Physically, it is a natural phenomenon, which may be called the loop-twist effect, with many annoying occurences. For example, you may have noticed that the cord connecting your telephone receiver to the telephone tends to twist around itself, making the telephone difficult to use. This happens because most right-handed people often make a right-handed loop on the cord when replacing the receiver. When iterated, these loops, having become twists because of the loop-twist effect, transform your telephone cord into a messy tangle.

2.4. The Whitney trick. This is also a belt trick, but it is best described (and demonstrated) by using a ribbon (whose extremities may be sewn together to form a flat annulus). The ribbon is lifted off the table and two opposite successive loops, an "inside" loop followed by an "outside" one, are made on it, and then the ribbon is replaced flat on the table (Fig. 2.5,a). The challenge is to undo the two loops without lifting the ribbon off the table.

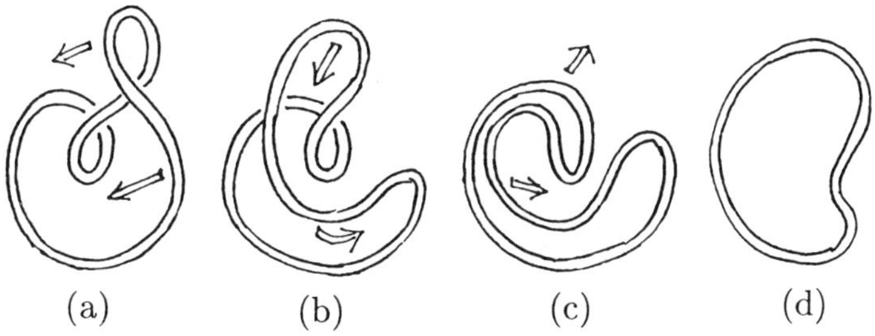

FIGURE 2.5. The Whitney trick

The solution (known to Whitney [**Whi**]) is simple enough, although most people fail to find it quickly by trial and error, and after a while usually reach the (erroneous) conclusion that the loops can't be undone without taking the ribbon off the table. But they can, as Fig.2.5,b-d shows.

Remark 1. If you intend to check how the Whitney trick works in practice by using a real ribbon, the choice of the latter is important: it should be long and flexible (paper won't do!). It might also help to use tacks or scotch tape to secure parts of the ribbon to the table when you are arranging or moving other parts of it.

Remark 2. In real life, there are actually four different ways to make a loop on an oriented ribbon, rather than two. In particular, there are two ways to make an outside loop on an (oriented) ribbon, depending on whether you make an overcrossing or an undercrossing when you complete the loop (Fig.2.6). Notice that in order to do the Whitney trick (see Fig.2.5,a), we used an inside loop with an underpass followed by an outside loop with an underpass.

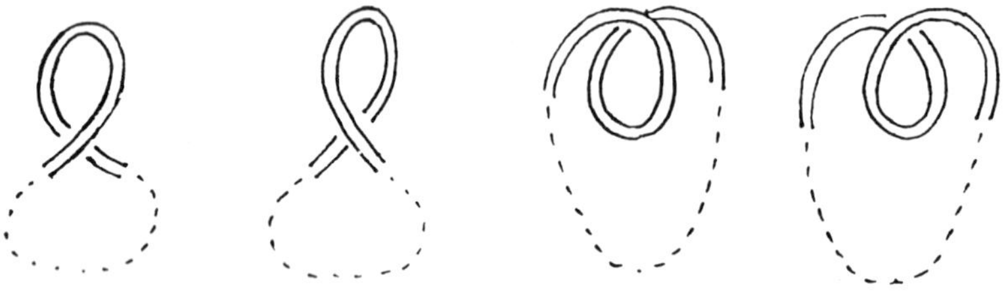

FIGURE 2.6. Four ways to loop a ribbon

Note that if you make two successive identical loops, you will not succeed in putting the ribbon back flat on the table: two twists appear on the ribbon (because of the loop-twist effect). Of course you can cut the ribbon, remove the double twist, glue back the extremities, and then replace the ribbon (with two right-hand loops on it) flat on the table. Such a ribbon cannot be transformed into a flat ribbon without loops by sliding it on the table; actually, this transformation can't be performed even if you are allowed to lift the ribbon off the table and manipulate it in space. But we are not ready to prove this fact: negative results require invariants!

Problem 2.1. What pairs of successive loops (of the 16 possible ones) on a ribbon lying flat on a table will cancel out each other as in the Whitney trick?

2.5. The sphere trick. Consider a flat ribbon with two successive inside loops, one with an overcrossing, the other with an undercrossing. In space, these two loops become opposite twists that cancel each other out. However, experiments with such a ribbon will convince us that the two loops will not cancel each other out if we are not allowed to lift any parts of the ribbon off the table's surface. Again we are not ready to prove this fact (no invariants!).

Now let us suppose that the ribbon described in the previous paragraph lies flat on the surface of a pretty large sphere (Fig.2.7,a). Can the loops be cancelled with each other without taking the ribbon off the sphere's surface? (Here we do not propose to carry out a real-life experiment, only an imaginary one; we also remind the reader that the ribbon was assumed very flexible and can therefore be stretched.)

Under these conditions, the surprising answer to the previous question is "yes" (Fig.2.7,b-d). The trick is to stretch the ribbon and take it "around infinity". This transforms the inside loop into an outside one, and the latter cancels out with the inside loop via a move similar to the move in Fig.2.5,b. This is what we call the sphere trick; it was also known to Whitney [**Whi**].

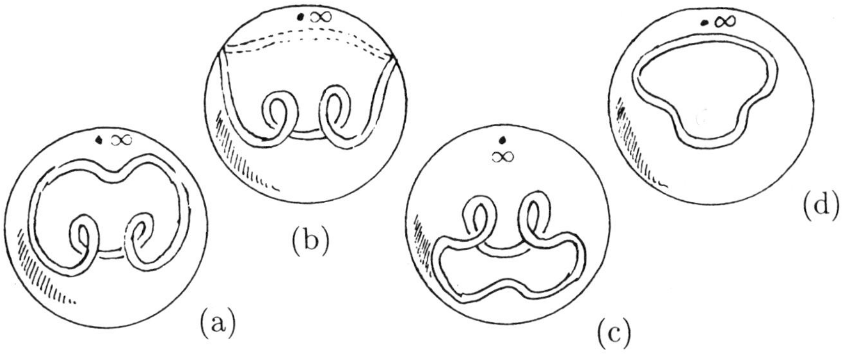

FIGURE 2.7. The sphere trick

Problem 2.2. What pairs of successive loops (of the 16 possible ones) on a ribbon lying flat on the sphere will cancel each other out by appropriately sliding the ribbon along the sphere's surface?

2.6. Games with twisted ribbons. The most famous twisted ribbon is the Möbius band. We shall not describe the classical tricks involving this object (like painting its surface or cutting it along its midline) since they are too well known. However, if the reader has never iterated the cutting procedure, we suggest the following series of experiments (to be carried first with a long paper Möbius band, and then in the reader's imagination). First cut the Möbius strip along its midline, then cut the resulting object along its midline, then keep cutting the things you get along their midlines. The aim is to predict what you will get after the third, fourth,..., nth step.

Now let us make a ribbon with a full twist (the Möbius band has only a half twist). What will happen if we cut this ribbon along its midline? What happens if we iterate this procedure? How is this related to the fact that the edge of this ribbon is a Hopf link? How is it related to the previous experiment?

Finally, let us make a ribbon with a twist and a half or, which is the same thing (by the belt trick), a half twist and a loop in the same direction (Fig.2.8). Observe the edge of this strip; clearly, it a trefoil knot. This observation should be useful for solving the concluding problem in this section. (We suggest that the reader attack this problem without preliminary experiments with paper strips, but once the results have been discovered, verify them experimentally.)

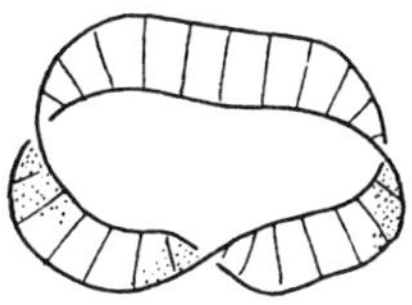

FIGURE 2.8. The ribbon with three half-twists

Problem 2.3. What do you get by cutting a twisted Möbius band (i.e., a ribbon with three half twists) along its midline? What happens if you iterate?

2.7. To conclude this section, let us note that the general theory of twisted ribbons in space is so complicated because it involves two different kinds of phenomena. One has to do with the intrinsic geometry of the objects considered, the other with their disposition in space. Thus, for example, the untwisted ribbon and the ribbon with one full twist model homeomorphic surfaces, but their embeddings in space are different.

Comments

The practice of tieing knots goes back to the very origins of civilization, if not beyond (some of the larger primates can make knots). The first treatises about knots that have reached us date back to the Renaissance and are, basically, handbooks for sailors. For a modern treatment of knots as practical objects, see [**Ash**]. However, the first scientific text on knot theory appeared much later, in 1796, and is an article by C. A. Vandermonde (better known for his determinant).

Nevertheless, the first substantial achievement of the theory concerns links and is due to Carl Friedrich Gauss (1833). In connection with his work in electrodynamics, Gauss wrote a beautiful integral (see [**Cou**], pp. 409–411, and [**Cal**]) whose value is an integer, the famous linking number of two curves (electrical circuits in the applications). The elementary topological definition of the linking number appears below in §15. It is remarkable that this beautiful piece of mathematics was totally ignored by topologists and its first generalization was found only 150 years later as the Kontsevich integral, explicitly defining Vassiliev invariants (see §4). Incidentally, the Gauss archives contain lovely pictures of knots drawn in 1792; it would seem that Gauss had begun to tabulate knots, but gave up the topic without having found a knot invariant.

The first important stage in the development of knot theory took place at the end of the 19th century, and did not involve mathematicians. The great physicist William Thomson (Lord Kelvin), speculating about the structure of matter, tried to reconcile the corpuscular and wave theories by imagining particles as little closed curves that he called vortex atoms ([**Kel**], 1867). The tabulation of knots, inspired by Kelvin and soon undertaken by P. G. Tait [**Tai**], was then viewed as a classification of atoms (from this viewpoint, molecules could be links). The results of Tait's titanic labors were the first knot tables and the famous Tait conjectures (see the survey [**Thi**]), but no meaningful theory (in particular, no effectively computable invariant) was developed. Kelvin's vortex atom theory, although it received the ideological support of J. C. Maxwell, was soon abandoned and forgotten, perhaps because of the success of another tabulation of atoms, more arithmetical than geometric in nature (Mendeleev's periodic table of elements).

Mathematicians started doing knot theory much later, but with much more success (due to the application of the fundamental group and some fairly sophisticated algebra to the study of knots). Most of the geometric material of this chapter, including the basic definitions, the Reidemeister theorem, and prime knots, was known to mathematicians in the twenties and thirties and appears in Reidemeister's beautiful book *Knotentheorie*, published in 1932 (for the English translation, see [**Rei**]). More recent are the examples of wild knots ([**AF**], 1948) as well as the proof of the theorem on the uniqueness of prime knot decompositions, due to

H. Schubert ([**Sch**], 1949). Newer monographs on knot theory include [**CF**], [**Rol**], [**BZ**], and [**Kau1**].

There are still many unsolved geometric problems in classical knot theory (for example, see [**Mor1**], [**Prz**], [**Thi**]). A number of the really fundamental problems in knot theory have been settled. Thus the prime knot decomposition theorem (see 1.8 above) has been proved, the most interesting Tait conjectures have been established ([**Mur1-2**], [**MT**]), W.Haken has found an unknotting algorithm [**Hak1-2**], a (complicated) comparison algorithm for knots has been proposed in [**Hem**], and there are extensive knot tables [**Thi**] (most of these questions are discussed in Chapter II).

Perhaps the most natural general open problem in classical knot theory is to find a comparison algorithm for knots (in particular, an unknotting algorithm) simple enough to be implemented on a computer. However, if this problem has a solution, we believe that the algorithm will be more algebraic than geometric in nature, involving some knot invariants. But this is the topic of the next chapter.

CHAPTER II

Knot and Link Invariants

In the previous chapter we pointed out that in order to prove "negative" results about links (and in particular about knots) some invariants are required. For example, at this stage we cannot rigorously prove such simple and intuitively obvious statements as the fact that the trefoil knot is nontrivial or, say, nonisotopic to the figure eight knot. In this chapter we present the most recent and powerful invariants of knots and links, due to Vaughan Jones (§3) and Victor Vassiliev (§4). Their original constructions were rather sophisticated, but extensive work by several other researchers has led to crucial simplifications, so that now the subject can be almost entirely developed along elementary lines. The only remaining difficult part (the existence of Vassiliev invariants), although it has three proofs based on totally different ideas ([**Vas1**], [**Kon1-2**] and [**BN**], [**Car**]), has resisted simplification. So in §4 the existence proof is omitted.

§3. The Jones polynomial

One of the most natural questions that arises in knot theory is, given two knot diagrams, to determine if they represent the same knot (this is the *knot comparison problem*). A particular case is the *unknotting problem*: to determine if a given knot diagram represents the unknot. Now if, say, the unknotting problem has a solution in some specific example, then we can prove this fact by finding the appropriate Reidemeister moves (which transform our diagram into the round circle). In the case when we are given a knot diagram that we suspect represents a nontrivial knot, in order to prove its nontriviality, the standard method is to use an *invariant*, i.e., an assignment to each knot diagram of some algebraic object (e.g. a number or a polynomial) that depends only on the knot isotopy class, and verify that the value of this invariant for the given knot differs from that for the unknot. This method will be effective if there is a simple algorithm for computing the invariant from the knot diagram.

The famous Jones polynomial is such an invariant for links (and hence for knots); it will be described in this section. We begin its construction by defining the so-called bracket polynomial of nonoriented link diagrams, due to Louis Kauffman. (Links and knots are assumed nonoriented until further notice.)

3.1. The bracket polynomial. Let us attempt to assign to each nonoriented link diagram L a polynomial in the variables a, b, c, denoted by $\langle L \rangle$, so as to satisfy the following defining relations:

(1) 

(2) $$\langle L \sqcup \bigcirc \rangle = c\langle L \rangle;$$

(3) $$\langle \bigcirc \rangle = 1.$$

Here the little pictures in relation (1) denote three link diagrams L, L_A, L_B that are identical outside a small disk and are as shown in the pictures inside it. Let us denote these three diagrams by L, L_A, L_B (see Fig.3.1).

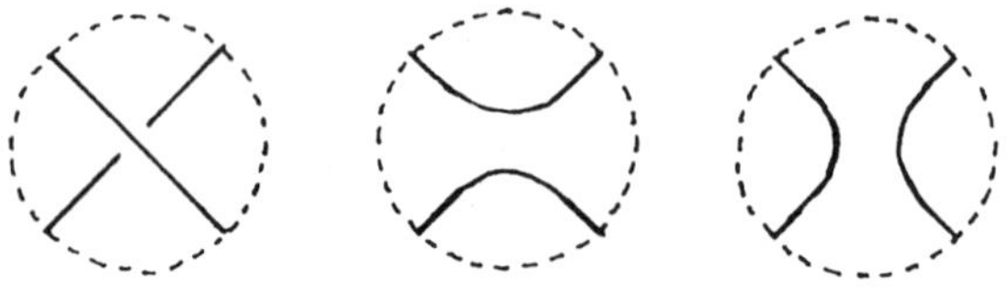

L L_A L_B

FIGURE 3.1. Eliminating a crossing point

In this notation, (1) may be rewritten in the form $\langle L \rangle = a\langle L_A \rangle + b\langle L_B \rangle$. Note that for the crossing contained in the small disk in the diagram L, no matter how it is rotated, the diagrams L_A and L_B are uniquely defined. Indeed, the arcs inside the small disks of the diagrams L_A and L_B are chosen in the regions A and B respectively (Fig.3.2); these regions are defined as follows: when we move along the upper branch of the crossing, we first see the region A to our left, and, after the crossing, to our right (and conversely for B); this definition clearly does not depend on the direction along which we move on the overcrossing.

FIGURE 3.2. A and B regions near a crossing point

Relation (2) means that the addition to the link L of a circular component that does not intersect L (and has no crossing points with L) results in the polynomial being multiplied by c. Further, relation (3) means that the polynomial assigned to the circle is equal to 1. Finally, we shall assume that the polynomial $\langle L \rangle$ does not change under plane isotopies of the diagram L.

Now let us try to impose relationships between the variables a, b, c so that the polynomial will be invariant with respect to the three Reidemeister moves. We begin with the move Ω_2. Using relation (1) repeatedly and (2) once, we obtain

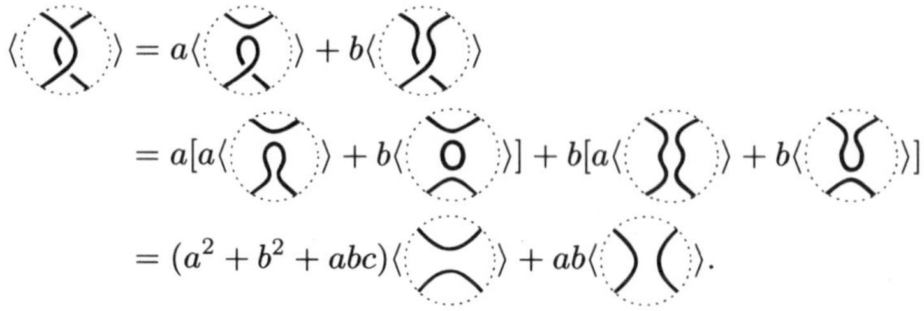

Now if we had $ab = 1$ and $a^2 + b^2 + abc = 0$, the polynomial would be invariant with respect to Ω_2. So we put $b = a^{-1}$ and $c = -a^2 - b^2$, thereby ensuring invariance with respect to Ω_2 of the bracket polynomial $\langle L \rangle$.

This does not leave us any room for further manœuvre, but luckily, the bracket polynomial as now defined turns out to be invariant with respect to Ω_3 as well. Indeed, by condition (1) we have the two following relations:

(3.1) 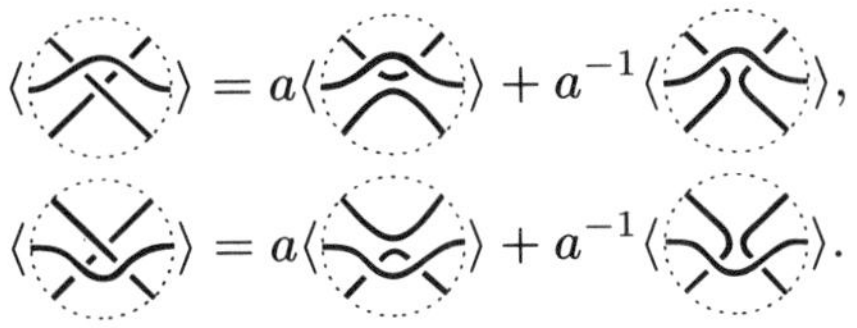

Clearly 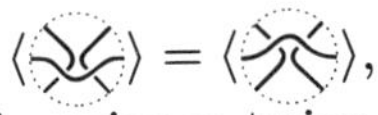, since the two diagrams are plane isotopic. Further, applying Ω_2-invariance twice, we get

Now comparing the right-hand sides of the two formulas (3.1), we see that they are equal term-by-term. Then so are the left-hand sides, which proves Ω_3-invariance of the bracket polynomial.

Now we turn to Ω_1, hoping that our luck will not change. By relations (1) and (2), we have

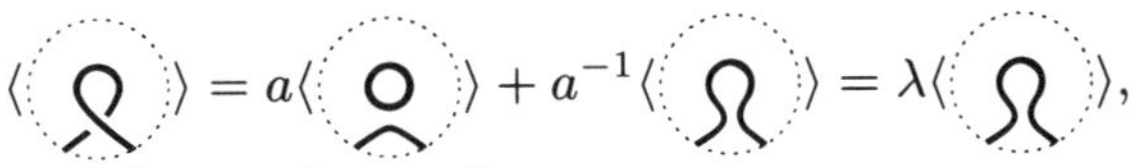

where $\lambda = a(-a^2 - a^{-2}) + a^{-1} = -a^3$.

A similar computation can be performed for the other type of little loop. Thus we have

(3.2) 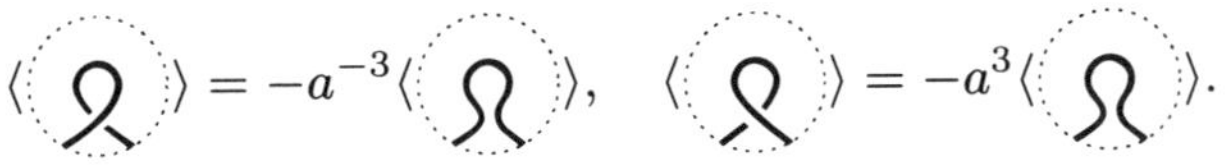

So our luck has turned: unless $a^3 = -1$, the bracket polynomial is not invariant with respect to Ω_1 and therefore is not an isotopy invariant of links. For example, we have

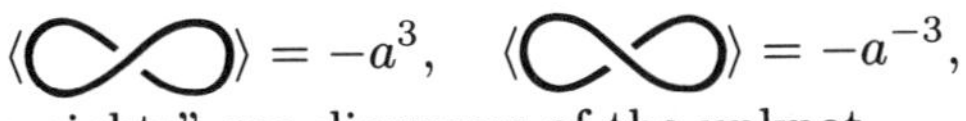

although both "figure eights" are diagrams of the unknot.

Nevertheless, we shall use the bracket polynomial as the main ingredient to construct the Jones polynomial, which will turn out to be invariant with respect to *all three* Reidemeister moves. But in order to do that, we must establish that the bracket polynomial exists and is unique. We begin with simple examples of its computation.

3.2. Computations of bracket polynomials. In the calculations that follow, we use results of the previous subsection, in particular relation (3.2).

Example 1. For the simplest diagrams with one and two crossings, we have

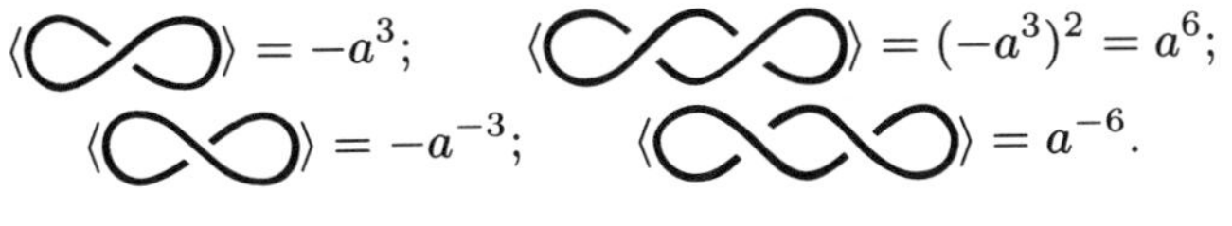

Example 2. For the Hopf links,

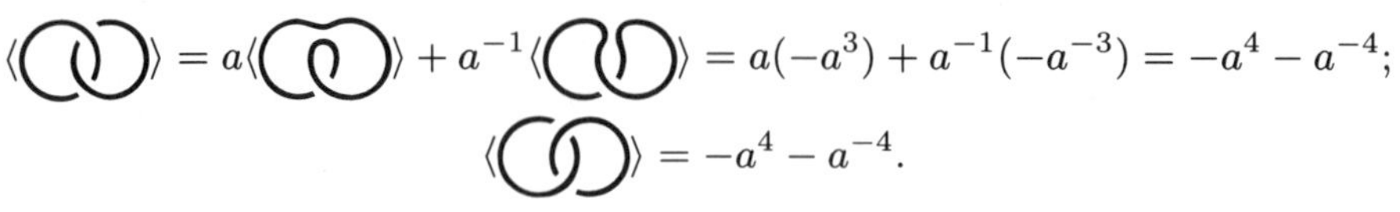

$$\langle\ \rangle = a\langle\ \rangle + a^{-1}\langle\ \rangle = a(-a^3) + a^{-1}(-a^{-3}) = -a^4 - a^{-4};$$

$$\langle\ \rangle = -a^4 - a^{-4}.$$

Example 3. For the trefoils

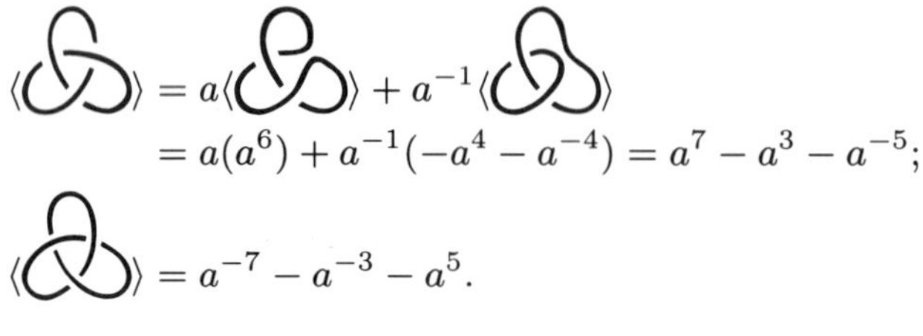

$$\langle\ \rangle = a\langle\ \rangle + a^{-1}\langle\ \rangle$$
$$= a(a^6) + a^{-1}(-a^4 - a^{-4}) = a^7 - a^3 - a^{-5};$$

$$\langle\ \rangle = a^{-7} - a^{-3} - a^5.$$

Problem 3.1. Compute the bracket polynomial for the knot and link shown in Fig.3.3.

FIGURE 3.3. Knot and link for the computation of $\langle \dots \rangle$

3.3. To simplify computations in Examples 1–3, we often used the results of previous calculations. This is not really necessary. If we have not saved previous results, we can still calculate the bracket polynomial of a link directly, successively eliminating all crossing points by using condition (1) (see Fig.3.1). After a crossing is destroyed, the number of diagrams is doubled. Hence for a diagram with n crossings, we finally get 2^n diagrams, each of which contains several nonintersecting circles. From each of the 2^n diagrams, we can recover the transformations that occured in its construction.

Let us illustrate this in the case of the trefoil (Fig.3.4,a). Suppose that the diagram shown in Fig.3.4,b was obtained from this trefoil by destroying the crossings. Then we can affirm that at the crossings marked 1 and 3 the arcs were chosen in the region A, while the arcs for the crossing marked 2 are in B.

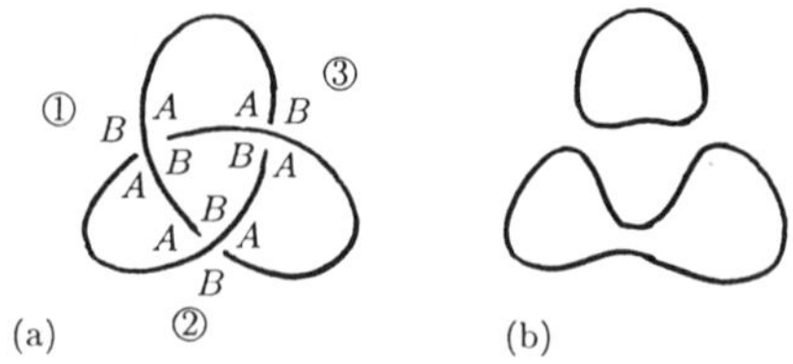

FIGURE 3.4. Destroying the crossing points of a trefoil

In the polynomial of this diagram, the coefficient $a^2b = a^2a^{-1} = a$ will then factor out; this coefficient does not depend on the order in which the crossings are destroyed. This is a general fact that can be used to prove existence and uniqueness of the bracket polynomial.

3.4. THEOREM. *There exists a unique polynomial $\langle L\rangle$ satisfying the defining relations* (1)–(3).

Proof. Let us number the crossings of the diagram L and assign to the ith crossing its *state*, i.e., either of the two formal "values" $x_i = A$ or $x_i = B$. A choice of states for all the crossings will be called a *state of the diagram* L. Let n be the number of crossings of L. Then the number of states of L will be 2^n. To each state s of the diagram L corresponds the system of nonintersecting circles obtained by destroying each crossing point in accordance with its state as shown in Fig.3.5.

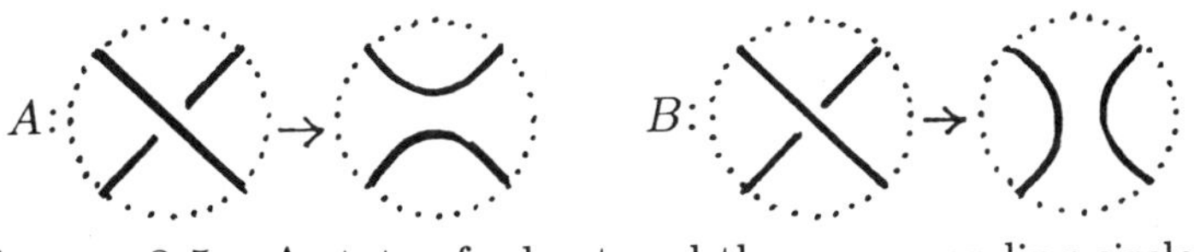

FIGURE 3.5. A state of a knot and the corresponding circles

Denote by $\alpha(s)$ and $\beta(s)$ the number of crossings in the state A and B, respectively, and by $\gamma(s)$ the number of circles obtained after all the transformations corresponding to s are performed. Now if we destroy all the crossings of L using condition (1), and then use conditions (2) and (3) to calculate the polynomial of the diagrams consisting of nonintersecting circles, we obtain

$$\langle L\rangle = \sum_s a^{\alpha(s)} b^{\beta(s)} c^{\gamma(s)-1} = \sum_s a^{\alpha(s)-\beta(s)} (-a^2 - a^{-2})^{\gamma(s)-1}, \tag{3.3}$$

where the sum is taken over all the 2^n states of L. Hence the polynomial $\langle L\rangle$ is uniquely defined.

Formula (3.3) establishes not only the uniqueness of the polynomial $\langle L\rangle$, but its existence as well. Indeed, let us *define* $\langle L\rangle$ by formula (3.3). Then conditions (2) and (3) are obviously satisfied. To prove condition (1) for the ith crossing, we need only split the sum over all s into two sums: one over those s for which $x_i = A$, the other over those for which $x_i = B$. □

3.5. The Kauffman polynomial. In the remaining part of this section, we consider *oriented* links, i.e., we assume that each component is supplied with an orientation (shown by arrows in the figures). For an oriented link diagram L, let us define its *writhe number* by setting

$$w(L) = \sum_i \varepsilon_i,$$

where the sum is taken over all crossing points and the numbers ε_i are equal to ± 1 depending on the *sign* of the ith crossing point, which is defined as shown in Fig.3.6.

$\varepsilon = 1$ $\varepsilon = -1$

FIGURE 3.6. Positive and negative crossing points

It is easy to see that if the orientations of all the components of the link L are reversed, then the writhe number $w(L)$ does not change. The computation of the writhe numbers for several knots and links is presented in Fig.3.7.

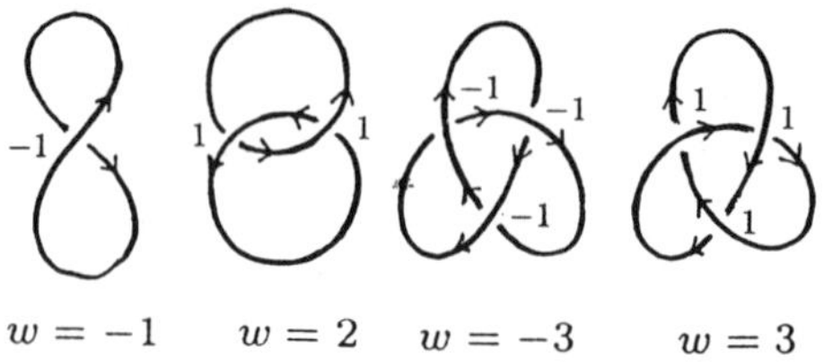

FIGURE 3.7. Computation of writhe numbers

The writhe number is invariant under the Reidemeister moves Ω_2 and Ω_3. Indeed, for the move Ω_2, the signs of the two new crossing points (or of the two disappearing ones) are opposite, while for Ω_3 the signs of the crossing points A, B, C (Fig.3.8) are respectively equal to those of A', B', C'.

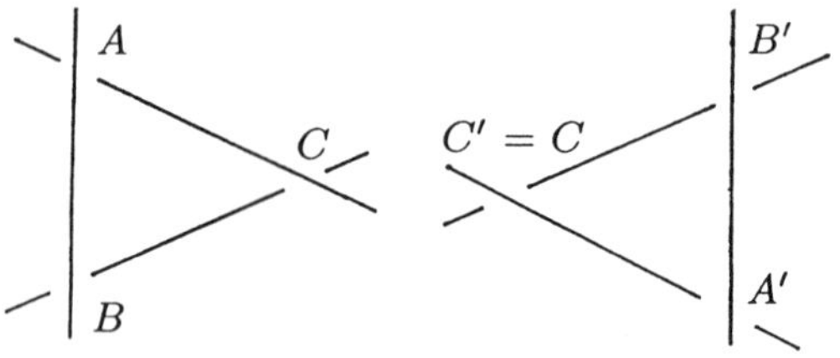

FIGURE 3.8. The writhe number is invariant under Ω_3

Under the move Ω_1, however, one new crossing point appears (or disappears), so that the writhe number is changed by ± 1.

Now let us define the polynomial $X(L)$ on any oriented link diagram L by setting

$$X(L) = (-a)^{-3w(L)} \langle |L| \rangle, \tag{3.4}$$

where the nonoriented diagram $|L|$ is obtained from L by forgetting the orientations of all the components (destroying the arrows).

3.6. THEOREM. *The polynomial $X(\cdot)$ is an isotopy invariant of oriented links.*

Proof. Since both $w(L)$ and $\langle |L| \rangle$ do not change under the Reidemeister moves Ω_2 and Ω_3, the same is true of $X(L)$. Further, both $w(L)$ and $\langle |L| \rangle$ change under Ω_1, but these changes cancel each other. For example,

$\langle$  $\rangle = -a^{-3} \langle$  $\rangle$ and $w($  $) = w($  $) - 1$,

so that by the definition of $X(L)$ and formula (3.2), we obtain

$X($  $) = (-a^{-3})(-a^{3})X($  $) = X($  $)$.

A similar calculation (using the other relation in (3.2)) for the move Ω_1 with the other type of little loop is left to the reader.

Being invariant with respect to all three Reidemeister moves, the polynomial $X(L)$ is an isotopy invariant by the Reidemeister Theorem 1.7. □

Using the polynomial $X(L)$, one can establish, among other things, that the left and right trefoils are different (nonisotopic) knots. Indeed, it follows from the result of Example 3 that

$X($  $) = -a^{16} + a^{12} + a^{4},$

$X($  $) = -a^{-16} + a^{-12} + a^{-4}.$

Recall that the polynomial $\langle \cdot \rangle$ was defined axiomatically, by means of relations (1),(2),(3). Then we had to prove (this turned out to be fairly easy) that such a polynomial exists and is unique (Theorem 3.4). The polynomial $X(L)$, on the other hand, was defined explicitly (by means of formula (3.4), involving $\langle |L| \rangle$). It turns out that the polynomial $X(L)$ also has an axiomatic description by simple and useful defining relations.

In order to obtain the main defining relation for $X(L)$, let us use condition (1) for the bracket polynomial twice:

$\langle$  $\rangle = a\langle$  $\rangle + a^{-1}\langle$  $\rangle,$

$\langle$  $\rangle = a^{-1}\langle$  $\rangle + a\langle$  $\rangle.$

Multiplying the first relation by $-a^{-1}$, the second by a, and adding, we obtain

$a\langle$  $\rangle - a^{-1}\langle$  $\rangle = (a^2 - a^{-2})\langle$  $\rangle.$

Choosing appropriate orientations, we obtain

$$a\langle L^+\rangle - a^{-1}\langle L^-\rangle = (a^2 - a^{-2})\langle L^\circ\rangle,$$

where $L^\pm$ and L° are the diagrams shown in Fig.3.9.

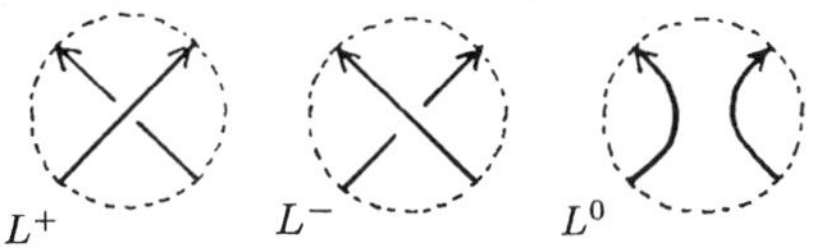

L^+ L^- L^0

FIGURE 3.9. Diagrams in the main defining relation for $X(L)$

It is also easy to check that $w(L^\pm) = w(L^\circ) \pm 1$. Hence

$$a(-a)^3X(L^+) - a^{-1}(-a)^{-3}X(L^-) = (a^2 - a^{-2})X(L^\circ). \tag{3.6}$$

We are finally ready to define the Jones polynomial.

3.7. Definition of the Jones polynomial. Let us substitute $a = q^{-1/4}$ into the Kauffman polynomial $X(L)$. We then obtain a polynomial in $q^{\pm 1/4}$, denoted by $V(L)$ and called the *Jones polynomial* of the oriented link L. The Jones polynomial satisfies the following relations:

$$(1') \qquad q^{-1}V(L^+) - qV(L^-) = (q^{1/2} - q^{-1/2})V(L^\circ);$$

$$(2') \qquad V(L \sqcup \bigcirc) = -(q^{-1/2} + q^{1/2})V(L);$$

$$(3') \qquad V(\bigcirc) = 1.$$

Here in condition $(1')$, which is known as the *skein relation*, L^+, L^-, L° are three link diagrams identical outside a small disk, inside which they are as shown in Fig.3.9. Further, in condition $(2')$, $L \sqcup \bigcirc$ stands for the link L with an added circle that does not intersect L (and has no crossing points with L). Finally, condition $(3')$ says that the Jones polynomial of the circle is 1. These conditions follow immediately from the corresponding properties of the polynomial $X(L)$, e.g. $(1')$ follows from (3.6).

Relations $(1'), (2'), (3')$ form the basis of a technique for computing the Jones polynomial that differs from the one used for the bracket polynomial or for $X(L)$ (which was based on destroying all crossing points by using the defining relation (1) for the bracket polynomial). Here the application of the skein relation involves not only destroying a crossing point, but also changing it to an opposite crossing. So the computational strategy consists in "unravelling" the link by making several crossing changes, which means that we apply the skein relation $(1')$ only to those crossing points where these crossing changes must be performed.

Before proving that this strategy always works, let us demonstrate how it allows us to find the Jones polynomial of the trefoil. Applying condition $(1')$ to (any) crossing point, we get the unknot and the Hopf link:

$q^{-1}V($$) - qV($$) = (q^{1/2} - q^{-1/2})V($$).$

Applying condition $(1')$ to the Hopf link, we obtain

$q^{-1}V($$) - qV($$) = (q^{1/2} - q^{-1/2})V($$).$

Hence, by relations $(2')$ and $(3')$,

$V($$) = -q^{-2}(q^{1/2} + q^{-1/2}) - q^{-1}(q^{1/2} - q^{-1/2})$

$= -q^{-1/2} - q^{-5/2},$

so that

$V($$) = q^{-2} + q^{-1}(q^{-1/2} + q^{-5/2})(q^{1/2} - q^{-1/2})$

$= q^{-1} + q^{-3} - q^{-4}.$

This concurs with our computation in the previous subsection,

$X($$) = -a^{16} + a^{12} + a^4,$

because $q = a^{-1/4}$.

Similarly one can compute the Jones polynomial of any m-component link L represented by a diagram with n crossings. Indeed, suppose that $L_1, ..., L_k$ is a sequence of links, beginning with the given link $L = L_1$ and ending with the trivial m-component link L_k, where each subsequent link L_{i+1} is obtained from the previous one L_i by one crossing change. Then we have one of the two following relations:

$$q^{-1}V(L_i) - qV(L_{i+1}) + (q^{-1/2} - q^{1/2})V(L_i') = 0,$$
$$q^{-1}V(L_{i+1}) - qV(L_i) + (q^{-1/2} - q^{1/2})V(L_i') = 0,$$

where the diagram of the link L_i' has $n - 1$ crossings. Since the value of the polynomial $V(L_k) = (-q^{-1/2} - q^{1/2})^{m-1}$ is known, we can successively compute the polynomials $V(L_{k-1}), ..., V(L_1) = V(L)$ provided we know how to compute the polynomial of the links L_i' with less than n crossings. And we can perform an induction on the number of crossings to take care of that.

Now to prove that the described algorithm for the computation of the Jones polynomial always works, it remains to establish that the required unravelling sequence $L_1, ..., L_k$ of links exists.

3.8. Theorem. *Any m-component link L can be transformed into the trivial m-component link by an appropriate sequence of crossing changes.*

Proof. First let us assume that $m = 1$, i.e., L is a knot. Instead of the knot diagram, consider its projection or *shadow*, i.e., the corresponding plane curve with transversal self-intersections. In the plane choose a straight line l that does not intersect the shadow. Let us move l parallel to itself until it touches the shadow at some point P. Choose two points A and B in space vertically above P, the point A being located above B. Consider the trajectory of a point that moves above the shadow of our knot starting at A, uniformly decreasing its distance from the plane so as to complete its motion around the knot at B (Fig.3.10). Joining the points A and B by a line segment, we obtain a trivial knot K (because the projection of K on a plane perpendicular to the line l has no self-intersection points). Now it remains to note that K can be obtained from the given knot by crossing changes.

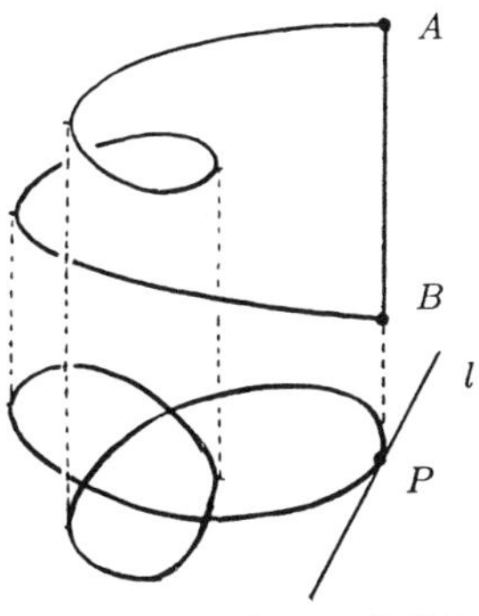

FIGURE 3.10. Unknotting the trefoil by crossing changes

Now in the case when L has more than one component, we apply the same construction to each component, taking care that each trivial knot K_i be located at a different altitude above the plane; this will be ensured if the projections of the segments A_iB_i on the vertical axis do not intersect. □

We have shown that the Jones polynomial of any oriented link can be computed from the Jones polynomials of trivial links by using condition (1′). Hence, the

Jones polynomial is a polynomial in $q^{\pm 1/2} = a^{\mp 2}$, i.e., the polynomial $X(L)$ of a link contains only even degrees of the variable a. It turns out that for knots K, the polynomial $X(K)$ contains only powers of a divisible by 4, so that the Jones polynomial is a Laurent polynomial in q and not only in $q^{1/2}$. Moreover, we have the following statement.

3.9. Theorem. a) *If the number of components of an oriented link L is odd, e.g. if L is a knot, then the Jones polynomial $V(L)$ contains only terms of the form q^k, $k \in \mathbb{Z}$.*

b) *If the number of components of an oriented link L is even, then the Jones polynomial $V(L)$ contains only terms of the form $q^{(2k+1)/2}$, $k \in \mathbb{Z}$.*

Proof. As we have seen previously, relations (2′) and (3′) imply that the Jones polynomial of the trivial m-component link is

$$(-q^{-1/2} - q^{1/2})^{m-1} = q^{(m-1)/2}(-q^{-1} - 1)^{m-1};$$

hence the statements of the theorem holds for trivial links. The Jones polynomial of any link may be computed from those of trivial links by using relation (1′). Hence it suffices to verify that if the theorem holds for any two of the three polynomials $V(L^+)$, $V(L^-)$, $V(L^\circ)$, then it holds for the third as well. If we look at the form of relation (1′), we see that it suffices to verify the following properties of the links L^+, L^-, L°:

1. the integers m_+ and m_- have the same parity,
2. the integers m_+ and $m_\circ$ have opposite parities,

where m_+,m_-,$m_\circ$ is the number of components of the link L^+, L^-, L° respectively.

We obviously have $m_+ = m_-$. It is also easy to check that $m_+ = m_\circ \pm 1$ (two cases must be considered, see Fig.3.11). □

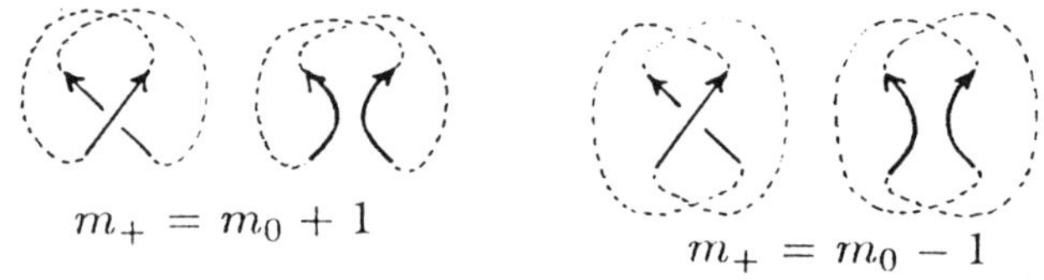

FIGURE 3.11. Proof of the fact that $m_+ = m_\circ \pm 1$

Problem 3.2. Compute the Jones polynomials of the knots shown in Fig.3.12 in two ways: first via the bracket polynomial, then by using conditions (1′)–(3′) directly; compare your answers.

FIGURE 3.12. Knots for computing Jones polynomials

3.10. Knot tabulation. We are now capable of rigorously classifying knots with a small number of crossings, say less than or equal to 7. To do that, we must first list all knot diagrams without repetitions (up to plane isotopy) with $0, 1, 2, ..., 7$ crossings. (This is a fairly easy, if tedious, combinatorial problem, whose solution we do not discuss here.) Once this is done, we can compute the Jones polynomials of all the knot diagrams in the list. If two diagrams have the same polynomial, we suspect that they are diagrams of the same knot, and attempt to prove that the diagrams are isotopic by means of Reidemeister moves. It turns out that in our case (7 crossings or less) these attempts are always successful. The result is the knot table of knots with 7 crossings or less (Fig.3.13), where we only present the so-called prime knots; recall that a *prime knot* is a knot that can't be represented as the connected sum of two nontrivial knots; knots that are not prime are called *composite*.

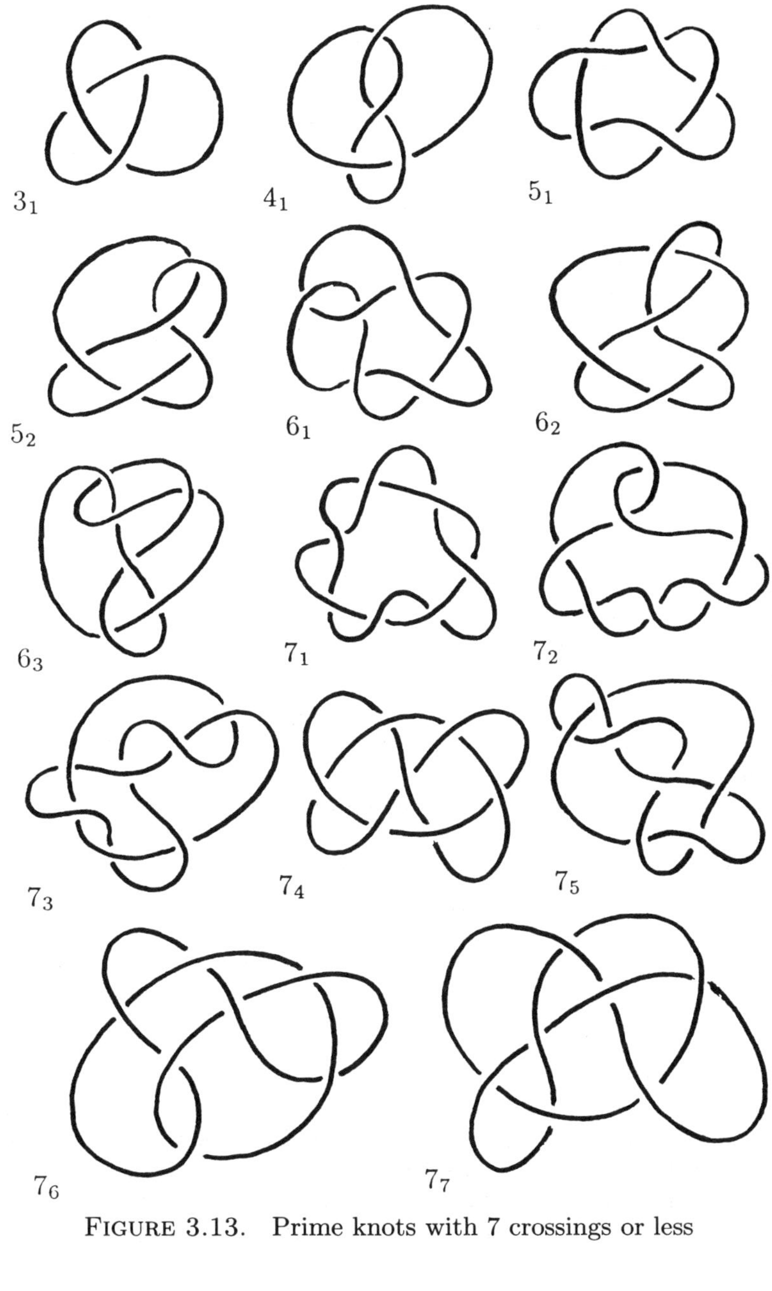

FIGURE 3.13. Prime knots with 7 crossings or less

Composite knots with 7 crossings or less are shown separately in Fig.3.14.

FIGURE 3.14. Composite knots with 7 crossings or less

Note that all the knots in Fig.3.13 are *alternating*, i.e., possess a diagram in which overcrossings alternate with undercrossings as we go around the knot projection. The simplest example of a prime nonalternating knot, 8_{19}, is shown in Fig.3.15.

8_{19}

FIGURE 3.15. A prime nonalternating knot

3.11. At this point it is natural to ask if the Jones polynomial always distinguishes nonisotopic knots (and links). The answer, unfortunately, is "no". It involves the connected sum operation.

Recall that the connected sum of knots was defined in 1.4. When links have more than one component, their connected sum is not well defined: its result may depend on the choice of components that we join together. We shall nevertheless use the notation $L_1 \# L_2$ in this case, having in mind that it may stand for several nonisotopic objects.

Problem 3.3. Prove that for any two links L_1 and L_2,

a) $V(L_1 \# L_2) = V(L_1) \cdot V(L_2)$;

b) $V(L_1 \sqcup L_2) = -(q^{-1/2} + q^{1/2}) V(L_1) \cdot V(L_2)$.

Here in formula a) the connected sum is arbitrary (any two components are joined).

Problem 3.4. Prove that the Jones polynomials of the two links shown in Fig.3.16 coincide.

It is easy to prove that the two links presented in Fig.3.16 are nonisotopic, so that we have an example of links not distinguished by their Jones polynomials. The cause of this unpleasant fact is the behavior of the Jones polynomial under the ambiguous (not well-defined) connected sum operation for links. Although this trick will not work in the same way for knots, we have a similar fact for knots.

FIGURE 3.16. Different links with identical Jones polynomials

Problem 3.5. Check that the Jones polynomials of the knots shown in Fig.3.17 coincide.

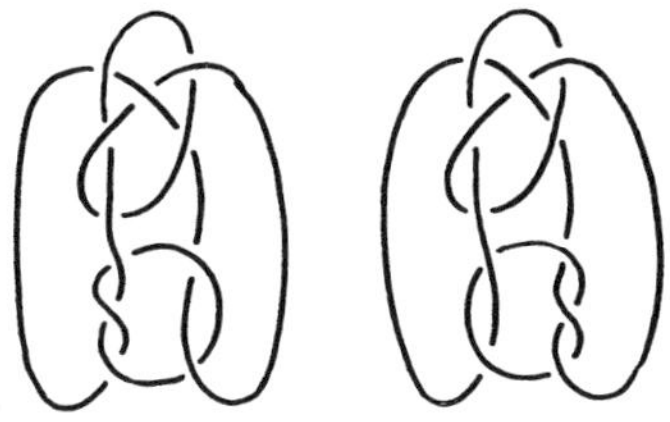

FIGURE 3.17. Different knots with identical Jones polynomials

It can be proved that these two knots are nonisotopic. So the Jones polynomial does not always distinguish knots either. This motivates the search for stronger invariants, if only to prove that the two knots in the last example are distinct. Such an invariant will be presented in the next section.

Remark. An attentive scrutiny of Fig.3.17 shows that a kind of tricky connected sum operation is hidden inside the knots, so the source of the phenomenon is the same as for links.

3.12. Other polynomial invariants. The Jones polynomial is not the only important polynomial invariant of knots and links, nor was it the first one to appear historically (the Alexander polynomial was defined a half century earlier). Here we list the most important known polynomials. All of them are link isotopy invariants, take the value 1 on the trivial knot and satisfy a specific form of the skein relation. Below we only give their names and specify the skein relation.

Conway polynomial ∇ in one variable z

Alexander polynomial Δ in one variable t

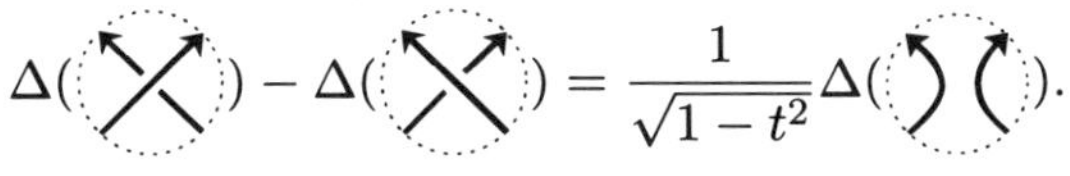

Jones polynomial V in one variable q

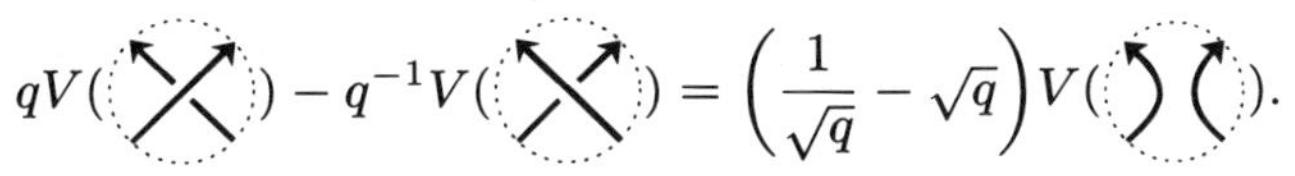

HOMFLY (or *LYMPHTOFU*) *polynomial* $\mathcal{P}$ in two variables x, t

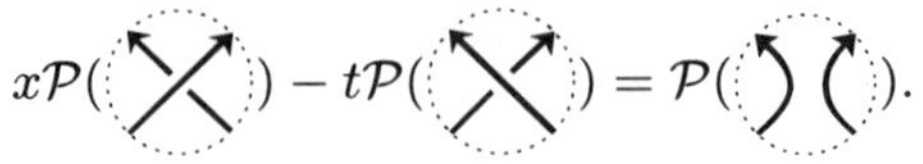

Jones polynomial $\mathcal{X}$ in two variables λ, q

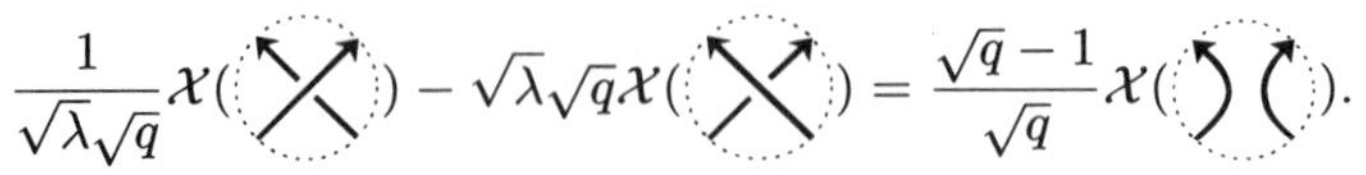

Note that any of the polynomials in this list can be obtained from the Jones two-variable polynomial by an appropriate change of variables.

§4. Vassiliev invariants

4.1. According to Theorem 3.8, any knot can be unraveled (i.e., modified so as to become the unknot) by performing a finite number of crossing changes. If we think of each crossing change as being a continuous process, the crucial moment is when the two branches meet, forming a point of self-intersection, where the two branches intersect transversally. The numerical knot invariants introduced by Vassiliev – the subject of this section – are based on a simple relation reflecting these crossing changes. But they require an extension of the class of objects with which we work: besides ordinary knots, we shall consider singular knots, i.e., curves whose only singularities are self-intersection points. More precisely, a *singular knot* k is a piecewise linear or smooth map $k : S^1 \to \mathbb{R}^3$ with no singularities except a finite number of transversal self-intersections, called the *double points* of k. (Note that the positive orientation of S^1 induces an orientation on its image $k(S^1)$; thus all singular and nonsingular knots in this section are always oriented.)

A numerical invariant v is said to be a *Vassiliev invariant* if it satisfies the relation

(4.1)

As in the previous section, formulas containing pictures in little dotted circles mean that we consider several (here, three) objects (here, singular knots) that are identical except within the circles, where they are as shown in the pictures. By a numerical invariant we mean an isotopy invariant of singular knots assuming values in some commutative ring with unit (which will in fact be the field of real numbers $\mathbb{R}$, unless stated otherwise). Note that an isotopy of singular knots must respect the order structure at the double points, i.e., the curve at the double points is assumed locally planar, and this property must be preserved in the process of the isotopy.

As compared to the previous section, we have enlarged the set of objects under study to singular knots, but it should be noted that links with two components or more are now excluded from our considerations.

We should also stress that, again in contrast to the previous section, Vassiliev invariants are cleary not unique (e.g., multiplication of v by any constant does not affect the validity of (4.1)). At present we leave aside the question of their existence, but we shall proceed by deriving some properties of an arbitrary Vassiliev invariant v.

4.2. The one-term relation

(4.2) 

This is an immediate consequence of definition (4.1). Indeed,

which is zero, because the two knots appearing in the right-hand side are isotopic (e.g. by two applications of the Reidemeister move Ω_1), and the function v is an isotopy invariant.

Problem 4.1. Prove that any Vassiliev invariant vanishes on any singular knot of the form shown in Fig.4.1 (the structure of the knot inside the two "boxes" in the picture is immaterial).

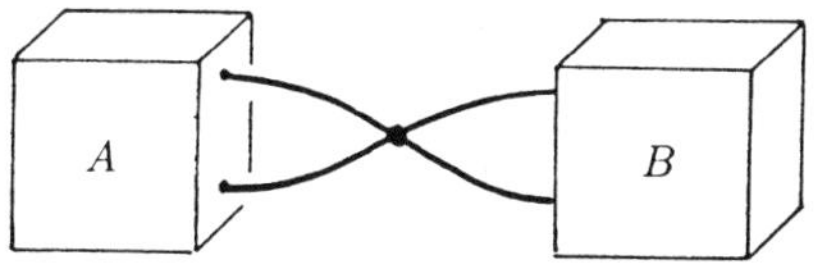

FIGURE 4.1. Singular knot living in two boxes

The relation

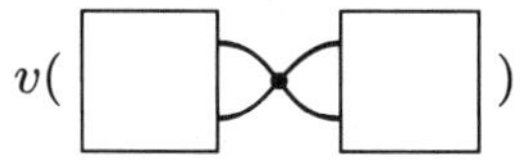

will be called the *general one-term relation.*

4.3. The four-term relation

(4.3) 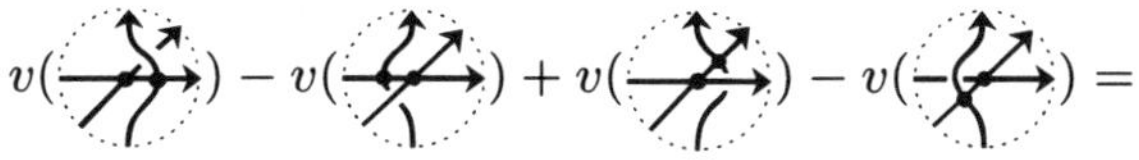

Proof. Using (4.1) four times, we get

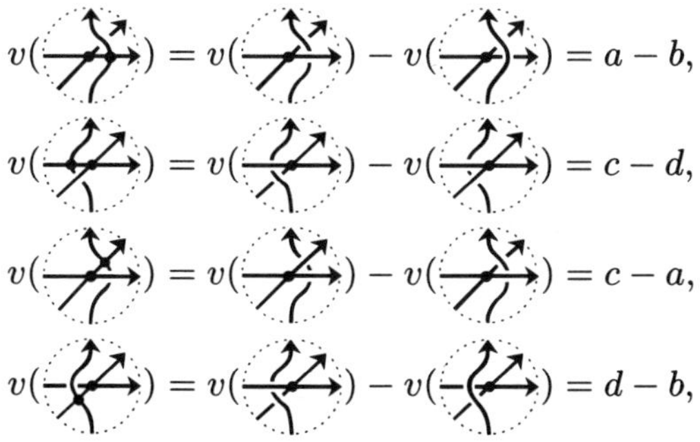

where the same letters denote identical numerical values of v on isotopic knots. Obviously,

$$(a-b)-(c-d)+(c-a)-(d-b)=0. \quad \square$$

4.4. Vassiliev invariants form a linear space over the field $\mathbb{R}$ (or over any other field chosen as the range of the invariants), because any linear combination of Vassiliev invariants obviously satisfies the (linear) relation (4.1). This space, however, is infinite-dimensional, and it would be more convenient to have a sequence of finite-dimensional spaces that increases so as to include sufficiently many Vassiliev invariants. We shall say that a Vassiliev invariant v is of *order* no greater than n (notation $\operatorname{ord} v \leqslant n$) if it vanishes on any singular knot with more than n double points. Denote $V_n = \{v \mid \operatorname{ord} v \leqslant n\}$. Then each V_n is (obviously) a linear space, and we have the sequence of inclusions

$$V_0 \subset V_1 \subset V_2 \subset V_3 \ldots .$$

4.5. Invariants of order 0 and 1. Any Vassiliev invariant v_0 of order 0 vanishes on any knot containing at least one double point. If v_0 assumes some value q on a specific knot without double points, then it assumes the same value on any other knot without double points. Indeed, formula (4.1) tells us that if two knots differ by one crossing change, then the values of v_0 on these knots differ by the value of v_0 on the corresponding singular knot (with a double point in place of the crossing point); but this value is zero. Since any two nonsingular knots can be taken to each other by a finite number of crossing changes (see Theorem 3.8), this means that v_0 is identically equal to q on all nonsingular knots. Thus $V_0 = \mathbb{R}$ and $\dim V_0 = 1$. Invariants of order 0 do not distinguish knots.

Surprisingly, neither do invariants of order 1. As we shall see, $V_1 = V_0$, because any invariant v_1 of order 1 vanishes on any knot with at least one double point (altough the definition only says that it is zero on knots with *two* double points). Indeed, let k_1 be a knot with one double point, and let k_1' be a knot obtained from k_1 by one crossing change. Then (4.1) implies $v_1(k_1) - v_1(k_1') = v_1(k_2)$, where k_2 has two double points. Then by definition, $v_1(k_2) = 0$ and $v_1(k_1) = v_1(k_1')$. Therefore, we can make all the crossing changes we want in the knot k_1 without changing the value of v_1. If the number of double points of k_1 is more than 1, we have $v_1(k_1) = 0$ (by definition). It remains to consider the case when there is exactly one. The double point splits k_1 into two parts, each of which can be made trivial and unlinked with each other by appropriate crossing changes (see Fig.4.2). As the result, we obtain the knot  on which any Vassiliev invariant vanishes by the one-term relation (4.2). Thus we have established the uninspiring result

$$V_0 = V_1 = \mathbb{R}. \tag{4.4}$$

So order one invariants do not distinguish knots either. Fortunately, the theory becomes meaningful at the next step, i.e., for order $n = 2$, but before going on to that case, we need a few more basic definitions and remarks.

Let us note that the argument used in connection with Fig.4.2 for $n = 1$ actually proves the following more general statement:

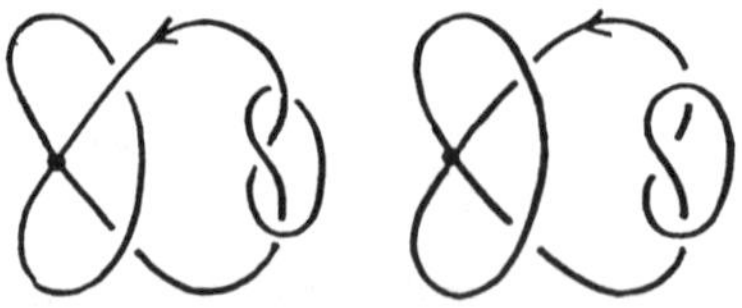

FIGURE 4.2. Crossing changes leaving $v \in V_1$ invariant

4.6. LEMMA. *The value of a Vassiliev invariant of order n on a knot with exactly n double points does not vary under crossing changes in the knot.* □

4.7. Symbols. Now, passing to the case of arbitrary order n, let us try to describe the quotient space V_n/V_{n-1}. When we are interested in an invariant of order n up to invariants of order $n-1$, it is convenient to consider the so-called symbol of the invariant rather than the invariant itself.

The *symbol* of a Vassiliev invariant v of order n is defined as the restriction of v to all singular knots with exactly n double points.

The importance of the notion of symbol is due to the fact that *two invariants of order n with the same symbol differ by an invariant of order no greater than $n-1$.* Indeed, their difference v vanishes on any knot with n double points. Formula (4.1) implies that the value of any Vassiliev invariant on a knot with $n+1$ double points is equal to the difference of its values on two knots with n double points. So v vanishes on all knots with $n+1$ double points. Similarly, it vanishes on any knot with $n+2$ double points, and so on. □

4.8. Gauss diagrams. Lemma 4.6 implies that the value of an invariant of order n on a knot with n double points (i.e., the value of the symbol) is not affected by crossing changes. So this value does not depend on knotting; it only depends on the sequence in which the double points appear as we go around the knot. It is convenient to encode this sequence in the following way. Consider the knot $k : S^1 \to \mathbb{R}^3$ with n double points. Moving around the circle S^1, let us mark all the points that are mapped to a double point, and then join each pair of marked points mapped to the same double point by a chord (Fig.4.3); what we obtain is called a *chord diagram* or *Gauss diagram* of order n of the knot k.

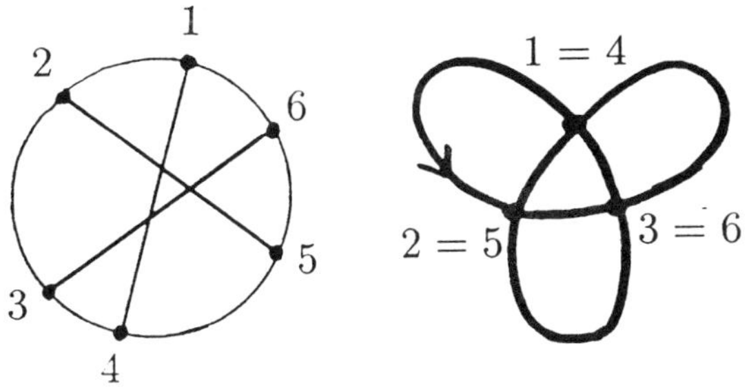

FIGURE 4.3. Gauss diagram of a singular knot

The two Gauss diagrams whose sets of chords differ only by an orientation-preserving diffeomorphism of S^1 are considered equivalent and are not distinguished. Fig.4.4 shows all (nonequivalent) Gauss diagrams of orders $n = 1, 2, 3$. Note that all nonsingular knots have the same chord diagram (which has an empty set of chords).

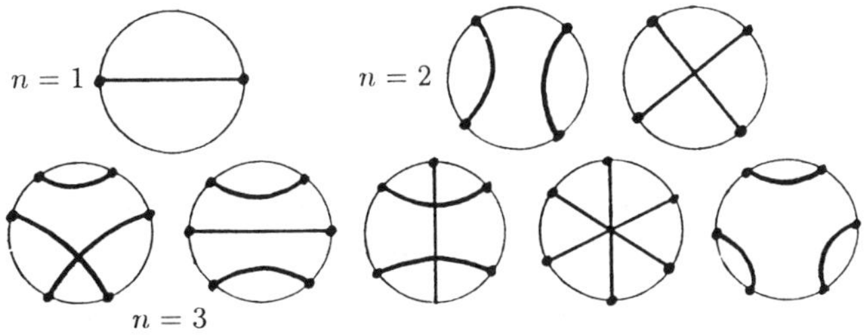

FIGURE 4.4. Gauss diagrams of orders $n \leqslant 3$

Problem 4.2. Verify that all diagrams of order $n \leqslant 3$ are equivalent to mirror symmetric ones, and give an example of a diagram of order 4 not equivalent to its mirror image.

It is easy to verify that any chord diagram is the Gauss diagram of some singular knot. Indeed, consider a point that traces a curve in $\mathbb{R}^3$ and mark the future double points on it in their order of appearance on the Gauss diagram; whenever a double point whose partner was already marked is to appear, direct the curve to the partner and make a transversal self-intersection there; continue in this way until you run out of double points, then take the curve back to the initial point; in 3-space there are no obstructions to this procedure. Fig.4.5 shows two knots with 2 double points corresponding to the two Gauss diagrams of order 2.

FIGURE 4.5. Singular knots with Gauss diagrams of order 2

Problem 4.3. Draw pictures of four singular knots corresponding to the four chord diagrams of order 3.

It will be useful for the sequel to restate Lemma 4.6 in the language of symbols and chord diagrams.

4.9. LEMMA. *The value of the symbol of a Vassiliev invariant of order n on a knot with n double points depends only on the Gauss diagram of the knot.* □

4.10. Let $\{e_1, \dots, e_k\}$ be the set of all nonequivalent chord diagrams of order n. Since the symbol depends only on Gauss diagrams and not on knotting (see Lemma 4.9), *any symbol of order n is entirely determined by its value on the diagrams* $\{e_1, \dots, e_k\}$. Hence to any element v_n of the space V_n/V_{n-1} we can assign the set of numbers $(v_n(e_1), \dots, v_n(e_k)) \in \mathbb{R}^k$, which determines a linear function in the space $\mathbb{R}^k$ with basis $\{e_1, \dots, e_k\}$. As the result, we get an injective map $V_n/V_{n-1} \to (\mathbb{R}^k)^*$, which is obviously linear. It is also easy to see that it is not surjective. For example, any Vassiliev invariant vanishes on the knot  corresponding to the diagram 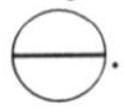. Hence no nontrivial element of the space V_1/V_0 corresponds to any linear function that does not vanish on the element 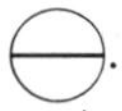.

In the language of chord diagrams, the one-term relation and the generalized one-term relation may be written as

(4.5)  $= 0,$  $= 0.$

These formulas mean that the symbol of any finite order invariant vanishes on any chord diagram possessing a chord that does not intersect any other chord.

As to the four-term relation, in this language it is expressed as the following diagrammatic formula:

(4.6)  $-$  $+$  $-$  $= 0.$

This formula means that the alternating sum of symbols of four chord diagrams that have $n-2$ identical chords (not shown in the picture), and have two more chords as pictured, vanishes provided the missing $n-2$ chords do not have endpoints in the forbidden segments of the circle (displayed by thickening the line). To prove this statement, we take the previous version of the four-term relation (4.3), complete the curve in the dotted circles ouside these circles (this can be done in two essentially different ways), and read off relation (4.6) from the resulting pictures (Fig.4.6). $\square$

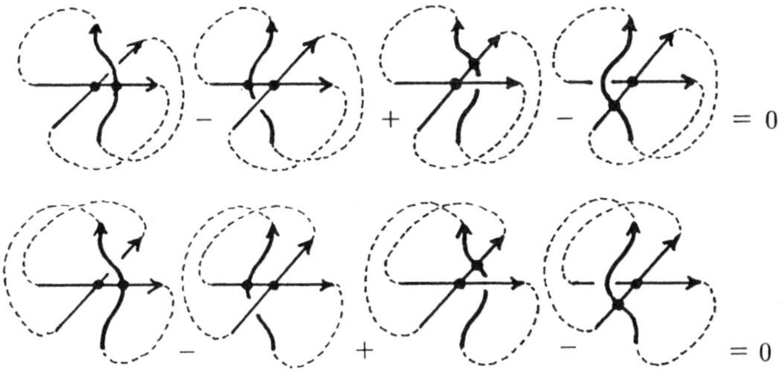

FIGURE 4.6. Two ways to read the four-term relation

4.11. Let U be a linear space with basis $e_1, \dots, e_k$, denote by U^* its dual, the space of linear functions on U, and let F be a subset of U^* such that any $f \in F$ satisfies the relations

$$\sum \lambda_{1i} f(e_i) = 0\,, \;\dots,\; \sum \lambda_{pi} f(e_i) = 0,$$

where λ_{ij} are fixed real numbers. Consider the subspace $U_1 \subset U$ generated by the vectors

$$w_1 = \sum \lambda_{1i} e_i\,, \;\dots,\; w_p = \sum \lambda_{pi} e_i.$$

Any element $f \in F$ vanishes on the space U_1; hence F can be identified with some subset in the space $(U/U_1)^*$.

Let us return to the space $\mathbb{R}^k$ whose basis $e_1, \dots, e_k$ is constituted by all the chord diagrams of order n. In this basis let us express all the equations that follow from the one-term relation (4.5) and the four-term relation (4.6), writing them in the form

$$v(e_p) = 0, \dots, v(e_s) = 0,$$
$$v(e_a) - v(e_b) + v(e_c) - v(e_d) = 0, \dots,$$
$$v(e_\alpha) - v(e_\beta) + v(e_\gamma) - v(e_\delta) = 0.$$

Then the symbol of an invariant of order n may be identified with a certain linear function on the space $\Delta_n = \mathbb{R}^k/U_1$, where the space U_1 is generated by the vectors

$$e_p, \dots, e_s, e_a - e_b + e_c - e_d, \dots, e_\alpha - e_\beta + e_\gamma - e_\delta.$$

The space Δ_n may be presented as the space generated by the vectors $e_1, \dots, e_k$ satisfying the equations

$$e_p = 0, \dots, e_s = 0, e_a + e_c = e_b + e_d, \dots, e_\alpha + e_\gamma = e_\beta + e_\delta.$$

We have proved that *there exists a natural monomorphism of the space V_n/V_{n-1} into Δ_n^*.*

This statement, however, may not assert anything at all: we have not established that nonzero finite order Vassiliev invariants actually exist. This is done in [**Vas1**], [**Car**], [**BN**]; in fact, we have the following stronger statement (see [**Kon1-2**]).

4.12. KONTSEVICH THEOREM. *Vassiliev invariants of order n exist, forming a linear space V_n such that V_n/V_{n-1} is isomorphic the space Δ_n^* of linear functions on n-chord diagrams modulo the one-term and the four-term relations.*

The rather difficult proof of this theorem is omitted (see [**BN**]).

4.13. In order to compute the values of Vassiliev invariants on specific knots, we must choose a basis in the space Δ_n of chord diagrams and express values of the chosen invariant on the basis elements. As n increases, the complexity of this problem grows very rapidly. Already for $n = 9$ this requires several days of work for a computer (work station). We shall limit ourselves to some specific examples in the cases $n = 2, 3$, and 4.

The space Δ_2 is generated by the diagram , the other 2-chord diagram  being trivial by the one-term relation.

The space Δ_3 possesses two nontrivial diagrams,  and . But these diagrams are not linearly independent, because the four-term relation (4.6) in this case has the form

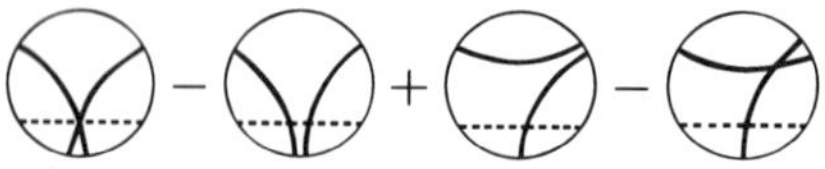 $= 0,$

and since the third diagram is trivial by the one-term relation, we get

(4.7) $v_3($$) = 2v_3($$)$

for any $v_3 \in V_3$.

Problem 4.4. (a) Prove the following relations:

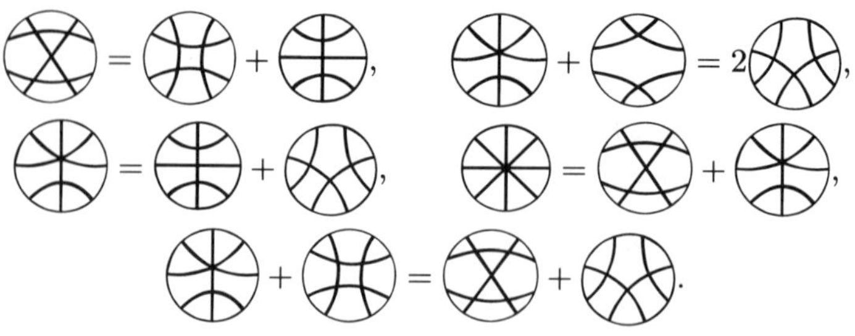

(b) Prove that $\dim \Delta_4 = 3$, and a basis of this space is constituted by

4.14. Now we can carry out some concrete computations of the values of Vassiliev invariants for the simplest knots. We have seen in 4.5 that invariants of orders 0 and 1 are constants, so we begin with $n = 2$. The symbols of order 2 invariants constitute a one-dimensional space (see the previous subsection). To fix a basis invariant $v_2 \in V_2$, we can put

(4.8) $v_2($  $) = 1, \quad v_2(\bigcirc) = 0.$

The first condition determines the symbol of v_2, while the second condition may be ensured by subtracting (if necessary) an invariant of order 0 (namely, if $v_2(\bigcirc) = r \neq 0$, we must subtract the order zero invariant identically equal to r on nonsingular knots).

We start with the trefoil. For brevity, instead of $v(k) = v(k_1) + v(k_2)$, we shall write $k = k_1 + k_2$, and omit v in analogous cases. Since any crossing change in the trefoil makes it the unknot, we have the relations shown in Fig.4.7.

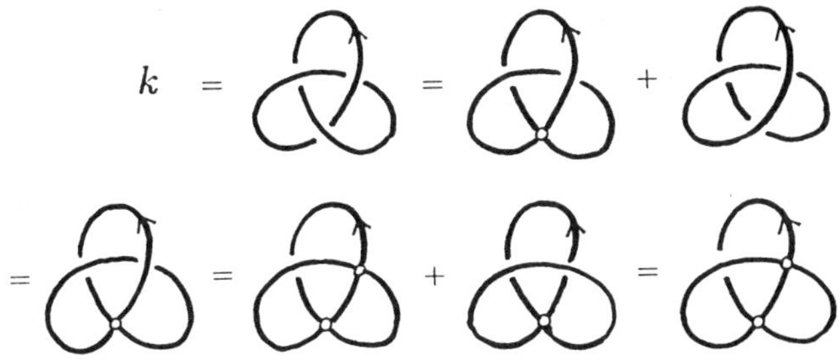

FIGURE 4.7. Calculating v_2 of the right trefoil

A similar calculation for the left trefoil yields the result shown in Fig.4.8. Since the knot on the right in that figure also has  for its Gauss diagram, the invariant v_2 has the same values on the left and right trefoils.

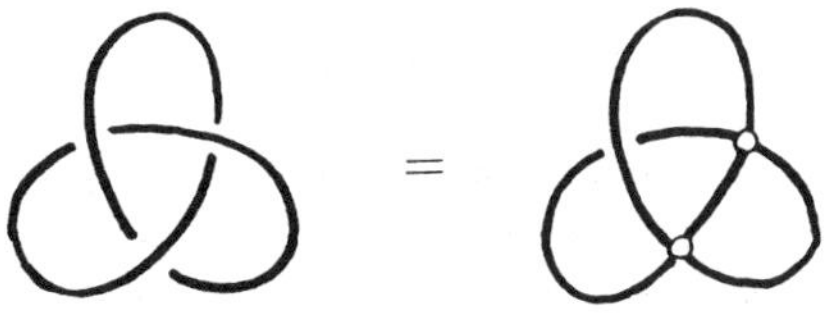

FIGURE 4.8. The invariant v_2 of the left trefoil

However, order three Vassiliev invariants do distinguish the left and right trefoils. Indeed, the calculation shown in Fig.4.9 yields a singular knot with Gauss diagram . Hence if we put $v_3($  $) = 1$, then the values of v_3 on the trefoils differ by 1.

FIGURE 4.9. The invariant v_3 distinguishes the two trefoils

Now consider the figure eight knot (Fig.4.10). It has two types of crossings, marked + and − in the figure. When we perform a crossing change at the points marked with a +, the corresponding diagram must be added, and subtracted otherwise (for the points marked −). This follows from the main defining relation for Vassiliev invariants (4.1). In this way we obtain the equalities shown in Fig.4.10. Thus the value of v_2 on the figure eight knot equals −1.

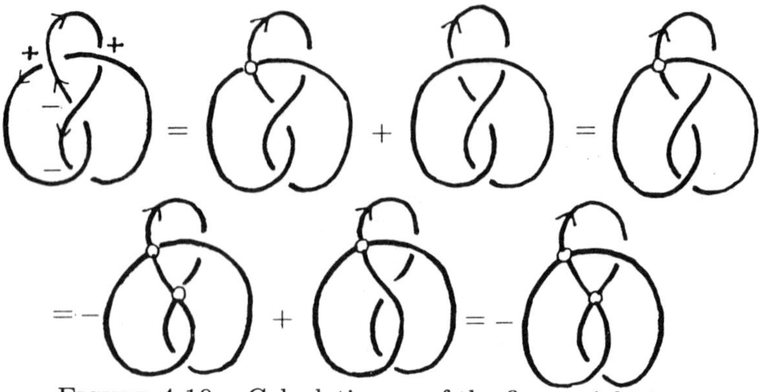

FIGURE 4.10. Calculating v_2 of the figure eight knot

Finally let us calculate the value of v_2 for the (2,5) torus knot (the third-from-last knot in Fig.4.11).

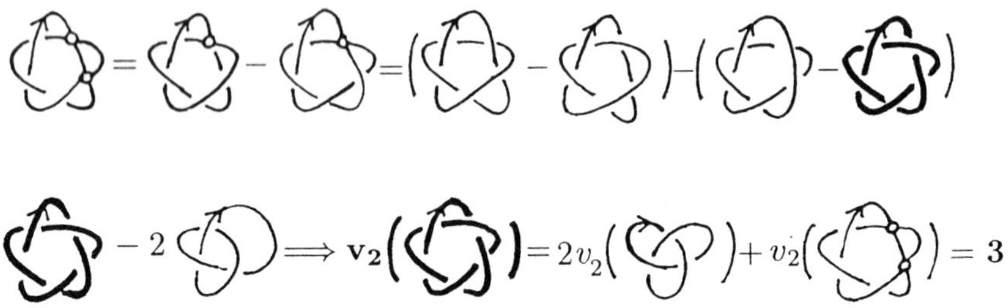

FIGURE 4.11. Calculating v_2 of the (2,5) torus knot

It is convenient to begin the computation by taking the knot with two double points (the first one on the figure), and then successively eliminate the double points via relation (4.1). This yields our torus knot, the unknot, and two trefoils. The value of v_2 for the unknot is 0 by the normalization condition (4.8). According to Lemma 4.9, the value of v_2 on a singular knot with two selfintersections is entirely determined by its Gauss diagram. The Gauss diagram of the first knot in Fig.4.11 is ; hence the value of v_2 on this knot is 1.

As the result, we see that the value of v_2 on the torus knot (2,5) is 3.

Comments

In our exposition of the knot (link) polynomials, they appear in reverse chronological order. This is because each later invention is more elementary than the previous one, but can be used to reconstruct the earlier version. In historical order, the Alexander polynomial came first ([**Ale2**], 1923). It was originally constructed, roughly speaking, as the determinant of the monodromy operator in the one-dimensional homology of the abelian covering of the complement to the given knot. For a detailed and very clear exposition of the Alexander polynomial, the reader is refered to the book [**CF**]. Although J. Alexander (and many other knot theorists) knew and used the skein relation for his polynomial, it was only much later that J. Conway noticed that this relation could be used as the main ingredient of an axiomatic approach to polynomial invariants, and invented his own version of the Alexander polynomial ([**Con**], 1970; it is significant that this extremely important article was not accessible in the Soviet Union and became known to Russian topologists only in the late eighties, after [**Jon1**] appeared). After that, and after the appearance of the Jones polynomial, other polynomials based on different skein relations started to proliferate.

The most natural one, the two-variable polynomial with a "homogeneous" skein relation (see 3.12), was simultaneously discovered by about a dozen mathematicians. In the West, six of them (P. Freyd, J. Hoste, W. B. R. Lickorish, K. Millett, A. Ocneanu, D. Yetter) got together and published jointly ([**FH...**]), while the two Polish experts (J. Pzrytycki and P. Traczyk) published later ([**PT**]). After the first publication, the two-variable polynomial became known as the HOMFLY polynomial in honor of the first six authors. D. Bar–Natan ([**BN**]) has suggested calling it the LYMPHTOFU polynomial, the U standing for the unknown discoverers.

The discovery of the Jones polynomial was rather surprising and dramatic. Vaughan Jones, a mathematician from New Zealand working on the representation of von Neumann algebras, had been using the so-called Ocneanu trace for this purpose. After one of his talks, Joan Birman, the leading expert in braid theory, pointed out the striking formal similarity between the properties of this trace and of the Markov moves (which relate braids to links; we discuss them in the next chapter). After that, and possibly a struggle with a fairly tricky calculation (see the coefficients of the Jones two-variable polynomial in 3.12), Vaughan Jones was able to produce the original construction of his polynomial ([**Jon1**]). We have not included this construction in our book for two reasons. First, it is not really elementary (it involves, among other things, Hecke algebras, the Ocneanu trace, and the Markov braid theorem). Secondly, the detailed exposition of this construction later published by Jones himself is so clear and self-contained that we felt it more expedient to refer the interested reader to this article directly ([**Jon2**]).

Vaughan Jones' work was immediately noticed, in particular by M. Atiyah (at whose seminar V. Jones gave a talk), by "N. Bourbaki", at whose seminar the first reports were given by A. Connes and P. Cartier, by O. Viro and V. Turaev in Leningrad, and by one of the authors of this book in Moscow. The excitement among the mathematical community was not due so much to the purely knot-theoretical aspect of this work, but rather because of the deep connections with operator algebras (V. Jones' original paper), statistical physics (V. Jones [**Jon3**], L. Kauffman [**Kau2**], V. Turaev [**Tur1**]), Hopf algebras (V. Drinfeld [**Dri1-2**]), and quantum field theories (E. Witten [**Wit2**], M. Atiyah [**Ati1-3**]). Predictably, Vaughan Jones was awarded the Fields Medal in 1990; so were Drinfeld and Witten. A mathematical version of the Witten invariants (due to several authors, including V. Turaev, N. Reshetikhin, O. Viro, R. Kirby, P. Melvin, and W. B. R. Lickorish, see [**RT1-2**], [**TV**], [**Lic4**]) appears in Chapter IX of this book.

In this chapter, we use Louis Kauffman's beautiful elementary approach, inspired by statistical physics, to define the Jones polynomial.

In contrast to the Jones polynomial, it took quite a while before the Vassiliev invariants (sometimes called finite type invariants) became known by and large to the mathematical community, although his original work was published in English with complete proofs immediately after it was done ([**Vas1**],1990), and it was obvious from the outset that his invariants were stronger than the Jones one-variable polynomial and could be generalized to other objects. The general construction of the Vassiliev spectral sequence, which computes the values of the invariants of knots and other objects, soon appeared in book form [**Vas2**]. (M. Gusarov, independently of Vassiliev, found a different approach to finite-type invariants, but his preprint on this subject was published only much later.) A remarkable relationship between the Vassiliev and Jones invariants was soon discovered by J. Birman and X.-S. Lin, their preprint was widely circulated long before their joint article appeared ([**BL**]), and J. Birman's talks helped make this topic more popular. Other work instrumental in popularizing Vassiliev invariants were the preprint [**Sos**] and V. I. Arnold's closing plenary report at the First Mathematical Eurocongress. More important were the remarkable papers by M. Kontsevich ([**Kon1-2**]), where an extremely unusual integral explicitly defining the Vassiliev invariants appeared, and by D. Bar–Natan ([**BN**]), who presented a proof of the Kontsevich Theorem (4.12 above) and described a totally different approach to the invariants based on the representations of simple Lie algebras. Later P. Cartier found a "purely combinatorial" proof of the existence of Vassiliev invariants ([**Car**]), while S. Chmutov, S. Duzhin, and S. Lando ([**CDL**]) discovered some deep relationships between these invariants and the Tutte polynomials of suitable graphs (e.g. the well-known dichromatic polynomial).

As of this writing, it is still unknown whether Vassiliev invariants classify knots (the Vassiliev Conjecture), but their importance, we believe, does not depend on how well they actually perform this classification. It is due to the generality of what may be called the Arnold-Vassiliev approach to defining invariants (of many different objects). Besides knots, these now include links, braids, tangles, plane immersions of disjoint circles, certain higher dimensional links, plane immersed curves, embedded graphs, and even nilpotent groups. This list will surely increase by the time this book is published.

CHAPTER III

Braids

Braids are very beautiful and useful objects. We have already considered braids on an intuitive level in §2, and noted that braids (unlike knots and links) form a group. Originally invented by E. Artin as a mathematical model to be used in the textile industry, their applications are numerous both within mathematics (complex polynomials, knots and links, representation of functions in n variables by composition of functions in less than n variables) and in physics (classical mechanics, statistical physics, quantum field theory). In a certain sense (that can be made precise), it can be said that braids are the quantum analog of permutations. Besides, Artin's braid relation (see (5.1) below) turns out to be simply another way of writing the Yang-Baxter equation (a fundamental relation in two-dimensional statistical physics, which also describes the behavior of particles in one-dimensional quantum theory). This analogy is one of the major sources of the deep relationship between physics and knot theory.

It was the close-knit relationship between braids and links (which we shall discuss in §6 below) and the analogies with physics that stimulated E. Witten and his followers to generalize the Jones polynomial to links in arbitrary closed oriented three-dimensional manifolds. But we have a long way to go before we can attack this topic directly (Chapter IX).

We begin by looking at the braid group from the point of view of geometry, algebra, complex polynomials, topology, and classical mechanics.

§5. The braid group

5.1. Geometric definition. In the space $\mathbb{R}^3$, consider the points $A_i = (i, 0, 0)$ and $B_i = (i, 0, 1)$, where $i = 1, 2, \ldots, n$. A polygonal line joining one of the points A_i with one of the points B_j will be called *ascending* if in the motion of a point from A_i to B_j along the line its z-coordinate increases monotonically. In other words, each horizontal plane cuts an ascending line at exactly one point.

A *braid* in n strands is defined as a set of pairwise nonintersecting ascending polygonal lines (the *strands*) joining the points $A_1, \ldots, A_n$ to the points $B_1, \ldots, B_n$ (in any order). Examples of braids are shown in Fig.5.1.

The equivalence of braids is defined similarly to that of (polygonal) knots or links (Fig.5.2,a) with the additional requirement that the line ABC must be ascending, i.e., the transformation shown in Fig.5.2,b is forbidden.

We shall use the word 'braid' indiscriminatively to mean an equivalence class of braids or a concrete representative of such a class; the reader will have no trouble in deciding which (in accordance with the context).

One can also consider braids whose strands are ascending smooth lines (rather than polygonal ones); then it is natural to define equivalence as isotopy, i.e., as a

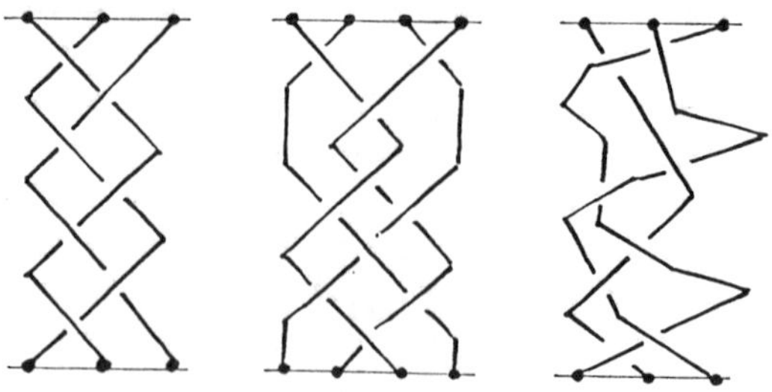

FIGURE 5.1. Examples of braids

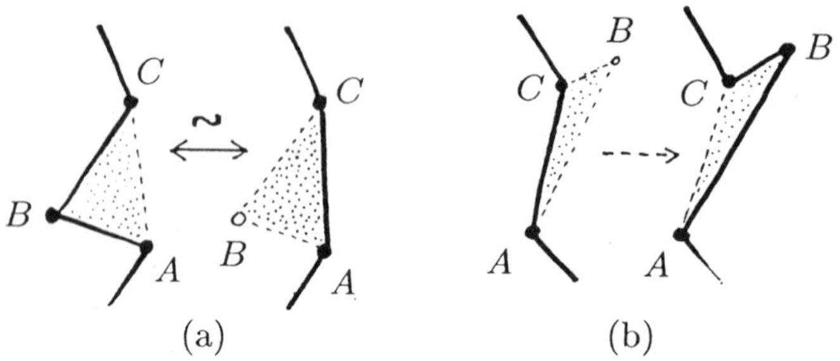

FIGURE 5.2. Equivalence of braids

smooth deformation in the class of braids. As in the case of links, this yields the same theory as the polygonal approach.

5.2. Group structure. The set of (equivalence classes of) braids in n strands has a natural group structure.

The product of two braids a and b is obtained by putting them end to end. More precisely, let us contract the braid a in the vertical direction to half its height, leaving its upper points in place, and contract b in the same way but with the lower points left in place; then the union of the contracted braids is by definition the *product* ab (Fig.5.3,a). It is clear from Fig.5.3,b that this operation is associative on the set of equivalence classes of braids, i.e., $(ab)c = a(bc)$ for any three classes a, b, c.

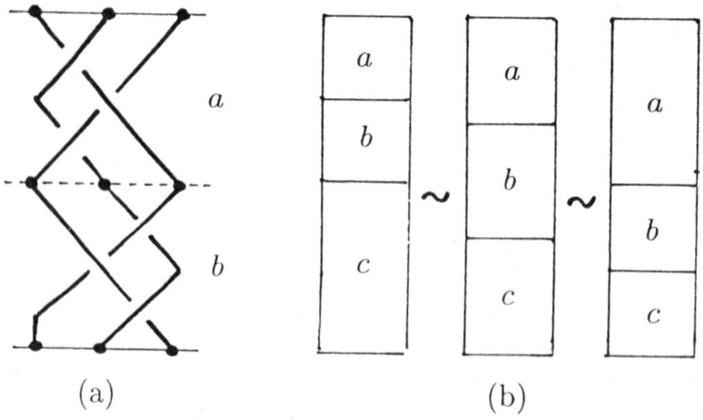

FIGURE 5.3. Product of braids

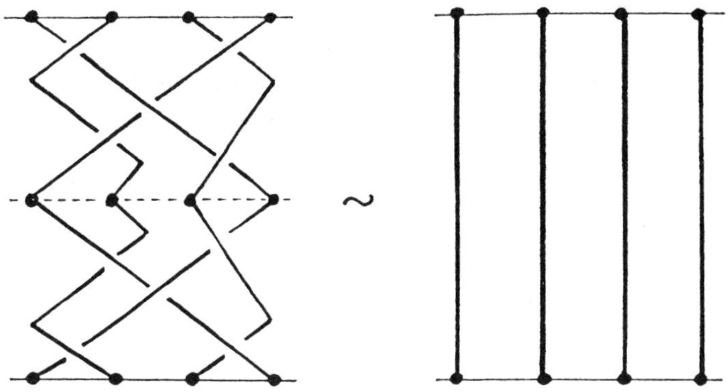

FIGURE 5.4. Inverse element in the braid group

The unit element is the braid consisting of n parallel vertical strands. It is easy to see (Fig.5.4) that the inverse braid b^{-1} to a given braid b is the mirror image of b in the plane $z = 1/2$ (recall the "braid trick" in §2).

The set of braids in n strands under this operation is called the *braid group* and is denoted by B_n.

5.3. Generators of the braid group. The braid group B_n may be presented by generators and defining relations between the generators. To do this, note that in the equivalence class of any braid there is a representative possessing a projection on the xz-plane with the following properties:

1. the projections of the strands are not tangent to each other;
2. no point of the xz-plane is the projection of three or more points from different strands;
3. all the crossings occur at different altitudes (above the xy-plane).

Then this representative will look like the braid in Fig.5.5,a. Hence any braid can be presented as the product of the elements $b_i^{\pm 1}$, where b_i is the braid shown in Fig.5.5,b. Thus the elements $b_1, \dots, b_{n-1}$ generate the group B_n.

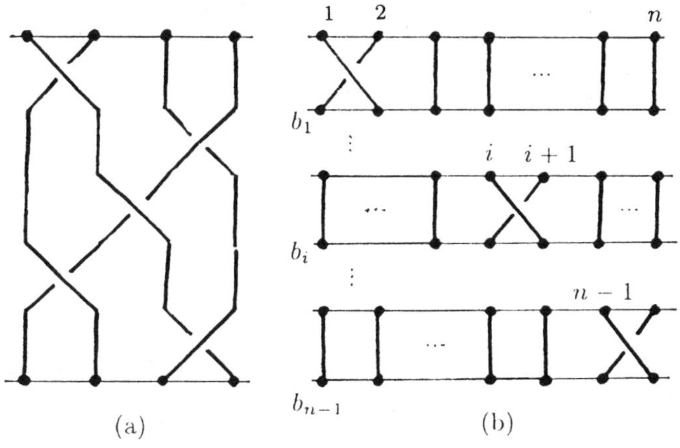

FIGURE 5.5. Braid generators

5.4. The Artin relations. In order to find the defining relations, we must consider the transformations of braids for which the properties (1)–(3) break down. Property (1) breaks down for the transformation shown in Fig.5.6. This transformation yields the relation $b_i b_i^{-1} = 1$, which holds trivially in any group.

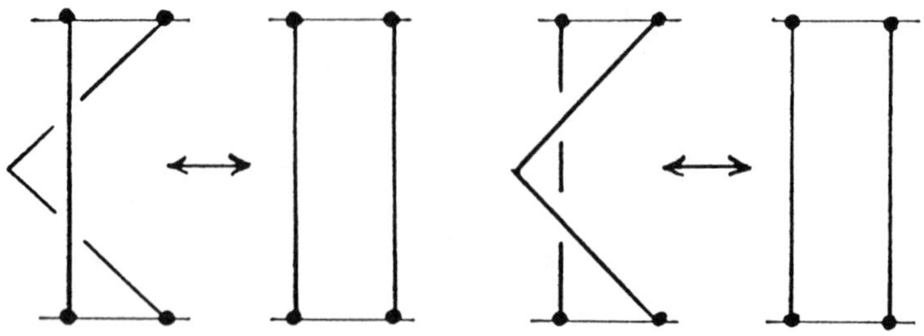

FIGURE 5.6. Braid transformations yielding the trivial relation

Property (2) breaks down for several different transformations. Three of them are shown in Fig.5.7. The first transformation yields the equation

$$b_i b_{i+1} b_i = b_{i+1} b_i b_{i+1}, \tag{5.1}$$

known as the *braid relation*. The second and third transformations give

$$b_i b_{i+1} b_i^{-1} = b_{i+1}^{-1} b_i b_{i+1}, \quad b_i^{-1} b_{i+1} b_i = b_{i+1} b_i b_{i+1}^{-1},$$

but these equations are equivalent to the braid relation (5.1). The other transformations related to the appearance of triple points can be obtained from the ones shown in Fig.5.7 by symmetry in the plane of the diagram. These transformations yield equations that also reduce to (5.1).

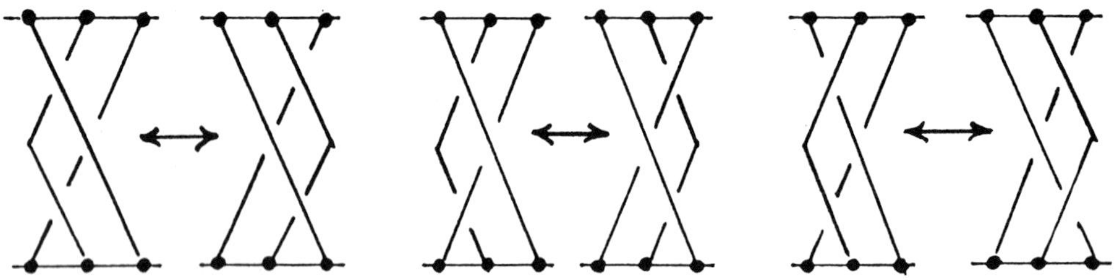

FIGURE 5.7. Transformations yielding the braid relation

Property (3) breaks down under the transformation shown in Fig.5.8. This transformation yields the relation

$$b_i b_j = b_j b_i \qquad \text{whenever} \quad |i-j| \geqslant 2, \tag{5.2}$$

sometimes called *far commutativity*, because it says that the generators commute pairwise when they are sufficiently far from each other, i.e., when their indices differ by 2 or more.

The relations that we have derived are in fact sufficient to characterize the braid group B_n algebraically.

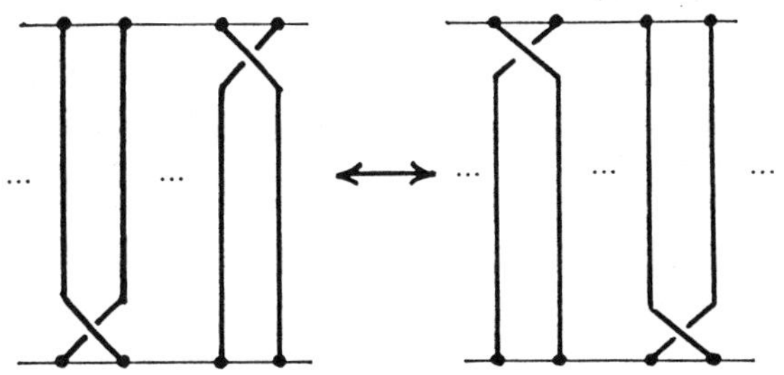

FIGURE 5.8. Transformation yielding far commutativity

5.5. ARTIN'S THEOREM. *The braid group B_n is isomorphic to the abstract group generated by the letters $b_1, \ldots, b_{n-1}$ that satisfies the braid relation* (5.1) *and far commutativity* (5.2).

Proof. We have already proved that braids are generated by elements $b_1, \ldots,$ b_{n-1} satisfying relations (5.1) and (5.2). It remains to show that any relation between braids follows from these two relations (and the trivial group relations $bb^{-1} = b^{-1}b = 1$, $b1 = 1b = b$). Geometrically, this means that if some braid is equivalent to the trivial braid, then this equivalence can be established by geometric operations corresponding to the relations indicated in the statement of the theorem.

Recalling the definition of the equivalence of braids, we see that it suffices to prove that any elementary braid equivalence in B_n, say $[AB] \mapsto [AC] \cup [CB]$, can be performed by using the relations of B_n and the trivial relations. The proof of this fact is similar to that of the Reidemeister Theorem 1.7. We begin by subdividing triangle $[ABC]$ into small triangles of types I–IV (see the begining of the proof in 1.7 and Fig.1.13) and replace our elementary move by a sequence of elementary moves on the small triangles, starting with the side $[AB]$ and working towards $[AC] \cup [CB]$.

The rest of the proof is quite similar to that of the Reidemeister theorem except for two points. The first is that we do not have to bother with "entering edges" (recall Fig.1.12), because such edges cannot occur in a braid (strands are ascending). The second is that we must take care of forbidden projections of type (3) (crossing points at the same level), which did not interest us in 1.7. Such a projection can occur when a move on a small triangle yields a plane isotopy changing the level (i.e., the z-coordinate) of a crossing point; but this is exactly the application of far commutativity (5.2). (As to the two other forbidden projections, (1) corresponds to Ω_2 and hence to one of the trivial relations, while (2) corresponds to Ω_3 and hence to a relation equivalent to the braid relation (5.1).)

Thus we see that the moves through the small triangles correspond to the relations listed above; the theorem is proved. □

Let us have a look at the groups B_n for small n. Obviously, B_1 is the trivial group, B_2 is isomorphic to $\mathbb{Z}$, while B_3 has two generators x and y and only one relation $xyx = yxy$.

Problem 5.1. Prove that a) the group B_3 is isomorphic to the group with generators a, b and relation $a^2 = b^3$; b) the element $(xy)^3$ lies in the center of B_3.

5.6. Mechanical interpretation. Consider the configuration space $C(n, \mathbb{R}^2)$ of the mechanical system of n identical distinct points in the plane. The space

$C(n, \mathbb{R}^2)$ consists of (unordered) sets of n points in $\mathbb{R}^2$ and is endowed with the natural topology. As the base point of this space, take $\omega := \{(1,0),(2,0),\ldots,(n,0)\}$. It is easy to verify that (homotopy classes of) loops based at ω correspond bijectively to (isotopy classes of) braids in n strands. Indeed, suppose that at the moment of time t_0, $0 \leqslant t_0 \leqslant 1$, the loop passes through the element $x(t_0) \in C(n, \mathbb{R}^2)$. The element $x(t_0)$ is the plane with n distinct points marked on it. Let us place this plane in 3-space by adding the third coordinate $z := t_0$ (Fig.5.9). Doing this for all $t_0 \in [0,1]$, we obtain a braid in n strands.

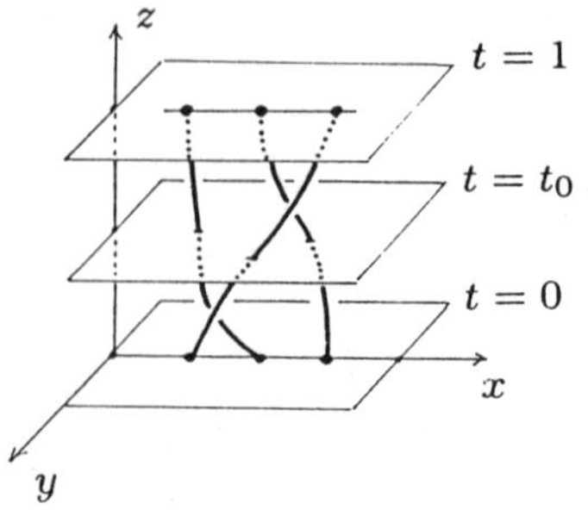

FIGURE 5.9. Braid corresponding to a loop in $C(n, \mathbb{R}^2)$

Clearly, isotopy of braids corresponds to the homotopy of loops, and the product of braids corresponds to the composition of loops. Thus we have proved that

$$B_n = \pi_1(C(n, \mathbb{R}^2)),$$

i.e., *the fundamental group of the configuration space of n identical particles in the plane is isomorphic to the braid group B_n.*

The mechanical interpretation presented above also has a purely geometric version, as the following problem shows.

Problem 5.2. In complex space $\mathbb{C}^n$ consider all points with pairwise distinct coordinates and identify all points that differ only by a permutation of their coordinates. Prove that the space thus obtained is homeomorphic to $C(n, \mathbb{R}^2)$.

5.7. Complex polynomial interpretation. Denote by $P(n, \mathbb{C})$ the topological space (in the natural topology) of degree n polynomials in one complex variable with leading coefficient 1 and without multiple roots.

Problem 5.3. Prove that $\pi_1(P(n, \mathbb{C})) \cong B_n$.

Problem 5.4. Suppose that S^3 is the 3-sphere, $K \subset S^3$ is the trefoil knot, and $\Sigma_3 \subset \mathbb{C}^3$ is the space of complex polynomials of degree 3 with leading coefficient 1 which have multiple roots (the 3 other coefficients determine the embedding into $\mathbb{C}^3$). Prove that a) $\mathbb{C}^3 - \Sigma_3$ is homotopy equivalent to $S^3 - K$; b) B_3 is isomorphic to $\pi_1(S^3 - K)$.

5.8. The word problem for braids. The *word problem* for a group presented by a system of generators and relations, in particular for the braid group, is to find an algorithm that determines, for any pair of words in the generators of the group, whether or not they present the same element (i.e., whether or not one can be taken to the other by applying the group's relations and the trivial relations). The word problem (which of course is always solvable for commutative groups) is solvable for the braid group B_n for any n. This is a theorem due to E. Artin (see [**Art2**]). However, Artin's algorithm, aptly called *combing*, is rather complicated and slow. Instead, we shall describe a more recent and more effective algorithm.

5.9. The Duhornoy algorithm. This algorithm, like Artin's combing algorithm, solves the word problem for braids. It can easily be implemented on a computer (even a weak PC). In order to describe how it works, we need a definition and some notation.

Let B_n, $n \geqslant 2$, be the braid group presented in the generators $b_1, \dots, b_{n-1}$. A word $w \in B_n$ is said to be *reduced* if, for any integer i, any occurence of the letter b_i is separated from any occurence of the letter b_i^{-1} by at least one occurence of a letter $b_j^{\pm 1}$ with $j < i$. For example $b_2 b_1 b_2^{-1}$ is reduced, while $b_2 b_3 b_2^{-1}$ is not.

Let us show how *any element $b \in B_n$ can be presented by a reduced word.* Let w be any word presenting b. If w is not reduced, then it is of the form

$$w = w_0 b_i^e w_1 b_i^{-e} w_2,$$

where $e = \pm 1$, the word w_0 is reduced, and w_1 contains only occurences of b_j and b_j^{-1} with $j > i$. Let us replace w by the word $R(w)$,

$$R(w) = w_0 b_{i+1}^{-e} b_{i+2}^{-e} \dots b_{j-1}^{-e} f(w_1) b_{j-1}^{e} \dots b_{i+2}^{e} b_{i+1}^{e} w_2,$$

where j is the least integer such that neither b_{j+s} nor b_{j+s}^{-1}, $s \geqslant 0$, occurs in w_1, and f is the transformation that takes $b_k^{\pm 1}$ to $b_{k-1}^{\pm 1}$ for $i < k < j$ and does not change the other generators. (If $j = i + 1$, then b_j^{-e} and b_j^{e} cancel and for f we take the identity.) Geometrically, this means that if w is as shown in Fig.5.10,a, then $R(w)$ is the braid shown in Fig.5.10,b. Let us call the operation $w \mapsto R(w)$ an *elementary reduction*. Clearly, w is isotopic to $R(w)$ and $R(w)$ no longer contains the forbidden combination $b_i^e w_1 b_i^{-e}$. Iterating this procedure, we obtain a reduced word equivalent to w.

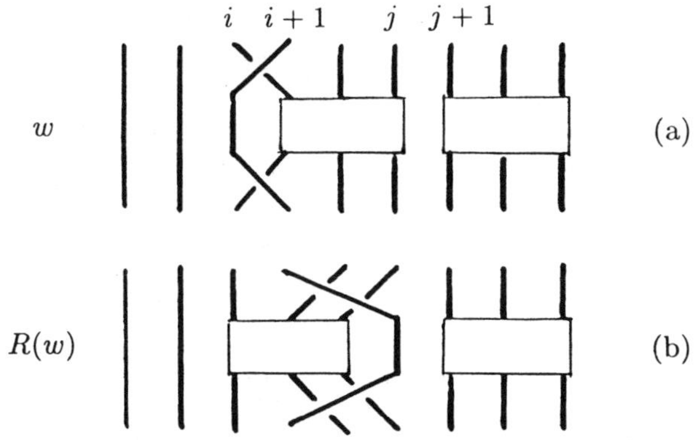

FIGURE 5.10. Elementary reduction

For example, if

$$w = b_1^{-1} b_2 b_1 b_3^{-1} b_1^{-1} b_3^{-1} b_1,$$

the corresponding sequence of elementary reductions (in which the forbidden combination to be replaced is shown in parentheses) will be

$$\begin{aligned}
w &= (b_1^{-1}b_2b_1)b_3^{-1}b_1^{-1}b_3^{-1}b_1,\\
R(w) &= b_2(b_1b_2^{-1}b_3^{-1}b_1^{-1})b_3^{-1}b_1,\\
R^2(w) &= (b_2b_2^{-1})b_3^{-1}b_1^{-1}b_2^{-1}b_3b_2b_3^{-1}b_1,\\
R^3(w) &= b_3^{-1}b_1^{-1}(b_2^{-1}b_3b_2)b_3^{-1}b_1,\\
R^4(w) &= b_3^{-1}(b_1^{-1}b_3b_2b_3^{-1}b_3^{-1}b_1),\\
R^5(w) &= b_3^{-1}b_2b_3b_2b_1b_2^{-1}b_2^{-1}b_3^{-1}b_2^{-1}.
\end{aligned}$$

Now we can describe the Duhornoy algorithm. Given two words $w_1, w_2 \in B_n$, apply elementary reductions to each of them until two reduced words w_1', w_2' are obtained; compare the two words w_1', w_2'; if they coincide (letter by letter), then the given words present the same element of the group B_n; if they do not coincide, they present different elements.

The validity of this algorithm is based on the following statements, due to P. Duhornoy:

1. *the reduction algorithm always terminates;*
2. *two words present the same element of the group B_n if and only if they have the same reduced form.*

We omit the rather complicated proof of these statements (see [**Duh**]). The algorithm itself, however, is quite simple (very easy to implement on a computer) and very fast: it is (conjecturally) quadratic in the length of the given words.

§6. The Alexander and Markov theorems

In this section we study the relationship between braids and links arising from the closure operation, which assigns a link to each braid in a natural way. We shall learn that this assignment is surjective (Alexander's theorem), but not injective (as simple examples show), and we shall algebraically characterize its "degree of noninjectivity" (Markov's theorem).

6.1. The *closure* of a braid b is defined as the link $\beta(b)$ obtained by joining the upper points of its strands to the lower ones as shown in Fig.6.1. The first question that arises is: for what braids does closure produce a knot, as in Fig.6.2,a, rather than a multicomponent link, as in Fig.6.2,b? To answer that question, we first establish an important relationship between the braid group and the permutation group.

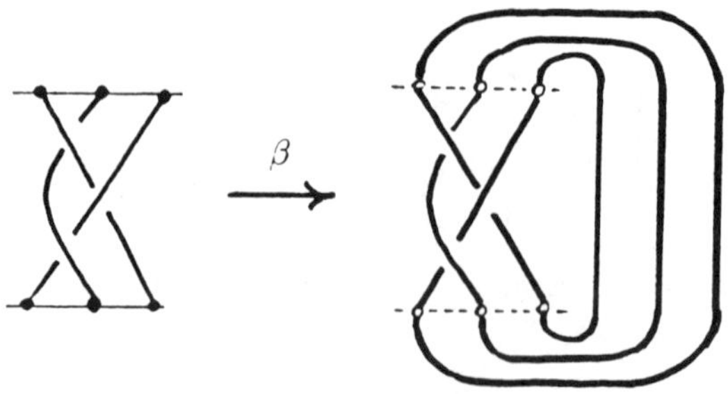

FIGURE 6.1. Closure of a braid

6.2. There is a *canonical epimorphism* $\sigma : B_n \to S_n$ of the braid group onto the permutation group. Geometrically, $\sigma(b)$ is the permutation of the endpoints of the braid induced by its strands, e.g. in Fig.6.2,c we have $\sigma(b) = (2,3,1)$. In terms of generators, σ is determined by the assignment $b_i \mapsto s_i$, where s_i denotes the transposition of the ith and the $(i+1)$st element. In terms of relations, the group S_n is obtained from B_n by adding the relation $b_i^2 = 1$. The fact that this assignment is a homomorphism and is onto follows immediately from the definitions.

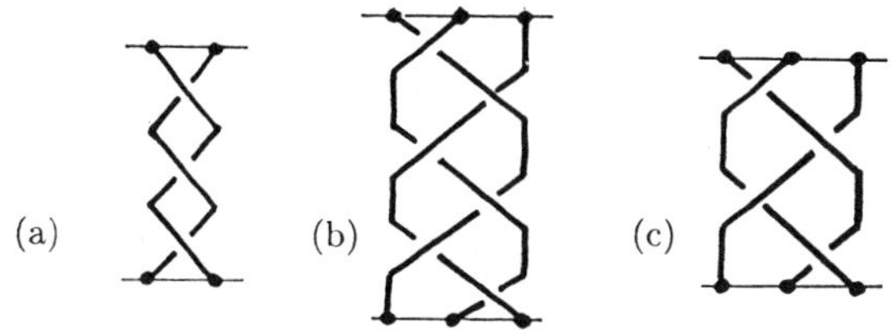

FIGURE 6.2. Three braids (whose closures are to be found)

6.3. Now we are ready to answer the question posed in 6.1: *the closure $\beta(b)$ of the braid b is a knot iff the permutation $\sigma(b)$ associated to the braid generates the cyclic subgroup of order n, $\mathbb{Z}/n\mathbb{Z}$, in the permutation group S_n.* The proof is obvious: we get a connected curve iff we visit all the endpoints of the braid, and this occurs iff the permutation is cyclic of order n.

6.4. Another natural question with a simple (positive) answer is whether the closure of different braids may yield the same link.

Problem 6.1. (a) Prove that the closures of the two braids $b_1, b_1^{-1} \in B_2$ coincide, as well as those of $b_1^2, b_1^{-2} \in B_2$.

(b) Prove that the closures of the two braids $b_1^3, b_1^{-3} \in B_3$ are distinct, as well as those of $b_1 b_2 b_1^{-1}$, $b_2^3 b_1^{-2}$.

Problem 6.2. What knots and links are obtained by taking the closure of the braids shown in Fig.6.2?

Now when we know that the closure map is not injective, the next natural question that we shall answer is whether it is surjective.

6.5. ALEXANDER'S BRAIDING THEOREM. *The closure map is surjective, i.e., any link (in particular, any knot) is the closure of some braid.*

Proof. Let a link diagram L be given. We shall say that the link L *winds around the point O* if every edge of L (we think of L as a polygonal object) is seen from O as *positive*, i.e., oriented from right to left (as in Fig.6.3,a).

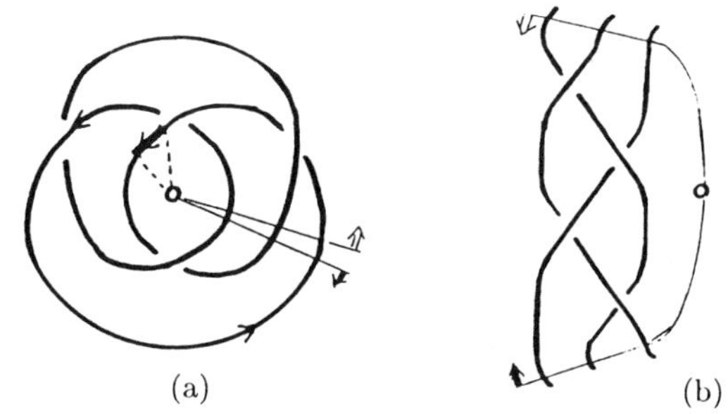

FIGURE 6.3. Braiding a knot that winds around a point

If the given link possesses a point around which it winds, the proof is immediate: cut L along a ray issuing from O and unwind it into a braid as shown in Fig.6.3,b.

Now to prove the theorem, it remains to show that any link is isotopic to a braided one. This can be achieved by repeated applications of the *Alexander trick*, which consists in replacing any negative edge of L (with respect to an arbitrarily chosen point O) by the two other sides (which will be positive!) of a triangle ABC whose new vertex C is just "behind" O (see Fig.6.4,a), until no negative edges remain, so that the new link winds around the point O.

Let us describe this construction in more detail. In the simplest case, when the negative edge AB has no crossing points, the application of the Alexander trick is straightforward (Fig.6.4,a). Indeed, moving the point C upward perpendicularly to the plane ABO, we can ensure that there will be no points of L inside the triangle ABC. By moving C upward along this perpendicular, we can ensure this same condition for the case in which the negative edge has exactly one crossing point and this point is an overpass of the edge (Fig.6.4,b). Or, if the negative edge still has exactly one double point, but this point is an underpass, then we can move C downward along the perpendicular until triangle ABC is freed from points of L (as in Fig.6.4,c). Finally, if the negative edge has several crossing points, then we subdivide the edge into several smaller edges, each containing exactly one crossing point, and apply the constructions used in the previous cases. □

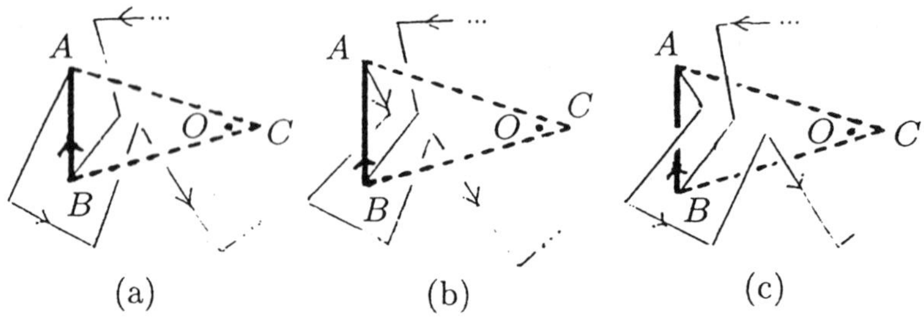

FIGURE 6.4. The Alexander trick

Note that the algorithm implicit in the proof of Alexander's theorem is not very economical: it may happen that a very large number of applications of the trick in Fig.6.4 will have to be performed to force the link to wind around the point O (especially if O is chosen inappropriately).

6.6. Vogel braiding algorithm. The proof of the Alexander theorem presented above is in principle based on an effective construction. However, the construction is not easy to implement on a computer. Neither is the proof presented in the book [**BZ**].

Here we present a simple braiding algorithm (due to the French mathematician P. Vogel) which is easy to program. (We omit the proof of its convergence.)

We shall say that the diagram of an oriented knot or link is *braided* if it possesses a point around which it winds. We have seen that from a braided link a braid whose closure is the given link can be read off immediately (recall Fig.6.3 above).

For example, the trefoil is braided, but the usual diagram of the figure eight knot is not. However, the eight has a braided diagram, obtained by taking an appropriate arc of the diagram "around infinity" (recall the sphere trick in §2). This operation

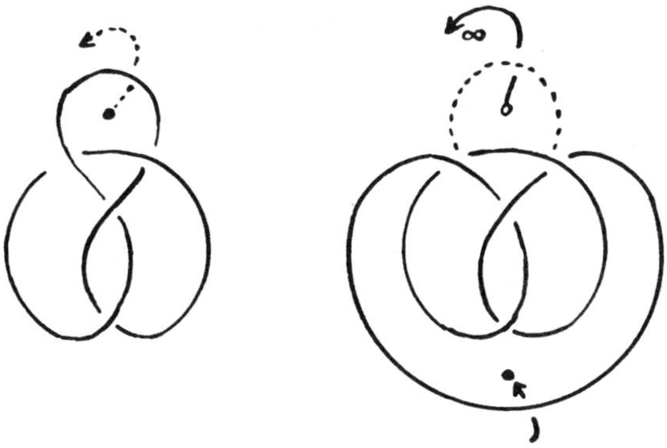

FIGURE 6.5. Change of infinity

(Fig.6.5), which we call *change of infinity*, can in fact be performed on the plane (not only on the 2-sphere) by means of the second and third Reidemeister moves.

On an oriented knot there is a well-defined operation of *destroying a crossing*, shown in Fig.6.6,a. If we destroy all crossings, our knot diagram falls apart into (oriented) closed curves called *Seifert circles* (Fig.6.6,d,e).

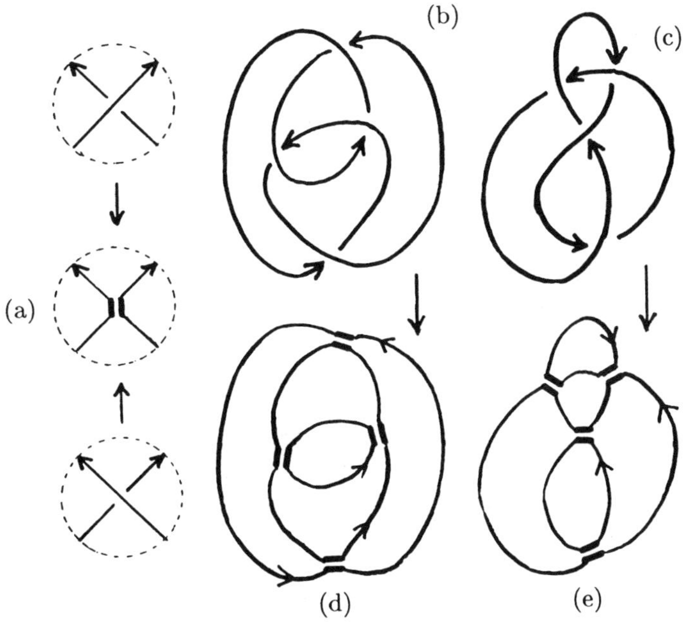

FIGURE 6.6. Seifert circles

Problem 6.3. Using Seifert circles, prove that any knot is the boundary of an orientable surface in $\mathbb{R}^3$.

The set of Seifert circles of a knot diagram is called *nested* if they induce the same positive orientation of the plane and bound an increasing system of disks (Fig.6.6,d). A knot diagram with a nested system of Seifert circles is obviously braided (look at Fig.6.6,b again).

Now consider the *shadow* of an oriented knot diagram, i.e., simply the knot projection (without overpass-underpass indications). The shadow may be regarded as a planar oriented graph, with vertices the crossing points, edges the oriented arcs joining them, and faces the regions bounded by the arcs. A region (face) will be called *troubled* if it has two *opposite* edges, i.e., edges not belonging to the same Seifert circle and having orientations that induce the same orientation of the region (Fig.6.7,a). To the opposite edges of any troubled region one may apply the second Reidemeister move Ω_2 (Fig.6.7,b). The result will be a rearrangement of regions: a new central region (which is not troubled) appears, as do several other regions (constituted by parts of the former troubled region and its neighbors).

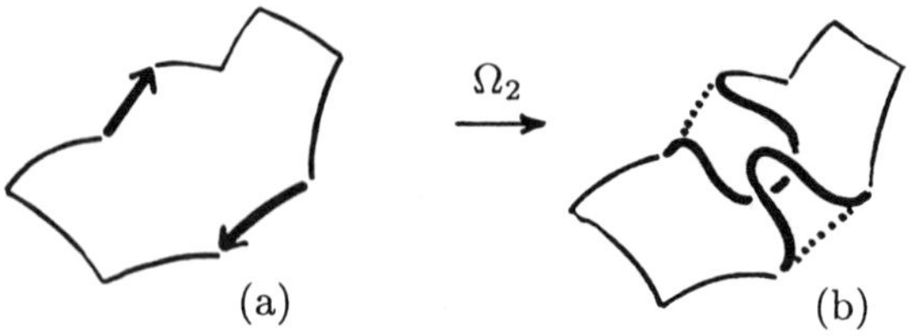

FIGURE 6.7. Applying Ω_2 to a troubled region

We now present the Vogel algorithm in the form of a "program" written in an informal Pascal-like language (which the reader should easily understand):

```
DO: DESTROY ALL CROSSINGS;
WHILE THERE EXISTS A TROUBLED REGION
   DO: Ω2;
   DO: DESTROY ALL CROSSINGS;
ENDWHILE
IF ALL SEIFERT CIRCLES ARE NESTED
   STOP;
ELSE
   DO: CHANGE INFINITY;
STOP
```

The conditions THERE EXISTS A TROUBLED REGION and ALL THE SEIFERT CIRCLES ARE NESTED were explained above, as well as the commands DESTROY ALL CROSSINGS, CHANGE INFINITY, and DO Ω_2 (applied to a troubled region). They are not difficult to program in a real programming language, provided we have a reasonable way of coding knot diagrams.

In Fig.6.8 we show what the program does when applied to the knot 5_2 from the knot tables (see §3).

Separately, in Fig.6.9, we present a more symmetric picture of the braided knot thus obtained, and of the corresponding braid.

In the general case it is not obvious that the Vogel algorithm will always stop, because it contains a potentially infinite loop WHILE...ENDWHILE. (In the example this loop was performed twice, after which the Seifert circles were not nested, so a change of infinity was performed.)

Problem 6.4. Prove that the Vogel algorithm always terminates. More precisely, show that (a) it terminates after not more than w operations Ω_2, where

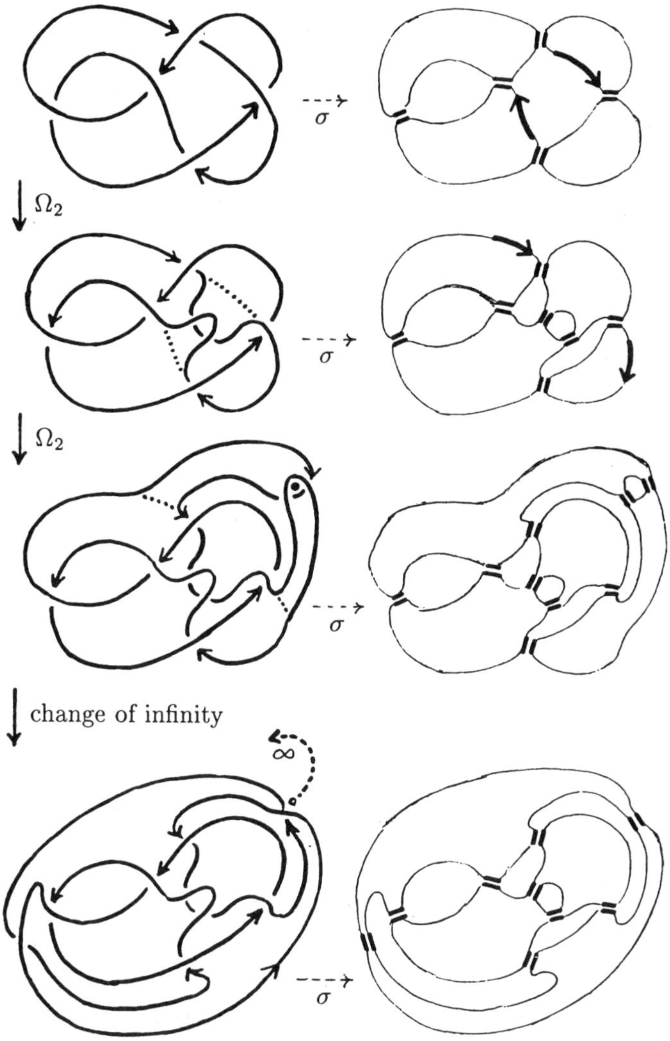

FIGURE 6.8. Applying the Vogel algorithm to the knot 5_2

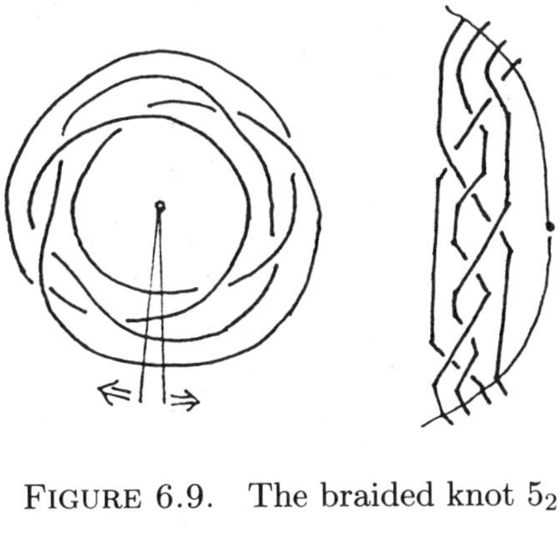

FIGURE 6.9. The braided knot 5_2

$w = (s-1)(s-2)/2$ and s is the number of Seifert circles for the given knot diagram; and (b) the resulting braid has s strands and is of length (i.e., number of generators in its algebraic expression) no greater than $n + (s-1)(s-2)$, where n is the number of crossing points.

6.7. Markov moves. Our next goal is to discuss in more detail how it comes about that different braids can have isotopic closures. We shall need the following algebraic transformations of braids presented in the standard generators b_i:

First Markov move. $b \leftrightarrow aba^{-1}$, where $a, b \in B_n$.

Second Markov move. $b \leftrightarrow bb_n^{\pm 1}$, where $b \in B_n$ and $bb_n^{\pm 1}$ is the element shown in Fig.6.10 (note that $b_n \notin B_n$, so that the notation bb_n makes sense algebraically only if we identify b with its image under the natural inclusion $B_n \hookrightarrow B_{n+1}$).

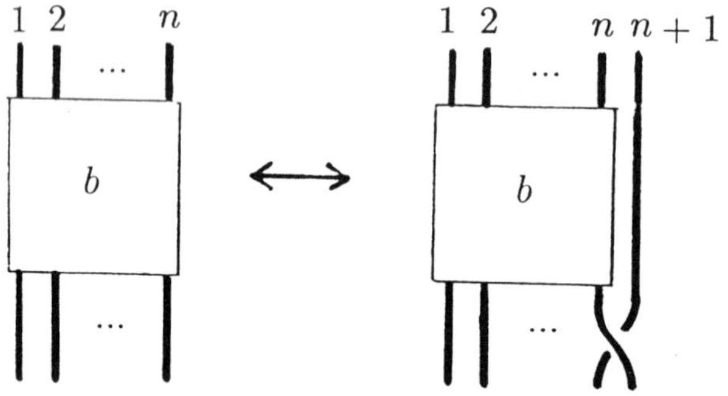

FIGURE 6.10. The element bb_n

6.8. MARKOV'S THEOREM. *The closures of two braids are isotopic if and only if one braid can be taken to another by a finite sequence of Markov moves.*

Proof. We shall not prove the rather difficult "only if" part of Markov's theorem (see [**Bir1**] or [**Mor2**]). To prove the easy "if" part, it suffices to check that the closure is not changed by the first Markov move, nor by the second one. For the first move, this is clear from Fig.6.11,a. For the second move, the proof is even more obvious: after we perform the closure operation, we get an extra loop that can be easily removed (e.g. by a Reidemeister Ω_1 move, see Fig.6.11,b). □

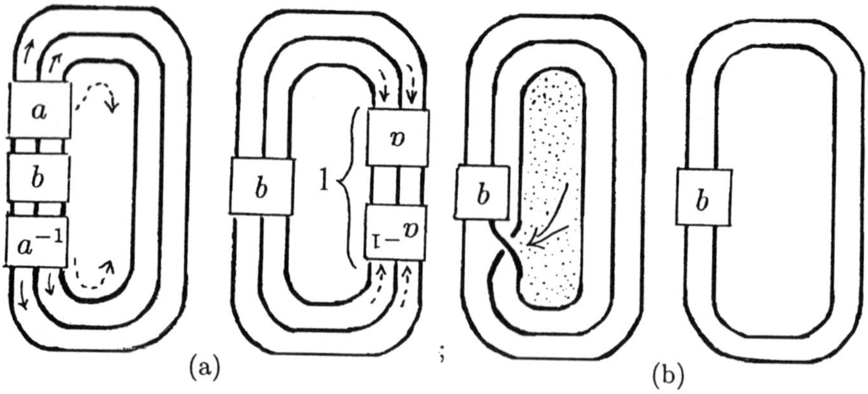

FIGURE 6.11. The closure is not affected by Markov moves

§7. Pure braids

7.1. Let the strand of a braid b join the point $A_i = (i, 0, 0)$ to the point $B_j = (j, 0, 1)$; the braid b is called *pure* (or sometimes *colored*) if each of its strands joins points with the same number, i.e., $i = j$ for all i. Then if we paint each strand in its own color, the upper and lower row of points on the braid will be painted in the same way. Two pure braids are considered *equivalent* if they are equivalent as ordinary braids. The product of two pure braids is a pure braid (because the strands of both can be colored consistently). Hence equivalence classes of pure braids constitute a group, called the *pure braid group* and denoted by K_n.

To a braid in n strands corresponds (see 5.6 above) a loop in the space of nonordered n-tuples of distinct points in the plane. Similarily, to a pure braid corresponds a loop in the space of *ordered* n-tuples in the plane, the isotopy of braids corresponding to the homotopy of loops. Hence $K_n = \pi_1(C - \Delta)$, where

$$\Delta = \bigcup_{i \neq j} \Delta_{ij}, \quad \Delta_{ij} = \{(z_1, \dots, z_n) \in \mathbb{C}^n | z_i = z_j\}.$$

7.2. THEOREM. *The pure braid group K_n possesses a finite family of generators, namely the braids b_{ij}, where $1 \leq i < j \leq n$* (Fig.7.1).

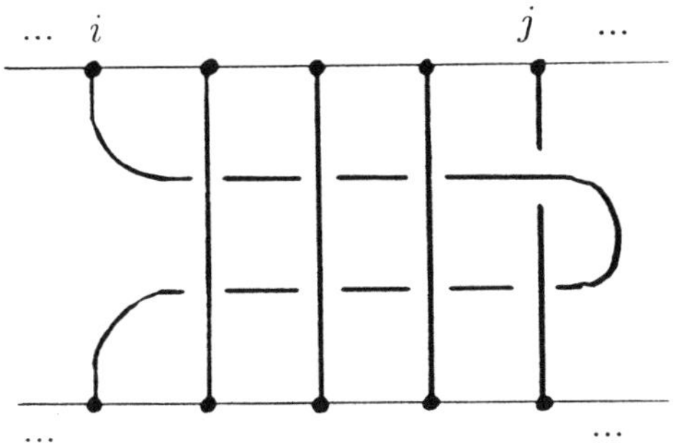

FIGURE 7.1. Generator of the pure braid group

Proof (by induction on n). For $n = 2$ it suffices to straighten out the second strand; then the first one will wind around the second and as the result we shall get the braid b_{12}^k, $k \in \mathbb{Z}$. Assume that the assertion of the theorem has been proved for pure braids in n strands, i.e., any braid $a_n \in K_n$ is representable in the form of a product of b_{ij}'s to the power ± 1. Now consider a pure braid $d_{n+1} \in K_{n+1}$. If we delete its first strand, we get a braid $a_n \in K_n$. To the braid a_n^{-1} add a vertical line (the new first strand) that does not link the other strands, obtaining the braid d'_{n+1}. There is an isotopy that takes the last n strands of the braid $c_{n+1} = d'_{n+1} d_{n+1}$ into parallel vertical segments, with the first strand winding around them. We claim that c_{n+1} can be presented as the product of the braids b_{ij}. Indeed, all we have to do is pull the first strand all the way to the left and then back again under all the strands each time it goes over one of the other strands (Fig.7.2).

By the inductive assumption, the braid a_n^{-1} (and hence d'_{n+1}) may be presented in the required form. Therefore so can the braid $d_{n+1} = (d'_{n+1})^{-1} c_{n+1}$. □

7.3. Denote by H_n the group of (isotopy classes of) homeomorphisms of the disk with n holes, where the homeomorphisms are identical on the boundary. It turns out that the pure braid group K_n is closely related to the group H_n. Before establishing this relationship in the general case, we consider the case $n = 0$.

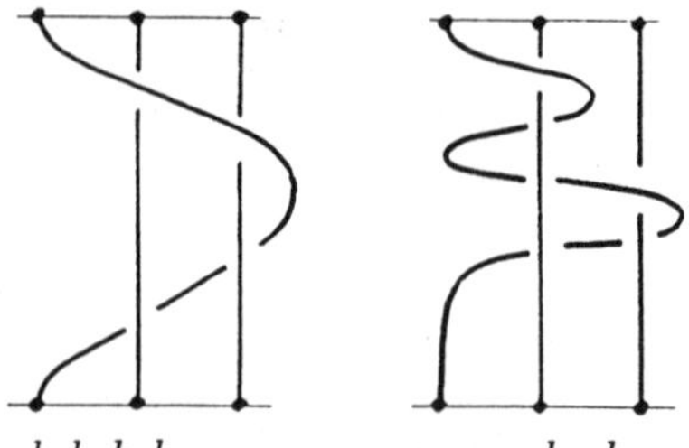

FIGURE 7.2. Presenting a pure braid in the generators b_{ij}

7.4. ALEXANDER'S HOMEOMORPHISM THEOREM. *The group H_0 is trivial, i.e., any homeomorphism of the disk identical on the boundary is isotopic to the identity.*

Proof. An isotopy h_t, $0 \leqslant t \leqslant 1$, that joins the homeomorphism $h = h_1$ with the identity $\mathrm{id} = h_0$ may be described as follows. The homeomorphism h_t is the identity on the annulus $t \leqslant |z| \leqslant 1$, and on the disk $|z| \leqslant t$ it is similar to the homeomorphism h on the disk $|z| \leqslant 1$ (Fig.7.3.). If we assume that $h(z) = z$ when $|z| \geqslant 1$, then for any $t \neq 0$, it suffices to set $h_t(z) = t(h(t^{-1}z))$. □

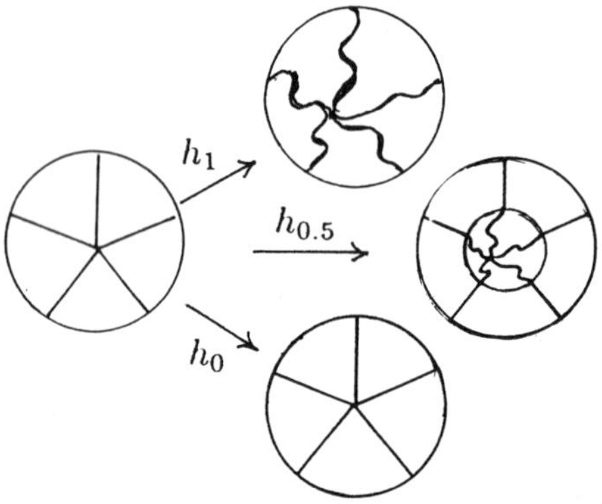

FIGURE 7.3. Isotopy joining h to the identity

7.5. The previous theorem may be used to compute the group H_n. Suppose $g \in H_n$ is a homeomorphism of the disk with n holes onto itself, identical on the boundary. The homeomorphism g may be extended to an auto-homeomorphism h of the entire disk, i.e., g can be extended from the boundary of each hole to the hole itself. This can be done simply by extending g by the identity (since g is the identity on the boundary by assumption). By Theorem 7.4, there exists an isotopy h_t joining $h = h_1$ to the identity $\mathrm{id} = h_0$. Suppose x_0 is a fixed point of the disk. The set of points

$$(t, h_t(x_0)), \qquad 0 \leqslant t \leqslant 1, \tag{7.1}$$

is an arc joining points of the upper and lower base of the cylinder $I \times D^2$, where I is the closed interval $[0, 1]$ and D^2 is the disk. If we consider such arcs for points $x_1, \dots, x_n$ belonging to the holes, we obtain a pure braid (Fig.7.4).

This braid encodes essential information about the homeomorphism h, but not enough to describe it completely up to isotopy. To obtain complete information, we must consider the arcs (7.1) for all the points of the boundary of each hole. For each

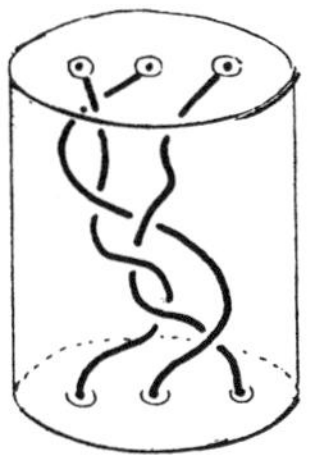

FIGURE 7.4. Pure braid corresponding to the isotopy h_t

hole, these arcs generate a cylindrical surface. To describe the homeomorphism, it suffices to indicate these cylindrical surfaces for all the holes and draw one of the arcs on each. The resulting picture will be called a *thickened braid.* Thickened braids form a group under the natural product operation (putting them end-to-end).

Now suppose γ is a closed curve in our disk with holes. Let us define a special type of homeomorphism related to γ. Suppose C is an ε-neighborhood of γ homeomorphic to $S^1 \times [0,1]$. Let γ_1 and γ_2 be the boundary components of C (Fig.7.5). Let us fix the curve γ_1 and start rotating the curve γ_2 along itself; here we think of the annulus C as being elastic, so that points very close to γ_2 follow the motion of points on γ_2, while points close to γ_1 barely move from their initial positions. When γ_2 performs a rotation by $360°$ along itself, its points return to their initial positions. Thus we obtain a homeomorphism of the disk with holes, identical on the boundary. Schematically, it is reprsented in Fig.7.5. We call this homeomorphism a *twist along the curve* γ.

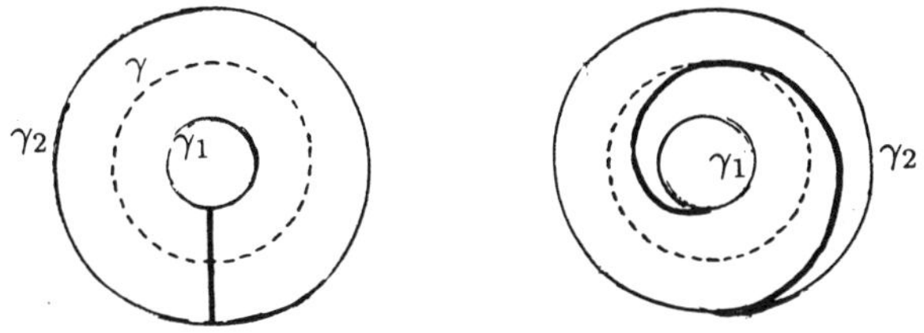

FIGURE 7.5. Twist along the curve γ

As an illustration, consider homeomorphisms of the disk with two holes. Twists along the curves shown by dotted lines in Fig.7.6,a,b correspond to the thickened braids appearing in Fig.7.6,c,d, respectively.

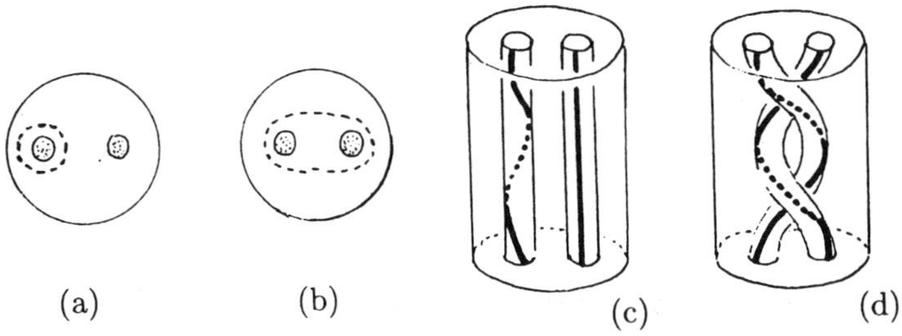

FIGURE 7.6. Twists and corresponding thickened braids

We have explained how to obtain a thickened braid from a homeomorphism. Conversely, let us show how to get a homeomorphism from a thickened braid. Without loss of generality, we can assume that our braid does not leave the cylinder over the disk. The upper base of the cylinder may be regarded as an elastic disk with n round holes, the outer boundary of which, as well as the boundaries of the holes, being rigid hoops. Let us think of each distinguished arc on the cylindrical surfaces as a narrow and shallow gully, and imagine that each of the hoops (except that of the outer boundary) has a short spike sticking out into the cavity of the corresponding gully (Fig.7.7). The hoops are attached to the elastic surface of the upper base, and in their motion pull nearby points along as the surface stretches.

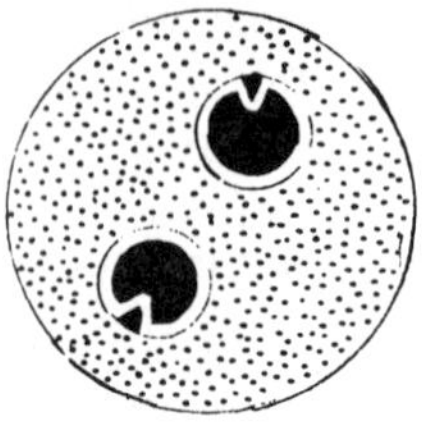

FIGURE 7.7. Hoops with spikes on the boundaries of holes

Let us begin to lower our construction parallel to itself so that points of the hoop on the outer boundary of our disk move vertically down parallel to the generatrix of the cylinder, while the inner hoops move downward in a winding motion (following the meanders of the cylindrical surfaces) and rotate (following the spikes sliding along the gullies twisting around these surfaces). When the disk reaches the lower base, we will have obtained the required homeomorphism.

The group of thickened braids is an extension of the group of pure braids K_n. More precisely, each thickened braid is characterized by a pure braid with an integer (indicating the number of twists) written on each strand. To each strand corresponds a complementary element of infinite order, contained in the center of the group. Therefore *the group of thickened braids is isomorphic to the direct sum* $K_n \oplus \mathbb{Z}^n$, where $\mathbb{Z}^n$ denotes the direct sum of n summands isomorphic to $\mathbb{Z}$.

We are now ready to prove a statement that will play a crucial role in §13, where we study the homeomorphism group of surfaces.

7.6. THEOREM. *The group H_n of homeomorphisms of the disk with n holes (up to isotopy) is generated by a finite number of twists along closed curves in this disk.*

Proof. The group H_n is isomorphic to the group of (isotopy classes of) thickened braids. Suppose that to the homeomorphism $h \in H_n$ corresponds (as explaned above in 7.5) the braid $\alpha'_n = \alpha_n + a$, where $\alpha_n \in K_n$ and $a \in \mathbb{Z}^n$. By Theorem 7.2, the pure braid α_n may be represented as the product of the generators b_{ij}. It is easy to see that the braid b_{ij} corresponds to the twist along the curve shown in Fig.7.8,a.

Taking the composition of these twists, we get a homeomorphism corresponding to the thickened braid $a_n + b$ for some value of $b \in \mathbb{Z}^n$. To obtain a instead of b, it suffices to perform the appropriate twists along the curves shown in Fig.7.8,b. □

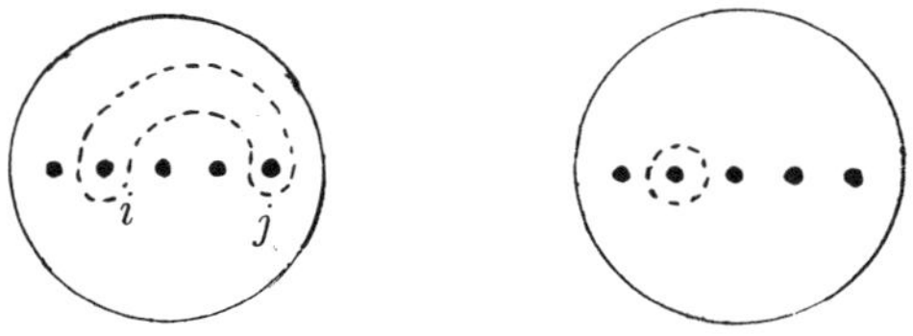

(a) (b)

FIGURE 7.8. Twists corresponding to b_{ij} and $(0, k, 0, 0, 0) \in \mathbb{Z}^5$

Comments

As we mentioned in the main text, the braid group is an invention of the German mathematician Emil Artin, who found its presentation and (positively) solved the word problem for it (the first exposition, in German, [**Art1**], 1925, was not flawless, but Artin wrote the definite text in English after he moved to the US, [**Art2**], 1947). The simple proof of Artin's theorem appearing here, however, is new (as far as we know). Many mathematicians have worked on braids since Artin, both from the algebraic viewpoint and the algorithmic one. Most of this work does not concern us here, so let us only mention the (positive) solution by F. A. Garside and G. S. Makanin of the conjugation problem ([**Gar**], [**Mak1**]) and refer the reader to the monograph by Joan Birman ([**Bir1**], 1974), where many other interesting results appear. Our own exposition of the solution of the word problem is based on recent work by Patrick Duhornoy (see [**Duh**]), made available to us in oral and preprint form by its author.

We don't know who invented the closure operation for braids (it was certainly known to J. W. Alexander). This operation establishes a fundamental connection between braids and links (in particular knots), a connection enhanced by the beautiful theorems of J. W. Alexander ([**Ale1**], 1923) and A. A. Markov ([**Mar1**], 1936, see also [**Mar2**]). The story of Markov's theorem is rather dramatic: although the author gave oral expositions of his proof, he never published it, possibly because of his change of interest from topology to "constructive mathematical logic", leaving this task to one of his pupils, N. Weinberg. The latter, however, was killed during the war, soon after his first contribution to the subject was published ([**Wei**], 1939). The first published proof of the Markov theorem is due to Joan Birman (see [**Bir1**]). A shorter proof was later published by H. R. Morton ([**Mor2**]).

Of course, from the algorithmic viewpoint, it is natural to try to apply the classification theorem of braids to the classification of knots via the Markov theorem. This tempting approach has not proved successful (many people have tried).

There are also several proofs of Alexander's braiding theorem (Theorem 6.5, above). The one we present, according to hearsay, is the one the author had in mind in [**Ale1**]; another simple geometric proof appears in [**BZ**], while the braiding algorithm described in this chapter is due to P. Vogel ([**Vog**]). The material on pure braids is traditional, and was apparently known to Artin, while the connection of pure braids with the homeomorphisms of disks with holes (Theorem 7.6 above), like so many unexpected and beautiful ideas in geometric topology, is due to J. W. Alexander. This last theorem is the major connection between this chapter and the subsequent chapters about 3-manifolds.

The theory of braids has numerous applications. Not all of its problems (in particular algorithmic ones) are solved. For a fairly recent account (including unsolved problems) the reader can consult the collection cited in the reference [**Mor1**].

CHAPTER IV

3-Manifolds

This chapter, as well as Chapters VI and VII below, has to do with various presentations of three-dimensional manifolds.

The importance of the topic is due to at least three circumstances. First of all, traditionally, we think of our world as being a three-dimensional space, and even the consideration of time as a fourth dimension by physicists does not make the study of 3-space less significant. Secondly, we are capable of visualizing our 3-space only locally, so that formal means of presenting the global structure of various "possible worlds" are needed. Finally, there is no practically usable classification theorem for 3-manifolds, and this open-ended subject is presently in a stage of rapid development.

We assume that the reader is familiar with the classification theorem for 2-manifolds (if not, the Appendix gives a brief description). We should also note that a famous theorem due to A. A. Markov [**Mar3**] asserts that the classification problem for 4-manifolds is algorithmically unsolvable. According to G. Hemion, there is an algorithm that classifies 3-manifolds, but its description requires a whole book [**Hem**], and we shall not discuss it here. Thus our main goal is to describe simple universal constructions used to present all (compact orientable) 3-manifolds. These constructions are given here in the historical order of their appearance. They are: Heegaard diagrams (this chapter), surgery presentations (Chapter VI), and branched covering presentations (Chapter VII). As for Chapter V, it is devoted to Dehn twists, which in a certain sense bridge the gap between homeomorphisms of two-dimensional surfaces and three-dimensional surgery.

In the following chapters (as in most of this book), we work in the context of the topological category. This is justified in dimensions less than or equal to three. In particular, the piecewise linear and the differentiable classifications of 3-manifolds are both equivalent to the topological one (see [**Moi1-2**], [**Mun**]), so that we shall not need either differentiable or PL structures in 3-manifolds. There is one exception: in order to construct a Heegaard decomposition in a 3-manifold, we must use the latter's triangulation. Readers who do not wish to believe that all 3-manifolds have such a triangulation can simply assume that we only deal with triangulable manifolds. (This is actually no loss in generality, by [**Moi1-2**], [**Mun**].)

§8. Heegaard splittings

8.1. By a 3-*manifold* we mean a compact connected Hausdorff space M^3 each point of which has a neighborhood homeomorphic to Euclidean space $\mathbb{R}^3$. A 3-*manifold with boundary* is defined similarly, except that besides neighborhoods homeomorphic to Euclidean space, neighborhoods homeomorphic to Euclidean half-space $\mathbb{R}^3_+ = \{(x, y, z) \in \mathbb{R}^3 | z \geqslant 0\}$ are also allowed. The set of points that have only neighborhoods of the second type is called the *boundary* of M^3 and is denoted by ∂M^3. It is easy to prove that ∂M^3 is a 2-manifold. A 3-manifold is said to be *triangulated*, if it is presented as the finite union of tetrahedra whose pairwise intersection is either a common face, or a common edge, or a common vertex, or is void. A triangulated manifold is said to be *oriented* if its tetrahedra are oriented (this means that a frame called "positive" is chosen in each) coherently (this means that the two orientations induced on each face from the two adjacent tetrahedra are opposite), and *orientable* if it can be oriented.

8.2. A 3-manifold can be obtained by means of the following construction. Consider the sphere with g handles N^2 standardly embedded in $\mathbb{R}^3$ (Fig.8.1). By a *handlebody* with g handles, we mean the compact subset of $\mathbb{R}^3$ bounded by N^2. Let us take two copies M_1^3 and M_2^3 of this handlebody. Let $f : \partial M_1^3 \to \partial M_2^3$ be an arbitrary homeomorphism of their boundaries. Attach the two copies of our handlebody to each other by this homeomorphism, i.e., in the disjoint union of M_1^3 and M_2^3 identify each point $x \in \partial M_1^3$ with the point $f(x) \in \partial M_2^3$.

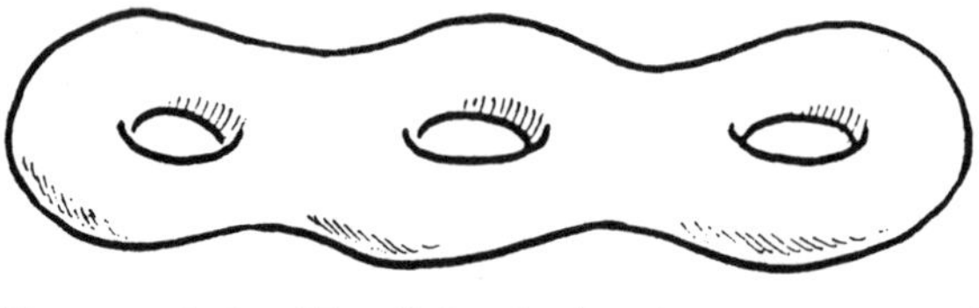

FIGURE 8.1. Handlebody (with three handles)

The space M^3 thus obtained is a 3-manifold. Indeed, a neighborhood of the point $x = f(x) \in M^3$ may be obtained by gluing together two half balls, the neighborhoods of the points x and $f(x)$ in M_1^3 and M_2^3 respectively (Fig.8.2). Thus we see that M^3 is a manifold without boundary.

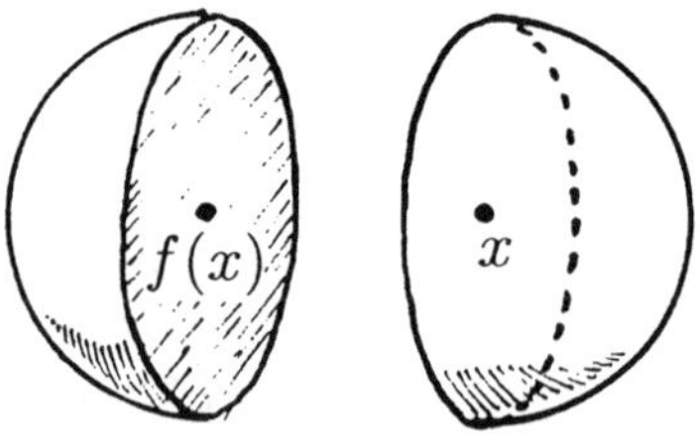

FIGURE 8.2. Neighborhood of identified points

Problem 8.1. Prove that the manifold M^3 thus obtained is orientable.

A *Heegaard splitting* of a 3-manifold M^3 is its presentation as the union of two handlebodies M_1^3 and M_2^3 with common boundary $\partial M_1^3 = \partial M_2^3$.

Examples of Heegaard splittings appear below in 8.4.

8.3. THEOREM. *Any orientable 3-manifold has a Heegaard splitting.*

Proof. Let K be a triangulation of M^3. Let us define the *barycentric subdivision* K' of K as follows. In each tetrahedron of our triangulation the medians of every face subdivide this face into 6 smaller triangles. Subdivide each tetrahedron into the

24 smaller tetrahedra obtained by taking the cones (with vertex at the barycenter of the tetrahedron) over the small triangles. These new tetrahedra constitute the first barycentric subdivision K' of K. Applying the same construction to K', we obtain the *second barycentric subdivision* K'' of K.

For M_1^3 let us take the union of all tetrahedra of the second subdivision K'' that have common points with the one-dimensional skeleton of K (i.e., with the set of edges of the tetrahedra of K), and for M_2^3 take the closure of the complement to M_1^3 in M^3. Let us show that M_1^3 and M_2^3 are homeomorphic handlebodies.

Suppose that Δ is a tetrahedron of K. The part of M_2^3 that lies in Δ is schematically shown in Fig.8.3,a. It has the shape of a "four-legged creature": four solid cylinders attached to a central solid sphere along one of their bases (Fig.8.3,b). These four-legged creatures have four free "soles" (one of them is shown in detail in Fig.8.3,c) along which they are attached to each other, forming M_2^3.

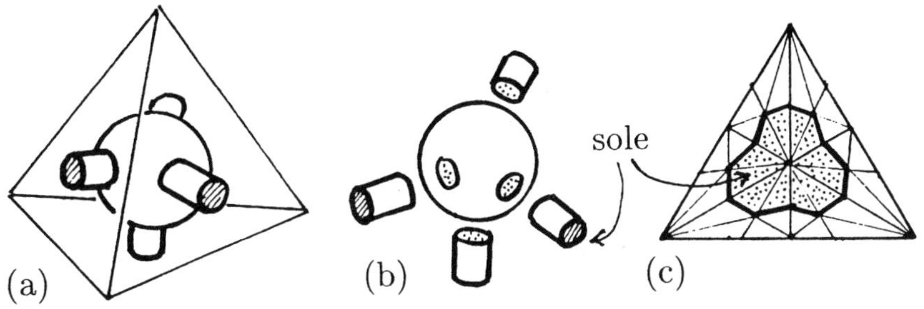

FIGURE 8.3. Four-legged creatures and their soles

To describe M_1^3, one can take the solid spheres centered at the vertices of the original triangulation K and join them by solid cylinders going along the edges of K (Fig.8.4).

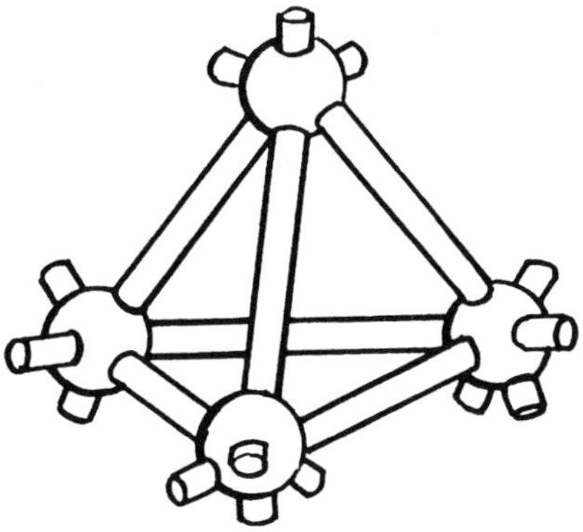

FIGURE 8.4. Complement to four-legged creatures

Thus the sets M_1^3 and M_2^3 may be obtained by gluing solid handles (cylinders) to solid spheres. Let us carry out this gluing procedure step by step, showing inductively that we obtain a handlebody (or several handlebodies) at each step. When the next solid cylinder is attached by its two soles to *different* handlebodies, no problems arise. But when the cylinder joins parts of the *same* handlebody, we

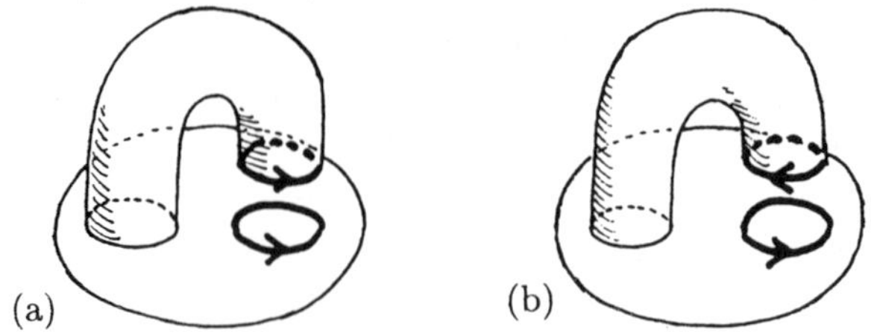

FIGURE 8.5. Oriented and nonoriented gluing

could *a priori* have some difficulties. This is because two types of gluing can occur in principle: oriented (Fig.8.5,a) and nonoriented ones (Fig.8.5,b).

However, if at some step of our process we had produced a nonoriented gluing, this would mean that M^3 is nonoriented, in contradiction with our assumption. So nonoriented gluing cannot occur, and inductively we have proved that M_1^3 and M_2^3 are both handlebodies.

The fact that they are homeomorphic is immediate: they have the same common boundary and hence the same number of handles. □

8.4. Examples. Any 3-manifold has many different Heegaard splittings. Let us consider some splittings of the sphere S^3.

The simplest one is obtained by cutting the 3-sphere along its equator, the 2-sphere, into two 3-disks.

A more interesting example is the classical splitting of the 3-sphere into two solid tori. Let us consider the 3-sphere as the following subset of complex 2-space:

$$S^3 = \{(z, w) \in \mathbb{C}^2 : |z|^2 + |w|^2 = 2\}.$$

Now let us put

$$M_1^3 = \{(z, w) \in S^3 : |z| \leqslant |w|\}, \qquad M_2^3 = \{(z, w) \in S^3 : |z| \geqslant |w|\}.$$

Note that in our situation the conditions $|z| \leqslant |w|$ and $|z| \geqslant |w|$ are equivalent to $|z| \leqslant 1$ and $|z| \geqslant 1$, respectively.

Now let us prove that M_1^3 and M_2^3 are solid tori. A point of the sphere S^3 may be presented in the form $(ae^{i\alpha}, be^{i\beta})$, where $a, b \geqslant 0$ and $a^2 + b^2 = 2$. The manifold M_1^3 is then determined by the condition $a \leqslant 1$. Consider the solid torus whose axial sections are pairs of 2-disks of radius 1. On it, introduce the coordinates (a, α, β) as shown in Fig.8.6.

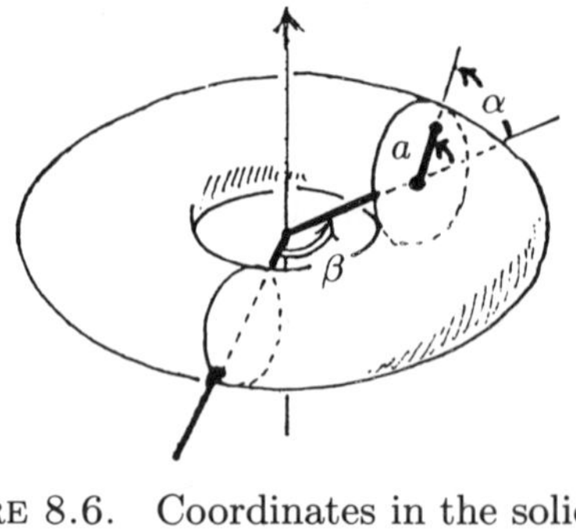

FIGURE 8.6. Coordinates in the solid torus

The assignment $(ae^{i\alpha}, be^{i\beta}) \mapsto (a, \alpha, \beta)$ defines a homeomorphism of the manifold M_1^3 onto our solid torus. (Note that $(0, \alpha, \beta)$ is the same point for all values of the parameter α; this point only depends on the parameter β.) In a similar way the assignment $(ae^{i\alpha}, be^{i\beta}) \mapsto (b, \beta, \alpha)$ defines a homeomorphism of M_2^3 onto the solid torus. Thus the 3-sphere has a Heegaard splitting into two solid tori.

Problem 8.2. Represent the 3-sphere S^3 as the one-point compactification of $\mathbb{R}^3$. Prove that if we remove a standardly embedded (in $\mathbb{R}^3$) open handlebody from S^3, we get a (closed) handlebody with the same number of handles.

8.5. By a *meridian* of the solid torus we mean a curve on its boundary given (in the coordinates described in the previous subsection) by the equation $\beta = \text{const}$; a *parallel* is given by $\alpha = \text{const}$. The common boundary of the manifolds M_1^3 and M_2^3 consists of points with coordinates $(e^{i\alpha}, e^{i\beta})$. The curve $\beta = \text{const}$ is at the same time a meridian of the solid torus M_1^3 and a parallel of M_2^3. This remark should help the reader visualize the 3-sphere as two solid tori glued together (Fig.8.7).

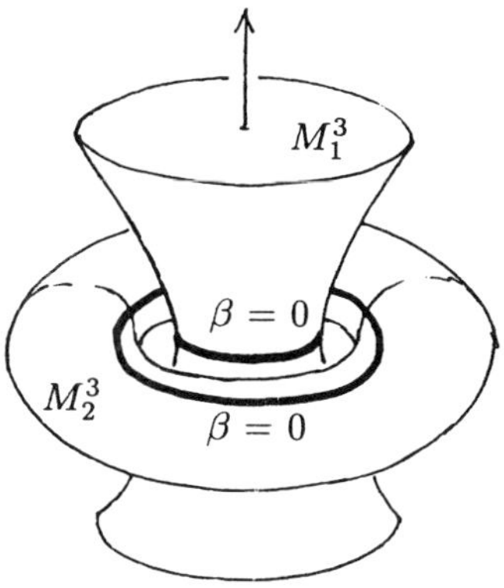

FIGURE 8.7. Heegaard splitting of S^3 into solid tori

Problem 8.3. What manifold is obtained by attaching a copy of the solid torus to another copy by a homeomorphism that takes meridians to meridians and parallels to parallels?

8.6. For the standard embedding of the solid torus in $\mathbb{R}^3$, there exists no isometry that takes parallels to meridians. (This is because the length of all the meridians is the same, while that of the parallels is not.) But in $\mathbb{R}^4 = \mathbb{C}^2$, the orthogonal map that interchanges the basis vectors e_1 and e_3, e_2 and e_4 takes the solid torus M_1^3 to M_2^3; hence, this map takes the common boundary torus T^2 to itself, interchanging parallels and meridians. This isometry was possible because the metric on the torus T^2 induced by the metric of $\mathbb{R}^4$ is locally Euclidean. Indeed, points of the torus $T^2 \subset \mathbb{R}^4$ have the coordinates

$$(x_1, x_2, x_3, x_4) = (\cos\alpha, \sin\alpha, \cos\beta, \sin\beta).$$

On the torus T^2 introduce the coordinates (α, β). Let dl be the distance in $\mathbb{R}^4$ between the points with coordinates (α, β) and $(\alpha + d\alpha, \beta + d\beta)$. Then

$$\begin{aligned} dl^2 &= dx_1^2 + dx_2^2 + dx_3^2 + dx_4^2 \\ &= (d\,\cos\alpha)^2 + (d\,\sin\alpha)^2 + (d\,\cos\beta)^2 + (d\,\sin\beta)^2 \\ &= (\sin^2\alpha + \cos^2\alpha)d\alpha^2 + (\sin^2\beta + \cos^2\beta)d\beta^2 = d\alpha^2 + d\beta^2, \end{aligned}$$

i.e., the induced metric on the torus $T^2 \subset \mathbb{R}^4$ is locally Euclidean. This means that in $\mathbb{R}^4$ it is possible to roll a torus from a square piece of paper. In contrast, in $\mathbb{R}^3$, although one can roll a cylinder from a square piece of paper, it is impossible to roll this cylinder into a torus.

The disposition of the torus in $\mathbb{R}^4$ described above is much more natural than its standard embedding in $\mathbb{R}^3$. Intuitively, one can say that the latter embedding distorts the torus, that we are in fact acustomed to a completely warped image of this object, whose true image can only be "seen" in its natural setting, i.e., in $\mathbb{R}^4$. In more rigorous terms, there is no isometric embedding of the torus with its (natural) flat metric into Euclidean 3-space.

8.7. Trefoils on the torus. In the 3-sphere

$$S^3 = \{(z,w) \in \mathbb{C}^2 : |z|^2 + |w|^2 = 2\},$$

one can position the trefoil knot so that it will lie on the torus $|z| = |w| = 1$. To do this, it suffices to take the intersection of the sphere S^3 with the complex surface $z^3 = w^2$. Indeed, $z^3 = w^2$ and $|z|^2 + |w|^2 = 2$ imply that $|z|^3 + |z|^2 = |w|^2 + |z|^2 = 2$. For $x > 0$ the function $x^3 + x^2 - 2$ is monotone increasing, so it only becomes equal to zero when $x = 1$. Therefore, the intersection of the sphere S^3 with the surface $z^3 = w^2$ consists of points (z,w) such that $z = e^{i\alpha}$ and $w = e^{i\beta}$, where $3\alpha \equiv 2\beta \pmod{2\pi}$. Fig.8.8,a shows the set of points (α, β) in the square of side 2π satisfying $3\alpha \equiv 2\beta \pmod{2\pi}$. One can glue a cylinder from this square in two ways (Fig.8.8,b), and from each of these cylinders glue a torus (Fig.8.8,c).

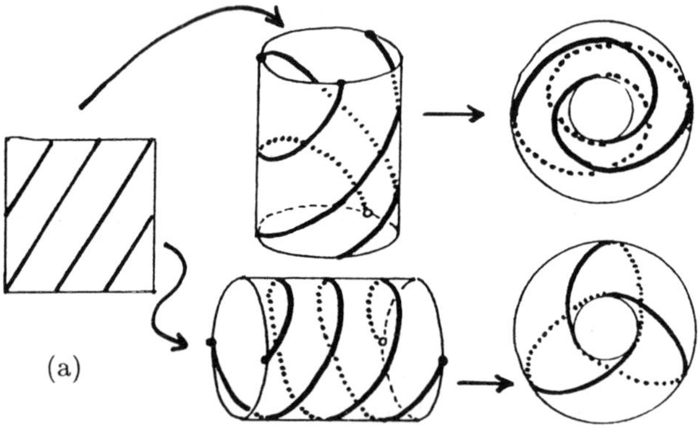

(a) (b) (c)

FIGURE 8.8. Two trefoils on the torus

In both cases we obtain a trefoil lying on the torus $|z| = |w| = 1$ in S^3. So we have proved that *the surface* $z^3 = w^2$ *cuts out a trefoil on the 3-sphere* $S^3 \subset \mathbb{C}^2$. This implies the following unexpected result.

8.8. THEOREM. *The space of complex cubic polynomials without multiple roots is homotopy equivalent to the complement* $S^3 - K$ *of the trefoil knot.*

Proof. Recall that two spaces X and Y are called homotopy equivalent if there exist continuous maps $F : X \to Y$ and $G : Y \to X$ such that $F \circ G \simeq \mathrm{id}_Y$ and $G \circ F \simeq \mathrm{id}_X$, where id_X and id_Y are the identity maps of the spaces X and Y, while $\simeq$ denotes the homotopy of maps. In the case when $X \subset Y$, it suffices to verify that there exists a map $G: Y \to X$ satisfying the following two conditions:

1) if $x \in X$, then $G(x) = x$;

2) the map G, regarded as a map from Y to Y, is homotopic to the identity.

First let us prove that the space X of polynomials of the form x^3+px+q without multiple roots is homotopy equivalent to the space Y of polynomials of the form x^3+ax^2+bx+c without multiple roots. To this end, consider the map $G: Y \to X$ taking the polynomial $f(x) = x^3+ax^2+bx+c$ to $f(x-a/3)$. The homotopy joining G to the identity map of Y is given by the formula $f_t(x) = f(x-at/3)$. Indeed, if f is a polynomial without multiple roots, then so is f_t for all $t \in [0,1]$, which means f_t is a homotopy of Y. So X is homotopy equivalent to Y.

Now it suffices to prove that X is homotopy equivalent to

$$Z := S^3 - K = \{(z,w) \in \mathbb{C}^2 : |z|^2+|w|^2 = 2,\ z^3 \neq w^2\},$$

since it was established in the previous subsection that K is the trefoil. To construct the required homotopy equivalence, first note that the polynomial x^3+px+q has multiple roots if and only if $4p^3+27q^2 = 0$; hence the polynomial $x^3-3zx+2w$ has multiple roots iff $z^3 = w^2$. Thus we can assume that

$$X = \{(z,w) \in \mathbb{C}^2 : z^3 \neq w^2\} \text{ and } Z \subset X.$$

Let us construct a map $G' : X \to Z$ as follows. Suppose that $(z_0, w_0) \in X$, i.e., $z_0^3 \neq w_0^2$. We claim that there exists a unique positive number λ_0 such that $(\lambda_0^2 z_0, \lambda_0^3 w_0) \in S^3$. Indeed, let $a = |z|^2$, $b = |w|^2$, and $\mu = \lambda^2$. Then the function

$$|\lambda^2 z|^2 + |\lambda^3 w|^2 = \mu^2 a + \mu^3 b = \mu^2(a+\mu b)$$

is monotone increasing for $\mu > 0$, so that λ_0 is unique as claimed. So now we can define

$$G'(z_0, w_0) := (\lambda_0^2 z_0, \lambda_0^3 w_0).$$

Clearly, the restriction of G' to $S^3 - K$ is the identity. Further, the map $G' : X \to Z$ is homotopic to the identity, say by the following homotopy:

$$F_t(z_0, w_0) = (\lambda^2 z_0, \lambda^3 w_0), \text{ where } \lambda = (1-t) + t\lambda_0.$$

Here $(\lambda^2 z_0)^3 - (\lambda^3 w_0)^2 = \lambda^6(z_0^3 - w_0^2) \neq 0$ for any $t \in [0,1]$, which means that F_t is indeed a homotopy of X. This establishes the required homotopy equivalence. □

Problem 8.4. Find a Heegaard splitting of real projective space $\mathbb{R}P^3$.

This problem, although it has a short solution, is not too easy. In general, it can be said that Heegaard splittings, being self-homeomorphisms of surfaces, are not too easy to describe. It would be nicer if we had a more effective, combinatorial description of the construction of 3-manifolds. Such a description exists, and will be presented in §10. Before that, however, we shall discuss the case of manifolds with boundary.

§9. Heegaard splittings for manifolds with boundary

9.1. Consider two handlebodies N and M standardly embedded in $\mathbb{R}^3$ and positioned so that the boundary of M divides N into two symmetric parts precisely

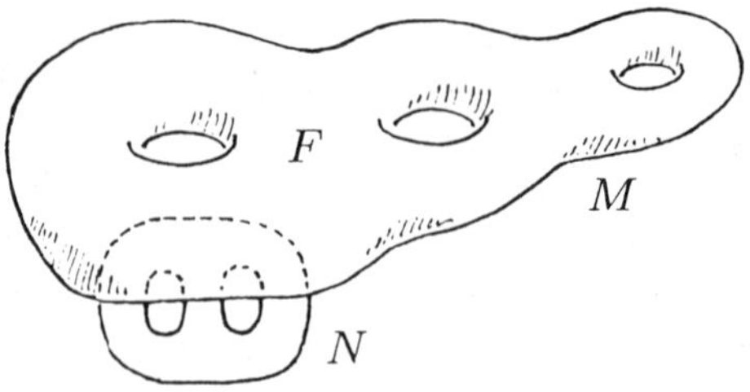

FIGURE 9.1. Two intersecting handlebodies N and M

as shown in Fig.9.1. Let F be the part of M that lies outside N and G the part of the boundary of M outside N.

Problem 9.1. Prove that if the handlebodies M and N have m and n handles respectively, then F is a handlebody with $m+n$ handles.

Consider two copies F_1 and F_2 of the handlebody F. Suppose G_1 and G_2 are the parts of the boundaries of F_1 and F_2 corresponding to G. Let us attach the handlebodies F_1 and F_2 to each other along some homeomorphism $h : G_1 \to G_2$. The result will clearly be an oriented 3-manifold M^3 with boundary whose boundary is (homeomorphic to) the boundary of N. Such a presentation is called a *Heegaard splitting of the manifold M^3 with boundary.*

9.2. THEOREM. *Any oriented 3-manifold M^3 with boundary has a Heegaard splitting.*

Proof. First we assume that M^3 has a connected boundary ∂M^3. Let g be the number of handles of ∂M^3, N^3 the handlebody with g handles, and T the *core* of N^3, i.e., the one-dimensional subcomplex of N^3 shown in Fig.9.2.

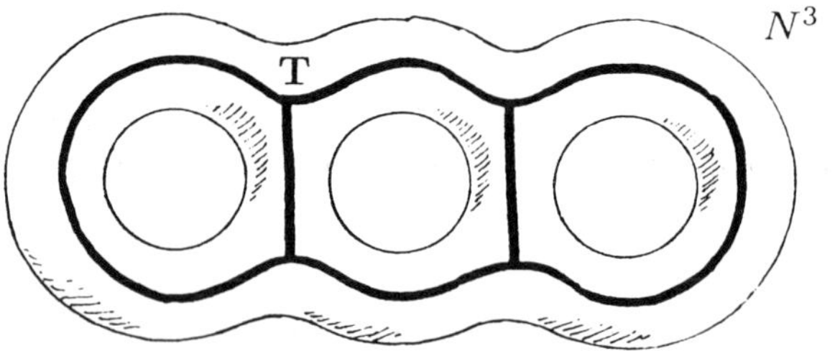

FIGURE 9.2. The handlebody N^3 with its core T

Let us attach the handlebody N^3 to our 3-manifold M^3 by an arbitrary homeomorphism of their boundaries. The result will be an orientable manifold L^3 without boundary. We can assume that L^3 has a triangulation K such that T is a subcomplex of K. Following the proof of Theorem 8.3, in L^3 consider the neighborhood L_1^3 (in the second barycentric subdivision K'') of the one-dimensional skeleton of K. Then L_1^3 is one of the handlebodies in the Heegaard splitting of L^3 (Fig.9.3,a). There exists an isotopy $f_t : L^3 \to L^3$ joining the identity map $f_0 = \mathrm{id}$ of L^3 with the map f_1 such that half of $f_1(T)$ is in L_1^3 while the other half is in L_2^3 (Fig.9.3,b). Now if we cut out the neighborhood of $f_1(T)$ from L^3, we get a manifold homeomorphic to the given manifold M^3. Cutting it along the common boundary of the handlebodies L_1^3 and L_2^3, we obtain the required splitting.

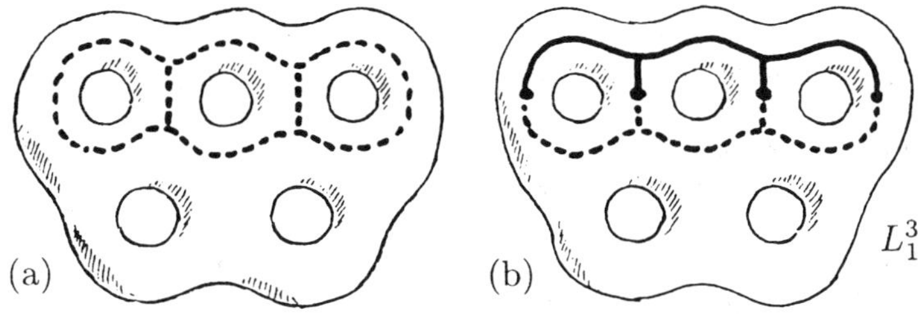

FIGURE 9.3. Pushing out half of T

In the case when the boundary of M^3 consists of several connected components, the above argument can be applied to each. □

§10. Heegaard diagrams

Any Heegaard splitting of an oriented manifold M^3 is entirely described by a homeomorphism of the sphere with g handles (specifying how two handlebodies M_1^3 and M_2^3 must be attached to each other to form M^3). This description would be practically useful if we had a method for the effective presentation of these homeomorphisms. Such a method is given in this section. It is based on the fact that for our aims all we need to know about our homeomorphism is what it does to a certain canonical system of curves on the sphere with g handles. So the attaching homeomorphism is encoded by the image of this system, in fact by a picture of a certain system of curves on a surface, called a Heegaard diagram.

10.1. To a given Heegaard splitting of a 3-manifold M^3 into two handlebodies M_1^3 and M_2^3 with g handles, let us assign a system of curves on the sphere with g handles N in the following way. Consider the standard embeddings of the handlebodies M_1^3 and M_2^3 in $\mathbb{R}^3$. On their boundaries draw the system of meridians $u_1, ..., u_g$ and $v_1, ..., v_g$ as shown in Fig.10.1.

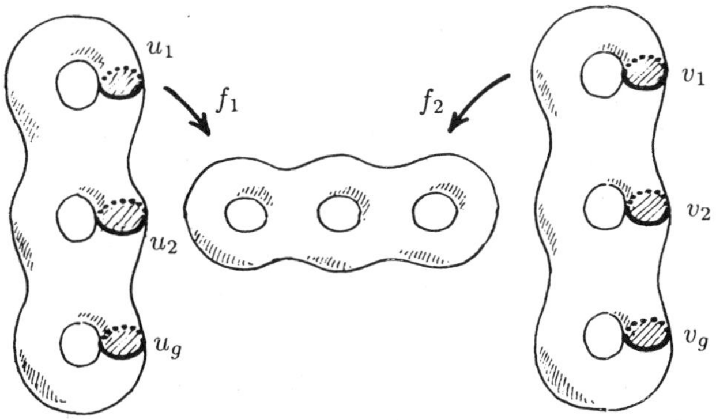

FIGURE 10.1. Canonical systems of meridians

For the sake of symmetry in our definitions, let us present the attaching homeomorphism $f : \partial M_1^3 \to \partial M_2^3$ as the composition of two homeomorphisms f_1 and f_2^{-1}, where f_i is the homeomorphism of ∂M_i^3 onto the sphere with g handles N. (We could have put $N = \partial M_1^3$ and $f_1 = \text{id}$ or $N = \partial M_2^3$ and $f_2 = \text{id}$, but either would mean giving preference to one of the handlebodies over the other.) On the surface N we now draw the curves $f_1(u_i)$ and $f_2(v_i)$, $i = 1, ..., g$.

Definition 1. The system of curves $\{f_1(u_i)\}$ and $\{f_2(u_i)\}$, $i = 1, ..., g$, on the sphere with g handles N is said to be a *Heegaard diagram* of the manifold M^3.

10.2. THEOREM. *If two manifolds M and M' have the same Heegaard diagram, then these manifolds are homeomorphic.*

Proof. We denote by $f_i : \partial M_i \to N$, $i = 1, 2$, the attaching homeomorphisms for the manifold M, and use similar notation but with primes for M'. We can assume (without loss of generality) that $M_2 = M_2'$ and $N = \partial M_2 = \partial M_2'$ (Fig.10.2).

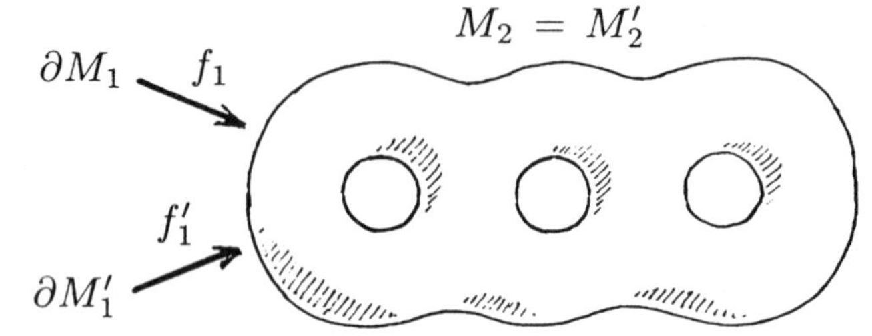

FIGURE 10.2. Manifolds with the same Heegaard diagrams

Since M and M' have the same Heegaard diagram, f_1 and f_1' are homeomorphisms of ∂M_1 and $\partial M_1'$ onto N such that the images of the meridians u_i and u_i', $i = 1, ..., g$, are the same. Let us prove that the identical homeomorphism $h_2 : M_2 \to M_2'$ can be extended to a homeomorphism $h : M \to M'$. The homeomorphism $(f_1')^{-1} \circ f_1 : \partial M_1 \to \partial M_1'$ takes each meridian of M_1 to the corresponding meridian of M_1'. Since any homeomorphism between two circles can be extended to the disks that they bound (along the radii), the homeomorphisms of the meridians can be extended to the disks that they bound inside the manifolds M_1 and M_1' (see Fig.10.1). Cutting M_1 and M_1' along these disks, we get manifolds D_1 and D_1' homeomorphic to the 3-disk (Fig.10.3).

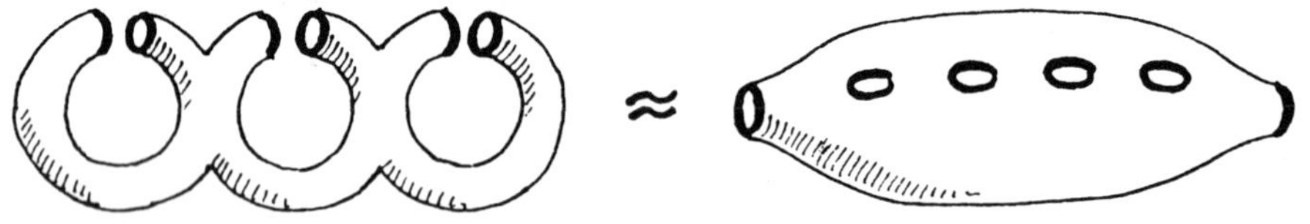

FIGURE 10.3. Cutting M_1 into a 3-disk

Since any homeomorphism between two 2-spheres can be extended to the 3-disks that they bound (along the radii), the homeomorphism between the boundaries of the 3-disks D_1 and D_1' can be extended to the disks themselves. As the result, we obtain the required homeomorphism $h : M \to M'$. □

10.3. The description of Heegaard splittings obtained in the previous subsection is not yet up to our requirements: the definition of Heegaard diagram given there still involves the homeomorphism f, and it does not explicitely specify what systems of curves can form a Heegaard diagram of a 3-manifold. The following definition is free of these drawbacks.

Definition 2. The system of closed curves $u_1, ..., u_g$ and $v_1, ..., v_g$ on the sphere with g handles N is said to constitute a *Heegaard diagram* if the two following conditons are satisfied:

1) the curves $u_1, ..., u_g$ are pairwise nonintersecting and the complement to their union is connected;
2) the curves $v_1, ..., v_g$ are pairwise nonintersecting and the complement to their union is connected.

Let us verify that *Definitions* 1 *and* 2 *are equivalent.* It is clear that the meridians of a handlebody are pairwise nonintersecting and do not disconnect its surface. Hence we need only prove that any Heegaard diagram in the sense of Definition 2 corresponds to the Heegaard splitting of some manifold.

First we claim that if the sphere N with g handles is cut along g nonintersecting circles that do not split N, a sphere S^2 from which $2g$ disks have been deleted is obtained. Suppose that instead we get a surface H with h handles and $2g$ deleted disks. The removal of one disk decreases the Euler characteristic[1] by 1. Hence the Euler characteristic of H is $(2-2h)-2g$. On the other hand, cutting along a circle does not change the Euler characteristic, so that $2-2h-2g = 2-2g$, whence $h = 0$ as claimed.

Now take two copies of the surface N. Cut one copy along the circles u_i (Fig.10.4,c) and the other along v_i (Fig.10.4,d). In both cases we get a sphere with $2g$ holes (Fig.10.4,b and e). These spheres may be homeomorphically deformed so that the boundary circles correspond to the canonical meridians of the handlebodies with g handles (Fig.10.4,a and f). Now it is easy to construct two handlebodies with g handles M_1^3 and M_2^3 together with homeomorphisms of their boundaries onto N taking the meridians of M_1^3 and M_2^3 onto the circles u_i and v_i on N respectively. This gives the required Heegaard splitting. □

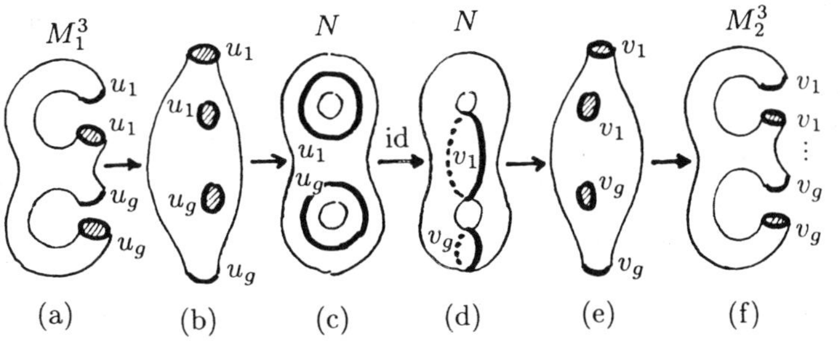

FIGURE 10.4. Heegaard splitting corresponding to Heegaard diagram

§11. Lens spaces

So far, we have demonstrated only a few examples of 3-manifolds presented by Heegaard splittings or diagrams. In this section we consider an infinite series of 3-manifolds that may be conveniently presented in this way: the classical lens spaces.

11.1. The only 3-manifold that can be obtained by gluing two 3-disks by a homeomorphism of their boundaries is the 3-sphere S^3. From two solid tori, as we

[1]The reader not familiar with this notion is referred to 12.6 or 21.2 below.

have seen above, besides the sphere S^3 (see 8.4), one can get projective space $\mathbb{R}P^3$ (Problem 8.4). But S^3 and $\mathbb{R}P^3$ are not the only 3-manifolds obtainable by gluing two solid tori by a homeomorphism of their boundaries. We shall begin by giving a geometrical definition of such manifolds, based on a discrete group action on S^3, and only then discuss their Heegaard presentation.

Suppose p and q are coprime positive integers, and $p \geqslant 3$. On the unit sphere $S^3 \subset \mathbb{C}^2$ consider the action (without fixed points) of the group $\mathbb{Z}/p\mathbb{Z}$ with generator σ by setting

$$\sigma(z, w) = (\exp(2\pi i/p)z, \exp(2\pi iq/p)w).$$

Take the quotient of S^3 by this action of $\mathbb{Z}/p\mathbb{Z}$, i.e., identify each point $x \in S^3$ with the points $\sigma x, \dots, \sigma^{p-1}x$. Since the action has no fixed points, it is easy to see that the quotient space is a 3-manifold; it is called a *lens space* and is denoted by $L(p, q)$.

11.2. Let us show that the quotient under the action of $\mathbb{Z}/p\mathbb{Z}$ of each of the two solid tori $|z|^2 \leqslant 1/2$ and $|w|^2 \leqslant 1/2$ is a solid torus, so that the lens space $L(p, q)$ can be glued from two solid tori. Consider the following cellular decomposition of the sphere S^3:

0) zero-dimensional cells $(0, \exp(2\pi ik/p))$;
1) one-dimensional cells $(0, \exp(2\pi i\theta))$, $k/p < \theta < (k + 1)/p$;
2) two-dimensional cells $(\rho \exp(2\pi ik/p), w)$, $0 < \rho \leqslant 1$, $|w| = \sqrt{1 - \rho^2}$;
3) three-dimensional cells $(\rho \exp(2\pi i\theta), w)$, $0 < \rho \leqslant 1$; $k/p < \theta < (k + 1)/p$, $|w| = \sqrt{1 - \rho^2}$, $k = 0, 1, \dots, p - 1$.

There are p cells in each dimension (indexed by the letter k). Under the action of the group $\mathbb{Z}/p\mathbb{Z}$, the cells permute with each other, so that the above cell decomposition of the sphere S^3 induces a cell decomposition of the lens space $L(p, q)$ with one cell in each dimension from 0 to 3. So our lens space can be obtained by taking one of the 3-cells and performing the appropriate identifications on its boundary under the action of $\mathbb{Z}/p\mathbb{Z}$.

Unfortunately, the coordinate presentation of our 3-cells in four-dimensional space $\mathbb{C}^2$ is not very convenient to work with, so we begin by changing to the more natural system of coordinates $(x_1, x_2, x_3) \in \mathbb{R}^3$, where our 3-cell will just be the unit 3-disk.

To carry out this change, to the point

$$(\rho \exp(2\pi i\theta), w), \text{ where } 0 < \rho \leqslant 1, \ k/p < \theta < (k + 1)/p, \ |w|^2 + \rho^2 = 1,$$

let us assign the point

$$(x_1, x_2, x_3) \in \mathbb{R}^3, \text{ where } x_1 + ix_2 = w, \ x_3 = (2p\theta - 2k - 1)\rho;$$

here $|x_3| \leqslant \rho$ and $x_1^2 + x_2^2 + x_3^2 \leqslant 1$ (Fig.11.1). Points of the sphere $x_1^2 + x_2^2 + x_3^2 = 1$ for which $x_3 > 0$ (respectively $x_3 < 0$) are assigned to points for which $\theta = (k+1)/p$ (respectively $\theta = k/p$).

Points with $\theta = k/p$ are taken by the generator $\sigma \in \mathbb{Z}/p\mathbb{Z}$ to points for which we have $\theta = (k + 1)/p$. It follows from the definition of our assignment that on the coordinates x_1 and x_2, the element σ acts via rotations by the angle $2\pi q/p$ about the origin; it also identifies the 2-cells of the upper hemisphere with 2-cells of the lower one as shown in Fig.11.2. These identifications produce $L(p, q)$.

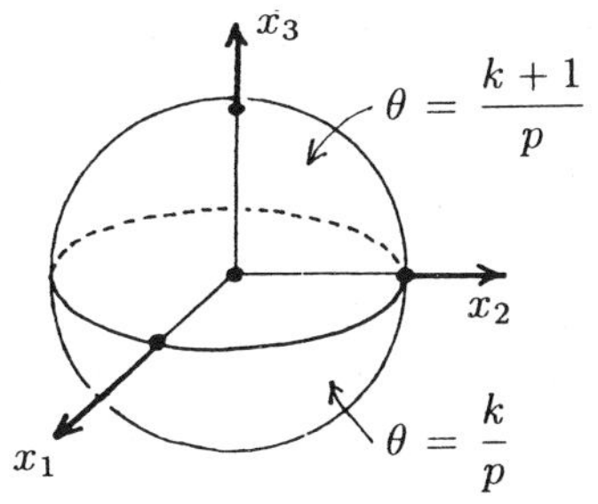

FIGURE 11.1. Points on the boundary of the 3-cell

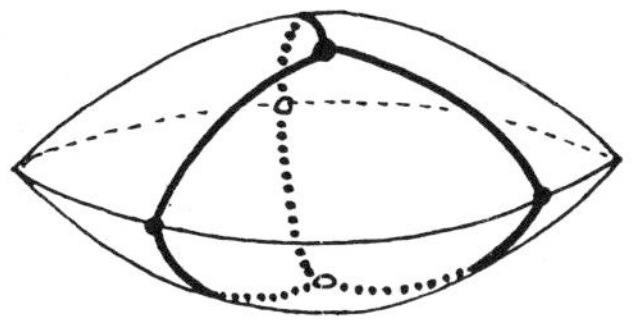

FIGURE 11.2. Identifying the boundary of a "lens"

Traditionally, the 3-cell with identified 2-cells on the boundary is pictured as a somewhat flattened 3-disk that resembles a lens, which is apparently the origin of the term "lens space".

The solid torus $|w|^2 \leqslant 1/2$ intersects the 3-cell from which we constructed the lens space $L(p,q)$ along the solid cylinder (with spherical bases) determined by the inequalities $x_1^2+x_2^2+x_3^2 \leqslant 1$ and $x_1^2+x_2^2 \leqslant 1/2$ (Fig.11.3). Under the identifications due to the action of the element σ, we must glue together the upper base of this cylinder with its lower base, after a rotation by $2\pi q/p$. The result will be the solid torus M_1^3. We leave to the reader the proof of the fact that its complement in the lens space $L(p,q)$ is also a solid torus. (This proof may be obtained by using the constructions described in 11.3.)

Problem 11.1. Find the fundamental group of the lens space $L(p,q)$.

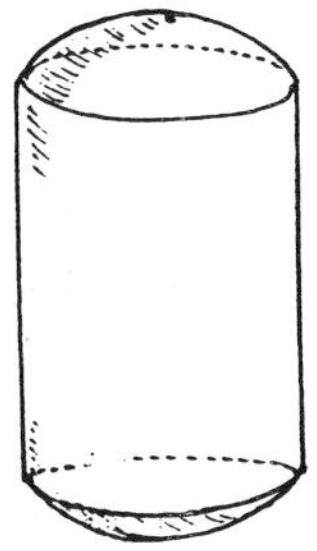

FIGURE 11.3. Cylinder with spherical bases

11.3. It follows from the result of the previous problem that *the lens spaces $L(p,q)$ and $L(p',q')$ are not homeomorphic if $p \neq p'$.* On the other hand, it is obvious from the construction that *the lens spaces $L(p,q)$ and $L(p,q')$ are homeomorphic provided $q \equiv q' \mod p$.*

Our next goal is to prove that *the lens spaces* $L(p, q)$ *and* $L(p, q')$ *are homeomorphic provided* $qq' \equiv +1 \mod p$. (Then they will also be homeomorphic when $qq' \equiv -1 \mod p$.)

To do this, cut the 3-cell into tetrahedra by means of p half-planes passing through the x_3-axis and the 0-cells (Fig.11.4). Denote the upper and lower faces of the first tetrahedron by T_1 and S_1, respectively, and the left and right lateral faces by A_1 and B_1, respectively. Denote the faces of the other tetrahedra in a similar way. Initially the faces B_i and A_{i+1} were identical, while the transformation σ identifies the faces S_i and T_{i+q} (here and below, we number the faces modulo p).

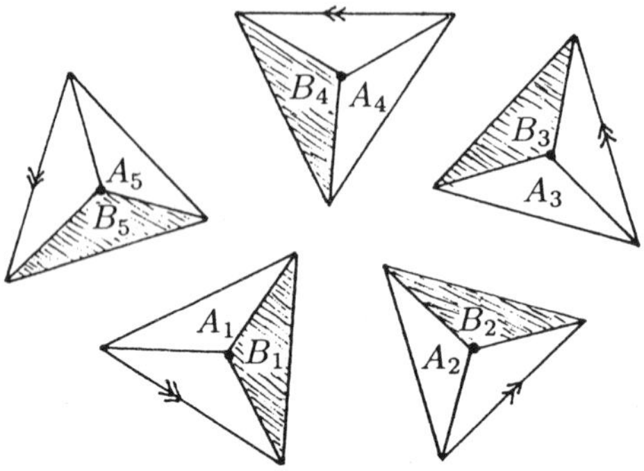

FIGURE 11.4. The lens cut into tetrahedra

Now instead of identifying the lateral faces first, let us begin by identifying the upper and lower faces, and only then identify the lateral ones. Then the adjacent tetrahedra will acquire the numbers $j, j + q, j + 2q, ..., j + q'q$. Hence the face B_j will be identified with the face A_{j+1}, which is the same as the face $A_{j+qq'}$, because by assumption the integer q' is the solution of the Diophantine equation $qq' \equiv +1 \mod p$. Since the end result does not depend on the order in which the identifications are performed, the two lens spaces are identical. □

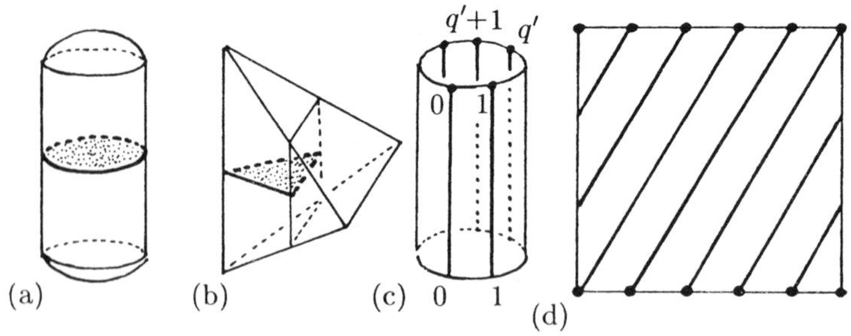

FIGURE 11.5. Heegaard diagram of $L(5, 3)$

11.4. Now we are ready to describe the Heegaard diagram of the lens space $L(p, q)$. For the meridional disk of the solid torus $|w|^2 \leqslant 1/2$, we take its section by the plane $x_3 = 0$ (Fig.11.5,a). On Fig.11.5,b we show the part of the meridional disk contained in one of the tetrahedra into which we have cut our "lens". This picture shows that the boundary circle of this disk corresponds to p segments on the other solid torus $|w|^2 \geqslant 1/2$ (Fig.11.5,c). But under our identifications, the lower point with number i is glued to the upper point $i+q'$, where $qq' \equiv 1 \mod p$. Since the integers p and q' are coprime, we obtain one closed curve on the torus (shown in Fig.11.5,d). This is the desired Heegaard diagram.

Comments

The idea of a manifold as a geometric entity not lying in some linear space, but possessing its own intrinsic geometry, is due to B. Riemann. It appeared before the notion of abstract topological space, and in those days was always supplied with a metric. We shall not attempt to describe the progressive "topologization" of the notion of 3-manifold that took place in the first third of the 20th century. The key names, besides Heegaard and Poincaré, are J. W. Alexander, M. Dehn, J. Nielsen, H. Seifert. The classical text that brings together the topological achievements of that period is the *Lehrbuch der Topologie* by H. Seifert and W. Threlfall ([**ST**], 1934).

Most of the material of this chapter is traditional and can be found in several texbooks, e.g. [**Rol**], [**MF**], [**ST**], except for §9, in which we follow the article [**Dow**], and Theorem 8.8, which is standard "folklore", but for which we were not able to find a proof in the literature.

CHAPTER V

Homeomorphisms of Surfaces

The main ingredient in a Heegaard splitting is a homeomorphism of two-dimensional surfaces. It would therefore be useful to have a more combinatorial description of these homeomorphisms, e.g. in terms of some kind of elementary operation. Such an operation, called a Dehn twist, was actually used in §7 in the particular case of a disk with holes. In the general case it is defined below and used to state and prove the Dehn-Lickorish theorem about the homeomorphisms of surfaces. This theorem has many important consequences that we discuss in this chapter. As we shall see, the Dehn-Lickorish theorem is not only a tool for encoding Heegaard presentations: it is in fact the basis of a more important and convenient way to present 3-manifolds, called surgery presentation and studied systematically in the next chapter. In this chapter, the main result for the sequel is Corollary 12.4, which asserts the existence of a certain surgery presentation of any 3-manifold.

§12. The Dehn–Lickorish theorem and its corollaries

12.1. In the previous chapter we have seen that any manifold M^3 can be obtained by gluing two handlebodies M_1^3 and M_2^3 along a certain homeomorphism $f : \partial M_1^3 \to \partial M_2^3$, and we can assume that $\partial M_1^3 = \partial M_2^3$.

Problem 12.1. Prove that in the above situation we can also assume without loss of generality that the homeomorphism $f : \partial M_1^3 \to \partial M_2^3$ is orientation preserving.

Now under a small change of the homeomorphism $f : \partial M_1^3 \to \partial M_2^3$, the manifold M^3 (considered up to topological equivalence) will not change. Therefore *to isotopic homeomorphisms in a Heegaard splitting corresponds one and the same 3-manifold.*

12.2. The simplest nontrivial example of an orientation-preserving homeomorphism (of a two-dimensional surface N) not isotopic to the identity may be obtained in the following way. Let us cut the surface N along a curve γ that does not split N into disjoint parts (Fig.12.1).

Let us fix one of the two copies of γ obtained by cutting and rotate the other copy along itself. Points located near this curve will follow its motion. More precisely, a one-sided neighborhood C of γ will be mapped homeomorphically so that this map is the given rotation on γ and the identity on the other side of the cylindrical neighborhood C. If we rotate the curve by 360°, then each point of the curve will return to its original position. Then we can glue the two copies of γ back together again, obtaining a homeomorphism of N onto itself, called a *Dehn twist* of N along the curve γ.

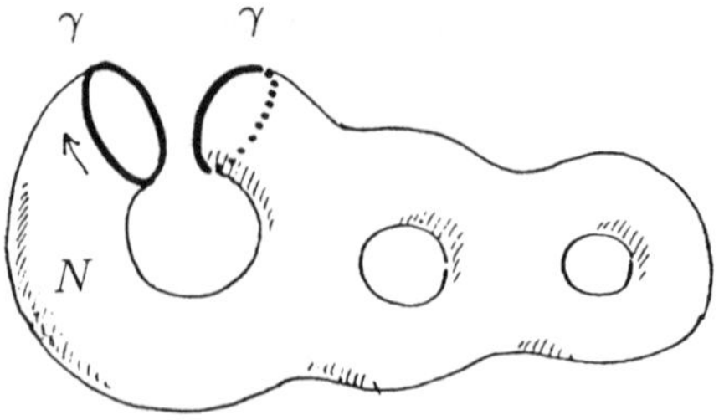

FIGURE 12.1. Surface cut along the curve γ

For example, the Dehn twist of the torus along the meridian α transforms the parallel β into the curve β' (Fig.12.2).

Problem 12.2. Prove that the Dehn twist along the meridian of the torus is not isotopic to the identity.

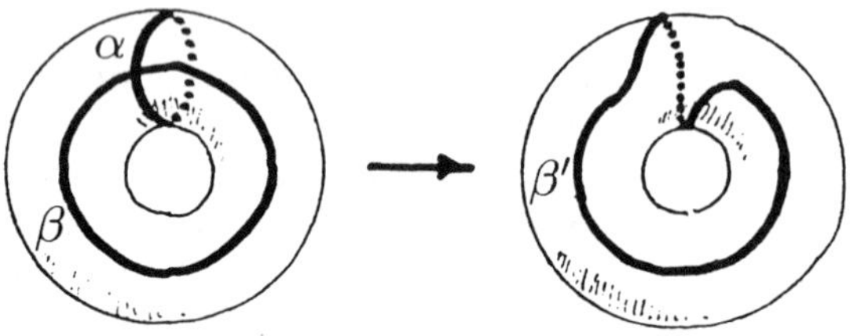

FIGURE 12.2. Dehn twist along the meridian of the torus

Dehn twists constitute a very particular and simple class of homeomorphisms. Nevertheless, it often suffices to consider this narrow class rather than generic homeomorphisms, because we have the following result.

12.3. DEHN–LICKORISH THEOREM. *Any orientation-preserving homeomorphism of an oriented 2-manifold (without boundary) can be presented as the composition of Dehn twists and homeomorphisms isotopic to the identity.*

Problem 12.3. Suppose the homeomorphisms $n_1, \dots, n_k$ of the 2-manifold N are isotopic to the identity, while the homeomorphisms $h_1, \dots, h_k$ of N are arbitrary. Prove that the homeomorphism $n_1 \circ h_1 \circ \cdots \circ n_k \circ h_k$ is isotopic to $h_1 \circ \cdots \circ h_k$.

The Dehn–Lickorish Theorem will be proved the next section. But first we shall derive two important consequences of this result.

12.4. COROLLARY. *Any orientable 3-manifold M^3 may be obtained by cutting out some solid tori from the 3-sphere S^3 and then pasting them back in, but along different homeomorphisms of their boundaries. Moreover, it can be assumed that all these solid tori in S^3 are unknotted.*

Proof. The manifold M^3 may be obtained by gluing together two handlebodies with g handles along some homeomorphism f. In the 3-sphere S^3, consider the standardly embedded handlebody with g handles; its complement is a homeomorphic handlebody, so the sphere S^3 can be obtained by gluing together two copies of the handlebody with g handles by a homeomorphism that we denote by f_0.

Consider the homeomorphism $f^{-1}f_0$. First we investigate the particular case when this homeomorphism is isotopic to a Dehn twist along a curve γ. Recall that the sphere S^3 was glued together from two copies M_1^3 and M_2^3 of the handlebody with g handles. On the boundary of M_1^3, draw the curve γ and then push it slightly inside M_1^3. Let T be a thin neighborhood of γ homeomorphic to the solid torus (Fig.12.3,a). Cut M_1^3 along the boundary of T and along the narrow ribbon surface swept through by γ in $M_1^3 - T$ (while this curve was being pushed inside M_1^3) and joining T to the boundary of M_1^3. Now we can construct a homeomorphism $j : M_1^3 - T \to M_1^3 - T$ which is identical outside the hatched region in Fig.12.3,b and is our twist along γ on ∂M_1^3.

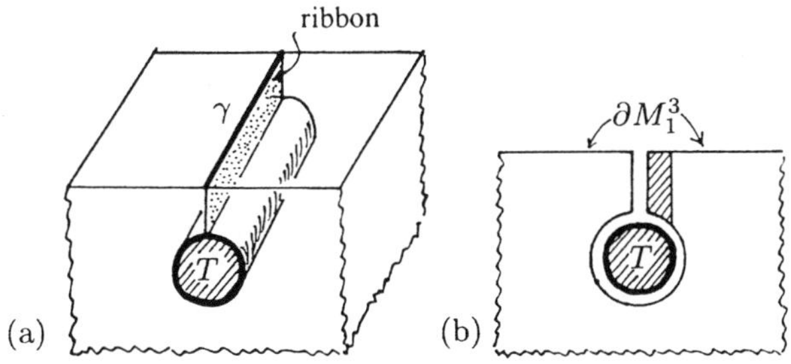

FIGURE 12.3. Homeomorphism of $M_1^3 - T$ induced by a twist

Now we claim that M^3 can be obtained from S^3 by removing the solid torus T and then gluing it back in, but along a different homeomorphism of its boundary. To prove that, it suffices to construct a homeomorphism of $S^3 - T$ onto $M^3 - T$. Assume that some manifold Z is obtained by gluing the manifolds X and Y along the homeomorphism $f_0 : \partial X \to \partial Y$. Suppose we are also given two homeomorphisms $j_1 : X \to X'$ and $j_2 : Y \to Y'$. In order to obtain a manifold Z' homeomorphic to Z from the manifolds X' and Y', one can perform the following identifications: if the points $x \in X$ and $f_0(x) \in Y$ were glued together, identify the points $j_1(x) \in X'$ and $j_2(f_0(x)) \in Y'$. Thus gluing along the homeomorphisms f_0 and $j_2 f_0 j_1^{-1}$ yields topologically equivalent manifolds. In the case at hand, $X = X' = M_1^3 - T$, $Y = Y' = M_2^3$, $j_1 = j$, and $j_2 = \mathrm{id}$ (Fig.12.4).

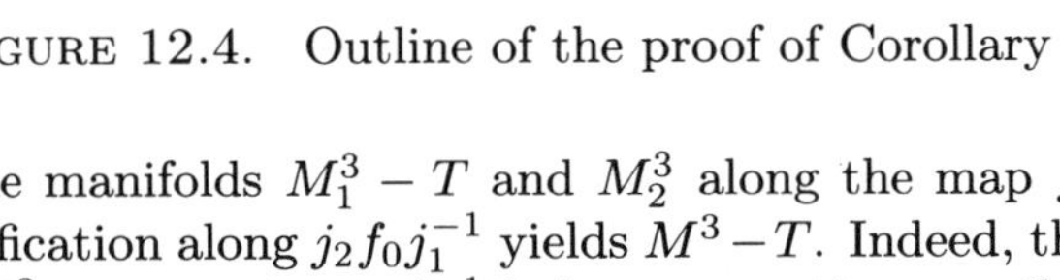

FIGURE 12.4. Outline of the proof of Corollary 12.4

By gluing the manifolds $M_1^3 - T$ and M_2^3 along the map f_0, we get $S^3 - T$, while their identification along $j_2 f_0 j_1^{-1}$ yields $M^3 - T$. Indeed, the homeomorphism $j : \partial M_1^3 \to \partial M_1^3$ is isotopic to $f^{-1} f_0$ by assumption, so the homeomorphism $f_0 j^{-1} : \partial M_1^3 \to \partial M_2^3$ is isotopic to $f_0(f^{-1} f_0)^{-1} = f$, and gluing along isotopic maps yields homeomorphic manifolds. So we have established our corollary in the case when the attaching homeomorphisms f and f_0 differ by a single twist.

In the general case of an arbitrary homeomorphism, we use the Dehn–Lickorish theorem to present it as a composition of twists. For these twists, we similarly choose the corresponding solid tori T_i and construct a homeomorphism of $S^3 - \bigcup_i T_i$ onto $M^3 - \bigcup_i T_i$ as above. Here we must take care that the T_i don't intersect, but this will be the case if the tubes T_i are thin enough and are pushed inside M_1^3 to different depths.

This completes the proof of the corollary, except for the assertion that the removed solid tori $T_i \subset S^3$ may be chosen so as to be unknotted. If some of them are knotted, we shall transform them into unknotted ones by removing and gluing back additional unknotted solid tori. To do this, recall that, according to Theorem 3.8, any knot can be untied by making the appropriate crossing changes. Therefore the last assertion of the corollary now follows from the following statement.

12.5. LEMMA. *Suppose $D^3 \subset \mathbb{R}^3$ is the 3-disk and $T \subset D^3$ is the solid torus shown in* Fig.12.5,a. *Then there exists a homeomorphism of the space $\mathbb{R}^3 - T$ onto itself taking the curves AB and CD shown in* Fig.12.5,a *to the curves AB and CD shown in* Fig.12.5,b, *and this homeomorphism is the identity outside D^3.*

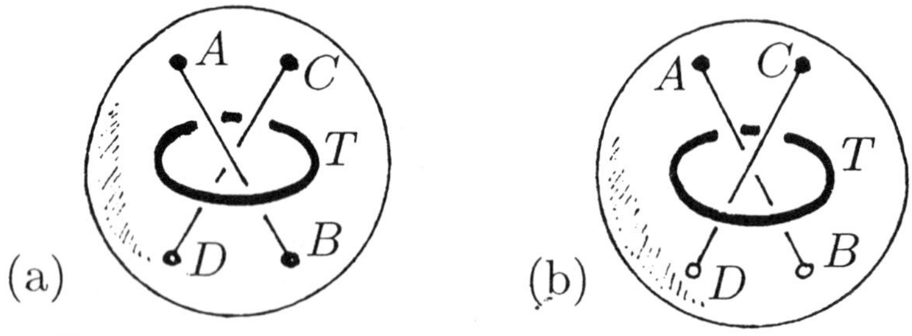

FIGURE 12.5. Crossing change with linking solid torus

Proof. The argument will be easier to visualize if we assume that the solid torus T is very thick, i.e., its inner diameter is much smaller than its outer diameter, so that the "doughnut hole" of the solid torus looks like a deep well (Fig.12.6).

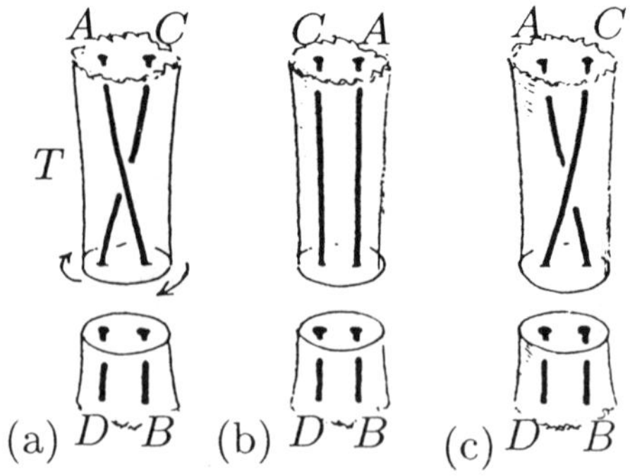

FIGURE 12.6. Homeomorphism producing a crossing change

Let us make a flat circular cut inside the 3-disk D^3 that slices through the well, as shown in Fig.12.6,a. Now we begin to rotate the upper copy of the cut (which is a 2-disk with two distinguished points) in the direction shown by the arrows; this is possible because T is removed. Turning through an angle of 180°, we get the picture shown in Fig.12.6,b, and continuing through 180° more, we get the configuration shown in Fig.12.6,c. As the result, each point of the cut 2-disk returns to its initial position, so we can glue back, obtaining a homeomorphism of $D^3 - T$ that can be extended by the identity outside of D^3 to a homeomorphism of $\mathbb{R}^3 - T$ producing the required crossing change. □ □

Problem 12.4. Prove that a) the links L_1 and L_2 in Fig.12.7 are not isotopic; b) their complements $\mathbb{R}^3 - L_1$ and $\mathbb{R}^3 - L_2$ are homeomorphic.

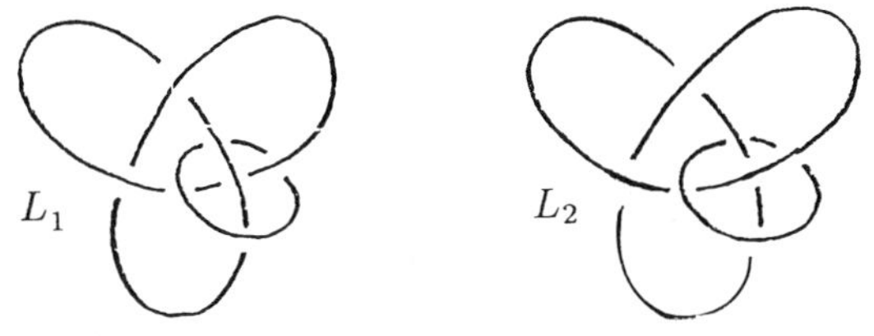

FIGURE 12.7. Nonisotopic links with homeomorphic complements

Problem 12.5. Prove that a) the knots K_1 and K_2 in the solid torus T (Fig.12.8) are nonisotopic, i.e., there exists no isotopy of T, beginning with the identity, that takes K_1 to K_2; b) the complements $T - K_1$ and $T - K_2$ to these knots are homeomorphic.

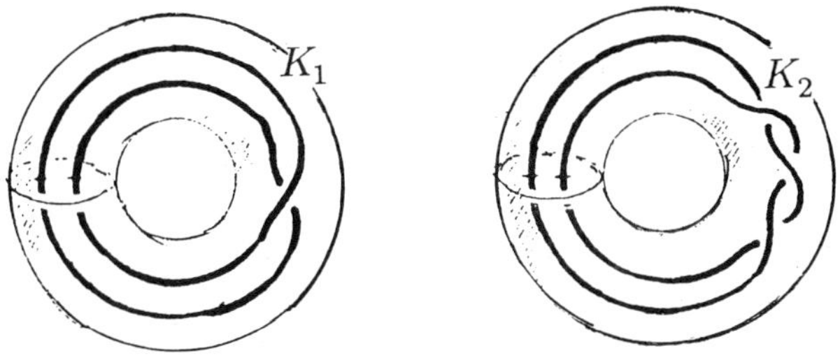

FIGURE 12.8. Knots in $D^2 \times S^1$ with homeomorphic complements

12.6. COROLLARY (ROKHLIN'S THEOREM). *Any orientable* 3*-manifold* (*without boundary*) *is the boundary of a* 4*-manifold.*

Before proving this corollary (in subsection 12.7 below), let us look at some examples of manifolds that are boundaries of other manifolds, as well as at some that are not. We shall say that a manifold (without boundary) M^n *bounds* if there exists a manifold (with boundary) W^{n+1} such that $\partial W^{n+1} = M^n$.

Problem 12.6. Find a manifold bounding the Klein bottle.

Recall that a nonidentical map $\sigma : X \to X$ is said to be an *involution* if we have $\sigma(\sigma(x)) = x$ for all $x \in X$.

Problem 12.7. Prove that if there is an involution without fixed points on a manifold M^n, then M^n bounds.

Problem 12.8. Prove that a) $\mathbb{R}P^{2n+1}$, b) $\mathbb{C}P^{2n+1}$, c) any compact Lie group, bounds by specifying involutions on them.

In order to establish that there exist manifolds that do not bound, let us recall the definition of the *Euler characteristic* of a triangulated manifold M^n: it is given by the formula

$$\chi(M^n) = a_0 - a_1 + a_2 - \cdots + (-1)^n a_n,$$

where a_i stands for the number of simplices in dimension i. We shall not prove the fact that the Euler characteristic does not depend on the triangulation of M^n.

Problem 12.9. Prove that the Euler characteristic of any odd-dimensional manifold (without boundary) is zero.

Problem 12.10. Prove that $M^n = \partial W^{n+1}$ implies $\chi(M^n) \equiv 0 \mod 2$.

Problem 12.11. Prove that neither of the manifolds $\mathbb{R}P^{2n}$ and $\mathbb{C}P^{2n}$ bounds.

12.7. Proof of Corollary 12.6. First we shall describe a "surgical operation" for manifolds of arbitrary dimension under which manifolds that bound are transformed into manifolds that still bound. Then we shall analyze the proof of Corollary 12.4 and establish that the manifold M^3 was obtained there from the sphere S^3 precisely by such surgical operations.

For arbitrary positive integers p and q, the manifold $S^p \times S^q$ bounds both manifolds $D^{p+1} \times S^q$ and $S^p \times D^{q+1}$. Therefore from any manifold M_1^n of dimension $n = p + q + 1$ we can cut out $D^{p+1} \times S^q$ and glue in $S^p \times D^{q+1}$ (along the natural homeomorphism of their boundaries). Denote the resulting manifold by M_2^n.

LEMMA. *There exists a manifold W^{n+1} whose boundary is the disjoint union of M_1^n and M_2^n.*

Proof. Consider the manifold $W_1^{n+1} = M_1^n \times I$, where $I = [0, 1]$. Its boundary consists of two copies of M_1^n, namely the manifolds $M_1^n \times \{0\}$ and $M_1^n \times \{1\}$. Consider the handle $D^{p+1} \times D^{q+1}$. Its boundary consists of $S^p \times D^{q+1}$ and $D^{p+1} \times S^q$, and these two manifolds intersect along $S^p \times S^q$. Let us identify the manifold $D^{p+1} \times S^q$ belonging to the handle $D^{p+1} \times D^{q+1}$ with the manifold $D^{p+1} \times S^q$ that was cut out from $M_1^n \times \{1\}$ (Fig.12.9). The result will be the required manifold W^{n+1}. Indeed, one of its boundary components is $M_1^n \times \{0\} = M_1^n$, while the other has the following structure: from the manifold $M_1^n \times \{1\}$ we have cut out $D^{p+1} \times S^q$ and pasted in $S^p \times D^{q+1}$ along the natural homeomorphism of their boundaries, thus obtaining M_2^n (up to a homeomorphism). □

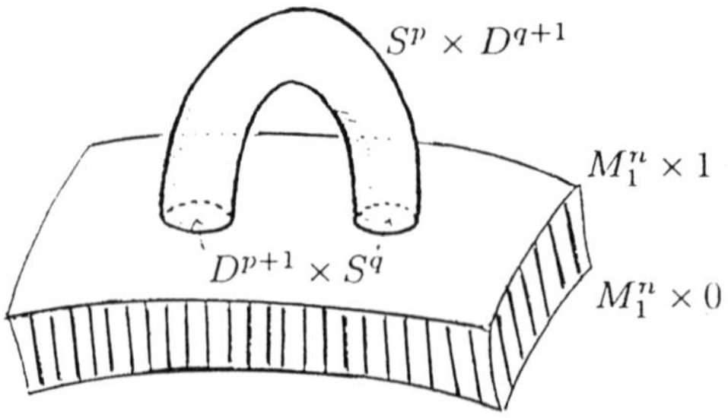

FIGURE 12.9. Adding a handle

In particular, if $M_1^n = \partial U^{n+1}$, then $M_2^n = \partial V^{n+1}$, where the manifold V^{n+1} is obtained from U^{n+1} and W^{n+1} by regarding these manifolds as glued together along their common boundary M_1^n.

For the surgical operation on three-dimensional manifolds, we must consider the case $p = q = 1$. In this case, we cut out the manifold $D^2 \times S^1$ and glue in $S^1 \times D^2$ in its place. Each of these manifolds is homeomorphic to the solid torus, and the natural homeomorphism of their boundary tori $\partial(D^2 \times S^1) \to \partial(S^1 \times D^2)$ identifies meridians with parallels. Therefore, in the case $p = q = 1$, the operation, which we call a *torus switch*, is carried out as follows: a solid torus is cut out of the given 3-manifold and then is pasted back in, along a homeomorphism that interchanges parallels and meridians.

In the proof of Corollary 12.4 we obtained M^3 from S^3 by cutting out solid tori from the latter and then pasting them back in along different attaching homeomorphisms. Let us look more closely at these homeomorphisms and verify that they are of the required type, i.e., it can be assumed that they are torus switches (they interchange parallels and meridians). Suppose that a is a meridian of the solid torus removed from S^3, while b is the meridian of the one that we have pasted in. Let us draw these curves on the boundary of $S^3 - T$ (Fig.12.10).

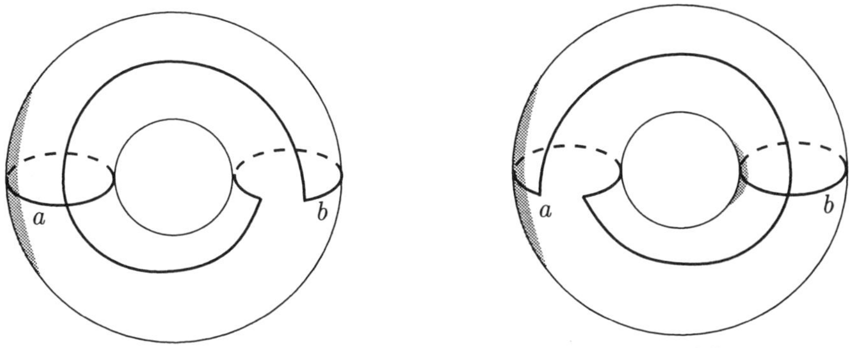

FIGURE 12.10. Twisted meridians and parallels

The result of gluing a solid torus to a boundary torus of a 3-manifold depends only on how the meridian of the solid torus is attached. Indeed, the passage from a solid torus with standard meridian and parallel to one with a standard meridian and an arbitrary parallel (Fig.12.11) can be carried out by isotopies and a finite number of twists of the solid torus along its meridional disk. But these transformations do not change the 3-manifold.

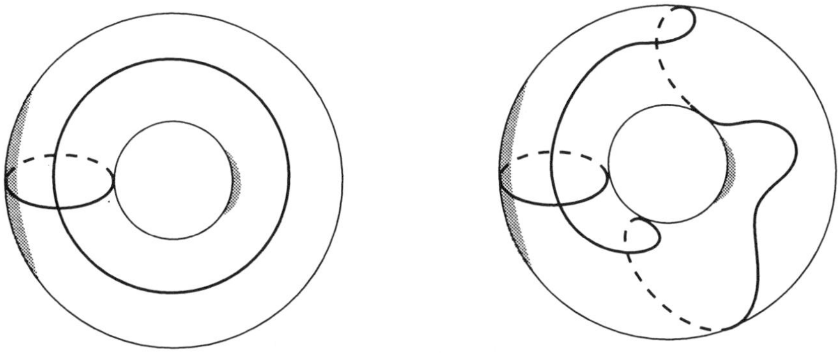

FIGURE 12.11. Standard and arbitrary parallel on the torus

Thus the gluing of the two solid tori shown in Fig.12.10 is equivalent to the gluing shown in Fig.12.12. Hence the regluing operation in the proof of Corollary 12.4 is equivalent to a regluing that interchanges parallels and meridians, i.e., to a torus switch, as required.

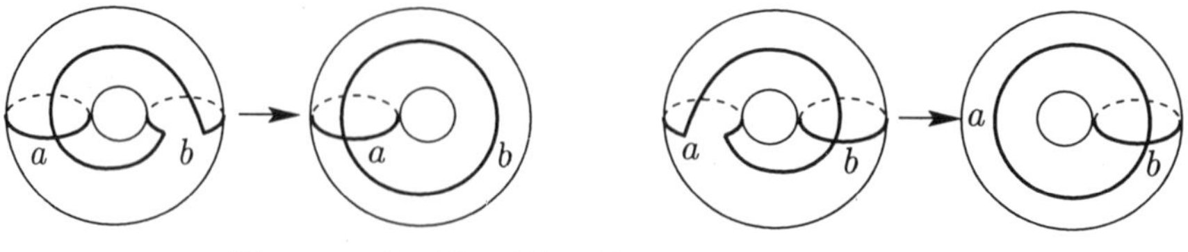

FIGURE 12.12. Identifications on the torus

When several regluing operations occur, the proof is similar. □

§13. Proof of the Dehn–Lickorish theorem

13.1. It is more convenient to prove this theorem in a more general form, by including 2-manifolds with boundary in our considerations. It is easy to check that any homeomorphism of an oriented 2-manifold with boundary which is the identity on a boundary component is orientation preserving. We shall prove the following statement.

13.2. THEOREM. *Suppose F is a compact oriented 2-manifold with boundary ∂F. Then any homeomorphism $h : F \to F$ identical on ∂F is isotopic to a composition of Dehn twists. (In the case when $\partial F = \varnothing$, we must additionally require that h be orientation preserving.)*

Let us call any composition of Dehn twists and homeomorphisms isotopic to the identity a *c-homeomorphism.* The proof of Theorem 13.2, which appears in subsection 13.4 below, is based on the following statement.

13.3. LEMMA. *Suppose that α and β are closed curves on the 2-manifold F (possibly with boundary), each of which does not disconnect F (i.e., its complement in F is connected). Then there exists a c-homeomorphism of F taking α to β.*

Proof. We shall consider three cases.

Case 1. The curves α and β intersect (transversally) at exactly one point. Then the composition of Dehn twists τ_α and τ_β along α and β transforms the curve α into a curve isotopic to β (Fig.13.1).

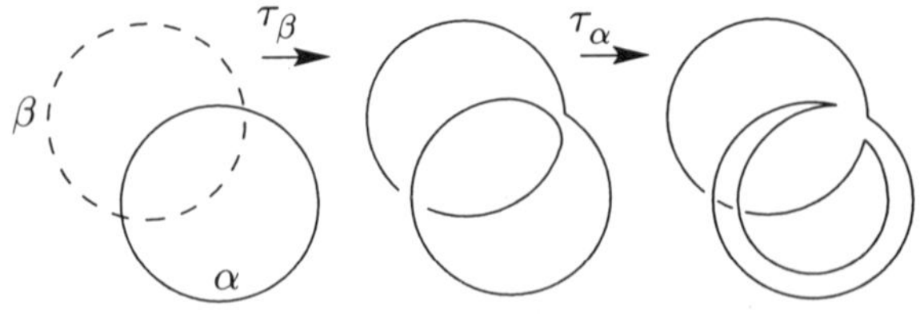

FIGURE 13.1. Case of two curves with one intersection point

Case 2. The curves α and β don't intersect. Let us show that in this case there exists an oriented curve γ that does not disconnect F and intersects each of the curves α and β transversally at one point. Then we will be able to use Case 1 and successively transform α into γ and γ into β.

Let us cut the surface F along α. In the case when $\alpha \cup \beta$ disconnects F, the needed curve γ is shown in Fig.13.2,a. If, on the contrary, $\alpha \cup \beta$ does not disconnect F, to the cut along α we add a cut along β; the needed curve γ is then shown in Fig.13.2,b.

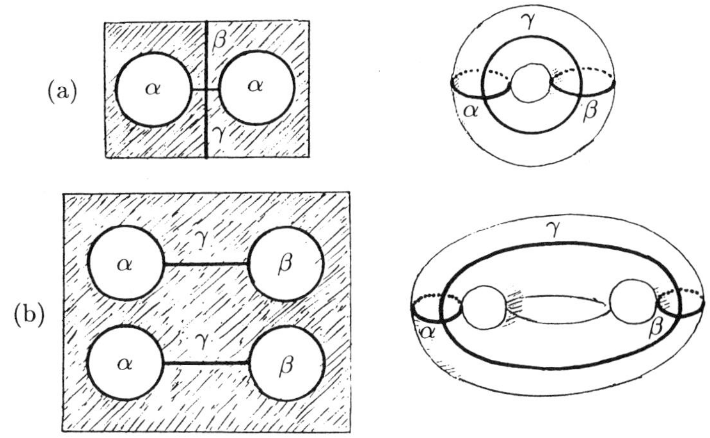

FIGURE 13.2. Case of nonintersecting curves

Case 3. The number of intersection points of α and β is greater than 1. If the curves α and β intersect at infinitely many points, we can replace (using a small isotopy) β by another curve β' such that α and β' intersect at a finite number of points. So we can assume that α and β intersect transversally at n points, where $n < \infty$. It now suffices to prove the existence of a curve γ such that:

1) γ does not disconnect the surface F;
2) γ intersects α at no more than one point;
3) γ intersects β at less than n points.

This is indeed sufficient, because if we have such a γ, we can transform α into γ by using the results of Cases 1 and 2, and then induction on n to transform γ into β.

To prove the existence of such a γ, consider two neighboring intersection points P and Q on β (this means that the arc l on β with endpoints P and Q intersects α only at the points P and Q). The points P and Q cut the curve α into two arcs α_1 and α_2. Consider the curves $\gamma_1 = \alpha_1 \cup l$ and $\gamma_2 = \alpha_2 \cup l$ (Fig.13.3). We claim that at least one of the curves γ_1 and γ_2 does not disconnect the surface F.

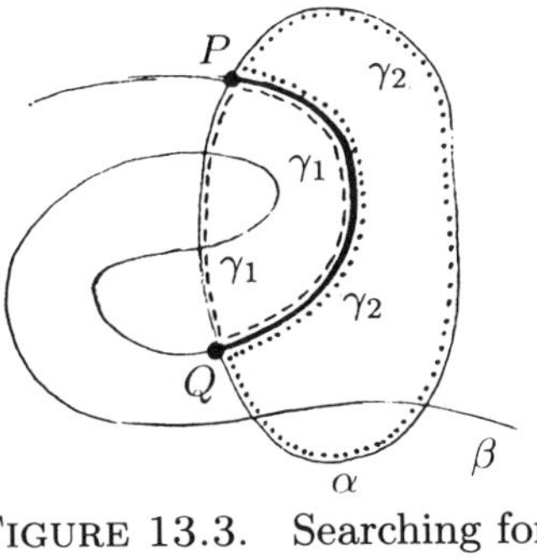

FIGURE 13.3. Searching for γ

If both do, it is impossible to get from the region marked 1 to region 2 (Fig.13.4) and from region 3 to region 2. Thus the region 2 is not connected to either region 1 or 3; so α disconnects F, contradicting the assumptions of the lemma and proving our claim.

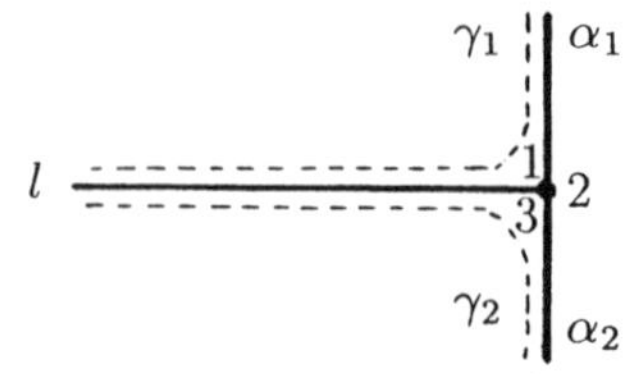

FIGURE 13.4. The curves γ_1 and γ_2

To be definite, assume that the curve γ_1 does not disconnect F. Pushing it off itself slightly, we obtain the needed curve γ (Fig.13.5), concluding the proof of Lemma 13.3. □

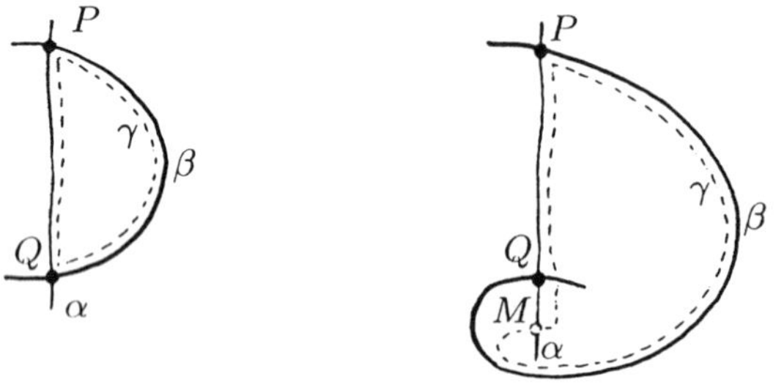

FIGURE 13.5. The curve γ (two possible dispositions)

13.4. Proof of Theorem 13.2. The surface F is, according to the classification theorem for 2-manifolds with boundary, a sphere with g handles and k holes. Cutting F along the meridians $m_1, \ldots, m_g$ (Fig.13.6), we get a disk with $k+2g-1$ little disks removed.

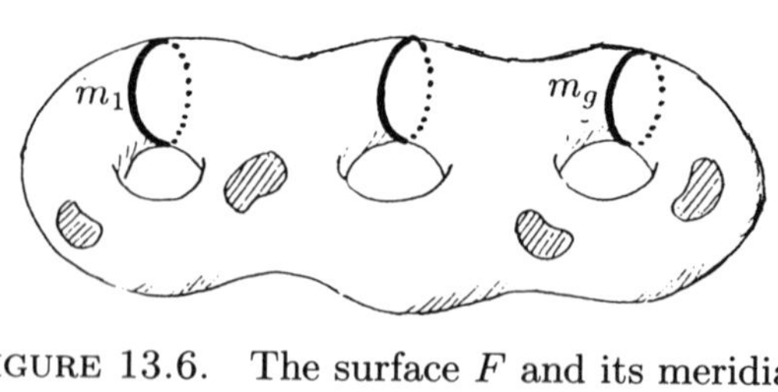

FIGURE 13.6. The surface F and its meridians

Under the homeomorphism $h : F \to F$, the meridian m_1 is transformed into a curve $h(m_1)$ that does not disconnect F. According to Lemma 13.3, there exists a c-homeomorphism f_1 taking $h(m_1)$ to m_1. First let us assume that the orientations of the curves $f_1 h(m_1)$ and m_1 coincide. Then there exists a homeomorphism f_1' isotopic to f_1 for which the map $f_1'h$ is the identity on the curve m_1. The curve m_1 splits its ε-neighborhood U into two parts U_1 and U_2. Since the homeomorphism $f_1'h$ is orientation preserving and is the identity on m_1, it maps points of U_i located near m_1 to points inside U_i. This means that if we cut F along m_1, then the homeomorphism $f_1'h$ will be a map of the resulting surface, identical on its boundary. After g such cuts we get a homeomorphism $f_g' \dots f_1'h$ of the disk with holes, identical on the boundary components. By Theorem 7.6, this homeomorphism is isotopic to a composition of Dehn twists, and therefore so is h.

It remains to figure out what to do when the curves $f_1h(m_1)$ and m_1 have opposite orientations. Since the curve $\alpha = m_1$ does not disconnect F, there exists a curve β intersecting α at exactly one point (Fig.13.7).

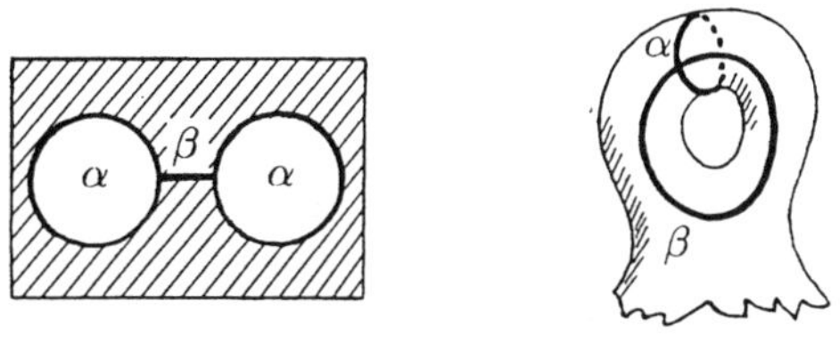

FIGURE 13.7. Finding the curve β

Let α^{-1} be a curve that coincides with α as a point set, but has the opposite orientation. The composition of twists $\tau_\beta\tau_\alpha\tau_\beta$ takes the curves α and β to curves isotopic to β and α^{-1} respectively (Fig.13.8).

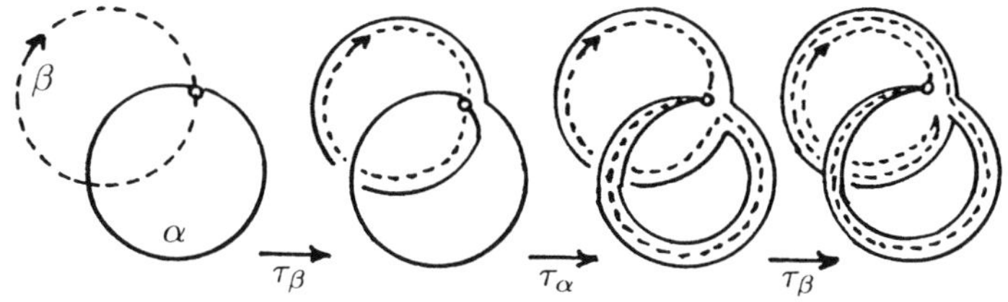

FIGURE 13.8. The composition of twists $\tau_\beta\tau_\alpha\tau_\beta$

Hence the homeomorphism $(\tau_\beta\tau_\alpha\tau_\beta)^2$ sends the curves α and β to curves isotopic to α^{-1} and β^{-1}. What is important for us is that it takes α to α^{-1}. This allows us to construct a c-homeomorphism f_1'' for which the curve $f_1''h(m_1)$ coincides with m_1 as a point set and has the same orientation, thereby reducing the case under consideration to the one already studied. □

Comments

Heegaard splittings are a bridge between two and three-dimensional topology, more precisely between homeomorphisms of surfaces and three-dimensional manifolds. Effective descriptions of the former yield effective descriptions of the latter. The main geometric ingredient in describing homeomorphisms of surfaces is the Dehn twist, invented by the German mathematician M. Dehn (see [**De1**]), who presented a proof of the theorem asserting that any homeomorphism of a surface is a composition of twists. However, this proof was not free of important errors. The definitive proof of Dehn's theorem was given by W. B. R. Lickorish (see [**Lic1**], 1962). The paper [**Rou**] and the book [**MF**] are the basis of our exposition.

It should be noted that much of the work in topology of Dehn and his contemporaries (e.g. Heegaard and Nielsen) was, from the modern point of view, technically rather imperfect, although the main ideas were sound. Only after World War II did this subject matter achieve a satisfactory degree of rigor. In particular, C. Papakyriakopoulos gave the first rigorous proof [**Pap**] of Dehn's famous lemma (on spanning disks for homotopy trivial curves) and H. Zieschang finally proved (see [**Zi1-2**]) Nielsen's theorem (which asserts, roughly speaking, that the fundamental group controls the homeomorphism group of surfaces).

The main consequence of the Dehn-Lickorish theorem was the fact that any 3-manifold can be obtained from the sphere by cutting out and pasting back solid tori. This fact, in its turn, was the starting point of much beautiful work on the topology of 3-manifolds by many authors, among whom V. A. Rokhlin, S. P. Novikov, R. Kirby, and W. Thurston are the most famous. For a modern survey of this subject matter (to which this chapter does not claim to do justice), emphasizing its algorithmic aspects, see the book [**MF**].

CHAPTER VI

Surgery of 3-Manifolds

In the previous chapter, we repeatedly transformed the 3-manifolds that we were working with by cutting out a solid torus and gluing it back in along a different homeomorphism. We proved that by iterating this operation, one can transform any orientable 3-manifold, say S^3, into any other orientable 3-manifold (Corollary 12.4). This approach to encoding 3-manifolds, called surgery presentation, is the main theme of the present chapter. In developing this theme, we shall study different types of surgery (§14 and §16), find convenient sets of operations sufficient to produce all manifolds, and finally discuss the beautiful Kirby calculus (§19), which tells us what different surgery presentations produce the same manifold. However, our exposition will not be very linear: we shall often stray from mainstream surgery presentation theory to wander among related topics, which become quite transparent when observed from the point of view of surgery. In particular, we shall finally introduce the Gauss linking numbers of two curves (§15), revisit lens spaces (§17), and have a look at another famous infinite series of 3-manifolds, the homology spheres (§18).

§14. Rational surgery along trivial knots

We saw in §12 that any compact orientable 3-manifold (without boundary) can be obtained from the sphere S^3 by several torus switches, i.e., by cutting out several unknotted solid tori and then pasting them back in 'inside-out' (interchanging the parallels and meridians). In many situations, however, it is more convenient to consider arbitrary reattaching homeomorphisms of the solid torus. The manifold obtained in this way is uniquely determined by the image J of the meridian α under the attaching homeomorphism of the boundray torus $S^1 \times S^1$, in fact by the isotopy class of J. Suppose that $J = p\alpha + q\beta$, i.e., J is the closed curve that winds around the boundary torus p times along the meridian and q times along the parallel; more precisely, this means that the curve represents the element $(p, q) \in \mathbb{Z} + \mathbb{Z} = \pi_1(S^1 \times S^1)$.

The result stated in Proposition 14.1(a) below asserts that in this case the integers p and q have no common divisors, while Proposition 14.1(b) asserts that the numbers p and q determine the curve J uniquely up to isotopy.

14.1. PROPOSITION. (a) *If the curve $p\alpha + q\beta$ is closed and has no self-intersections, then either the integers p and q are coprime, or one of them is 0 and the other is ± 1.*

(b) *If two closed curves without self-intersections on the torus are homotopic, then they are isotopic.*

Proof. (a) The torus can be obtained from the plane $\mathbb{R}^2$ by identifying the points (x, y) and $(x+m, y+n)$, where $m, n \in \mathbb{Z}$. Moreover, any curve on the torus can be obtained from a curve in the plane in this way. We shall prove the following statement: *if d is a natural number, then for any curve $\gamma \subset \mathbb{R}^2$ with endpoints A and B there exist points P and Q such that $PQ \| AB$ and $|PQ| = |AB|/d$.*

First let us show that this statement implies assertion (a) of the proposition. Indeed, to the curve $p\alpha + q\beta$ corresponds the plane curve whose extremities have coordinates differing by p and q. Hence if p and q are divisible by d, then the difference of the coordinates of the points P and Q are integers, and so one and the same point on the torus will correspond to P and Q. Thus the curve will be self-intersecting, contrary to the assumption of the proposition.

To prove the italicized statement, we rephrase it by using the following definition. We say that the distance n is *realized* on the plane curve joining the points A and B if there exists a segment of length n parallel to AB with endpoints on the curve. We must prove that *the distance $|AB|/d$ is realized on any curve joining the points A and B.*

First we shall prove that for any $0 < \delta < 1$ and on any curve joining A and B at least one of the distances $\delta|AB|$ and $(1-\delta)|AB|$ is realized. Without loss of generality we may assume that the segment AB lies on the x-axis and the x coordinates of its endpoints are 0 and 1. Assume that for some curve γ_0 joining the points A and B neither of the distances δ and $1-\delta$ is realized. Then when we translate this curve along the x-axis by the distances δ and $1-\delta$, we get curves without common points with γ_0. Let γ_δ and γ_1 be the curves obtained by translating γ_0 in the positive direction along the x-axis by the distance δ and 1, respectively. Then γ_δ does not intersect either γ_0 or γ_1. Construct the curve L as follows. On the curve γ_δ choose a point with maximal y-coordinate and draw a ray parallel to the y-axis from this point towards $+\infty$, then choose a point with minimal y-coordinate and draw a ray parallel to the same axis from that point towards $-\infty$ (Fig.14.1). The curve L consists of these two rays and the part of the curve γ_δ that goes from the origin of one ray to that of the other. The curves γ_0 and γ_1 do not intersect either the curve γ_δ or the two rays. Hence they do not intersect L. But L splits the plane into two disjoint components, the points of the curves γ_0 and γ_1 with maximal y-coordinates lying in different components. On the other hand, the curves γ_0 and γ_1 have a common point, namely $(1,0)$. This contradiction implies that at least one of the distances δ and $1-\delta$ is realized.

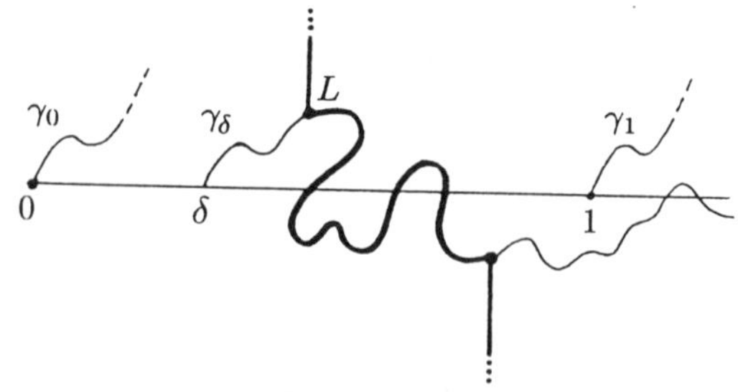

FIGURE 14.1. The curve L separating γ_0 from γ_1

Now let us prove, by induction on d, that the distance $|AB|/d$ is realized on any curve. For $d = 1$ the statement is obvious. The induction step goes as follows.

Let $\delta = 1/d$. Then $1 - \delta = (d-1)/d$, and so one of the distances $|AB|/d$ or $(d-1)|AB|/d$ is realized. If the first one is, we are done. But if the distance $(d-1)|AB|/d$ is realized, then for the curve joining the endpoints of the segment of length $(d-1)|AB|/d$, by the induction assumption, the distance

$$\frac{1}{d-1}\frac{(d-1)|AB|}{d} = \frac{|AB|}{d}$$

is realized as well. □

(b) The proof of statement (b) will be in four steps.

Step 1. *The given curves may be assumed homotopic to the meridian of the torus.* Indeed, suppose the given curves are of type (p,q). According to (a), the integers are coprime. Hence there exist integers r and s such that $ps - qr = 1$. The linear transformation

$$e_1 \mapsto pe_1 + qe_2, \qquad e_2 \mapsto re_1 + se_2$$

induces a homeomorphism of the torus onto itself which takes the curve of type $(1,0)$ to a curve of type (p,q).

Step 2. *It suffices to prove that a curve of type* $(1,0)$ *without self-intersections is isotopic to the meridian of the torus.* Indeed, having constructed isotopies of the curves γ_1 and γ_2 to the meridian, it is easy to construct an isotopy of γ_1 to γ_2.

Step 3. *It suffices to prove that a curve of type* $(1,0)$ *without self-intersections is isotopic to a curve that does not intersect the meridian.* Let us cut the torus along the meridian. As the result, we obtain an annulus containing a curve without self-intersections that is homotopic to the boundary components of the annulus (Fig.14.2). It obviously follows from the Jordan curve theorem that this curve is isotopic to the midline of the annulus.

FIGURE 14.2. Annulus with a nonselfintersecting curve

Step 4. *A curve of type* $(1,0)$ *without self-intersections that intersects the meridian in* $n > 0$ *points is isotopic to a curve that intersects the meridian in less than* n *points.* Cut the torus along the meridian, obtaining a cylinder. From an infinite collection of such cylinders, let us glue an infinite cylinder that covers the torus. Consider a lifting γ_0 of the given curve γ to this cylinder, i.e., a component of the inverse image of γ under the covering projection (Fig.14.3).

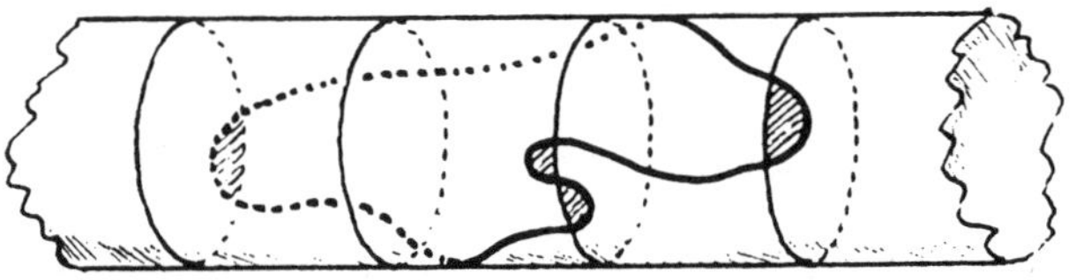

FIGURE 14.3. "Caps" on the infinite cylinder

The curve γ_0 (assumed to be in general position) cuts off several "caps" of the type displayed in the figure. We can assume that there is at least one such cap (otherwise we are done by Step 3). We must prove (working by induction) that one cap can always be destroyed by an isotopy. Note that in order to obtain a nonselfintersecting curve, during the isotopy we must not cross the other liftings γ_m of our curve. However, any cap of least height does not contain points of the curves γ_m, $m \neq 0$, as otherwise the curve γ would have selfintersections. Hence such a cap can be isotopically destroyed (by the Jordan curve theorem). This isotopy can be lowered to the torus. □

14.2. Definitions and conventions. Consider the surgery of the sphere S^3 in which a tubular ε-neighborhood of the trivial knot $J \subset S^3$ is removed and the meridian α of the repasted solid torus is identified with the curve $p\alpha+q\beta$. It is easy to verify that the surgery that glues α back onto $-\alpha$ (Fig.14.4,a) does not change our manifold. Hence surgery with the identification of the meridian and the curve $-p\alpha - q\beta$ is equivalent to surgery with the identification of the meridian and the curve $p\alpha + q\beta$.

We shall agree that the orientation of the meridian and parallel are chosen as shown in Fig.14.4,b.

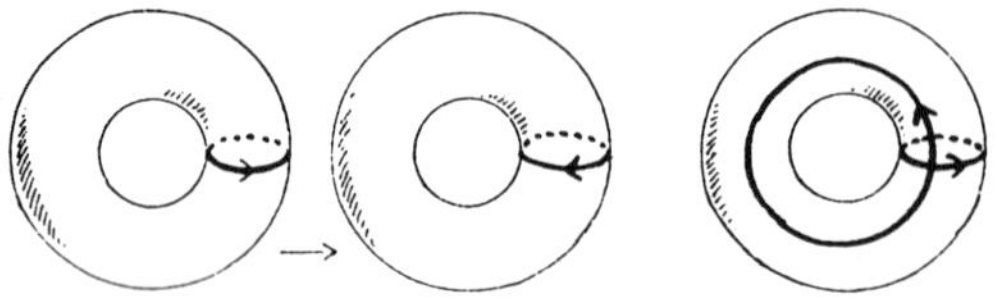

(a) (b)

FIGURE 14.4. Orienting the meridian and parallel

Now from Proposition 14.1 it follows that the surgery of the sphere S^3 considered above is entirely determined by the rational number $r = p/q$. We shall call this number the *framing index* or simply *framing* of the trivial knot J, and the corresponding operation, *rational surgery* with framing index r.

14.3. Note that for identical surgery, we have $r = 1/0 = \infty$, while rational surgery of index $r = 0$ (along the trivial knot) is a torus switch (interchanges parallels and meridians). This may seem strange at first glance; indeed it would appear more logical to take $r = q/p$ instead of $r = p/q$ for the framing index, so that identical surgery would have index 0. But, as we shall soon see, the crucial role in this theory is played by so-called integer surgery (for which $q = 1$), and for this reason it is better to take $r = p/q$ rather than $r = q/p$.

In the following subsections we consider some examples of manifolds obtained from the sphere S^3 by rational surgery along the framed trivial knot $\bigcirc^r$.

14.4. PROPOSITION. $\bigcirc^0 = S^1 \times S^2$.

Proof. The sphere S^3 can be obtained by gluing together two solid tori T_1 and T_2 along a homeomorphism $f : \partial T_1 \to \partial T_2$ interchanging parallels and meridians. Rational surgery with framing index $r = 0$ consists in cutting out the solid torus T_2 from S^3 and pasting it back in along a homeomorphism that exchanges parallels and meridians. The result is the gluing together of two solid tori T_1 and T_2 along the identical homeomorphism of their boundaries. Since $T_i \approx S^1 \times D^2$ and gluing together D^2 and D^2 along the identity map of their boundary circles produces S^2, gluing together T_1 and T_2 along the identity map of their boundaries produces $S^1 \times S^2$. □

14.5. PROPOSITION. $\bigcirc^{p/q} = L(p,q)$

Proof. The lens space $L(p,q)$ was obtained by gluing together two solid tori along the homeomorphism of their boundary that takes the meridian α of one torus to the curve $q\alpha + p\beta$ on the other torus (see §11). On the other hand, the sphere S^3 can be obtained by identifying the boundaries of these tori along the homeomorphism that takes α to β and β to α. Hence $L(p,q)$ can be obtained from S^3 by regluing a solid torus along the homeomorphism that takes the curve α to $p\alpha + q\beta$. □

14.6. PROPOSITION. $\bigcirc^{\pm 1/n} = S^3$

Proof. If we use the definition of $L(p,q)$ given in §11, then this proposition is a particular case of the previous one. Indeed, $L(p,q)$ is defined as the quotient of the unit 3-sphere in $\mathbb{C}^2$ by the equivalence

$$(z,w) \equiv (\exp(2\pi i/p)\, z, \exp(2\pi i q/p)\, w)\,.$$

This definition implies that for $p = \pm 1$ no identifications of points occur, i.e., $L(\pm 1, n) = S^3$. □

14.7. PROPOSITION. $\bigcirc^{r} \sim \bigcirc^{1/(\pm n + \frac{1}{r})}$.

Proof. It is clear from the definition of lens spaces that $L(p,q) \approx L(p, q \pm np)$, but this is exactly the statement of the proposition in a different notation. □

14.8. Note that the homeomorphism in the previous proposition is far from obvious if we use the definition of rational surgery directly. Nevertheless, this homeomorphism, which we denote by $f : M^3 \to N^3$, where the manifolds M^3 and N^3 are obtained from the sphere S^3 by rational surgery along the trivial knot with framing r and $1/(\pm n + \frac{1}{r})$ respectively, plays an important role in the surgery of 3-manifolds. Therefore we shall describe it in more detail.

The surgery of the sphere S^3 along the circle with framing $r = p/q$ produces a manifold that can be glued together from two solid tori T_1 and T_2 along a homeomorphism of their boundaries identifying the meridian α with the curve $q\alpha + p\beta$ (Fig.14.5).

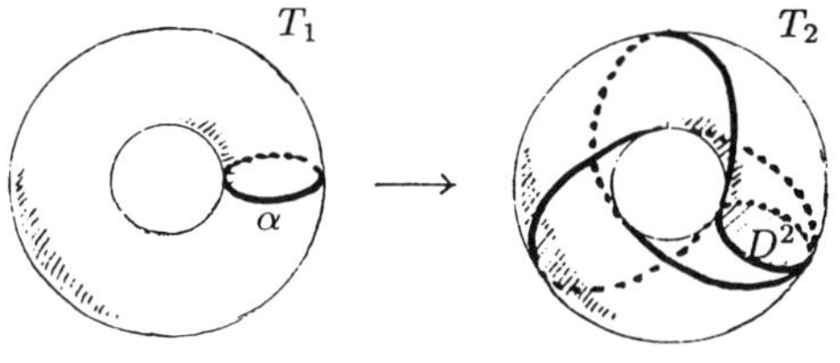

FIGURE 14.5. Rational surgery with framing 2/3

This curve intersects the meridional disk D at p points. Cut the solid torus T_2 along D and give T_2 a twist by n revolutions. Then the curve $q\alpha + p\beta$ will be transformed into $(q \pm np)\alpha + p\beta$ (the sign depends on the choice of the direction of twisting). If the curve $q\alpha + p\beta$ corresponds to surgery along J with framing $r = p/q$, then the curve $(q \pm np)\alpha + p\beta$ will correspond to surgery along J with

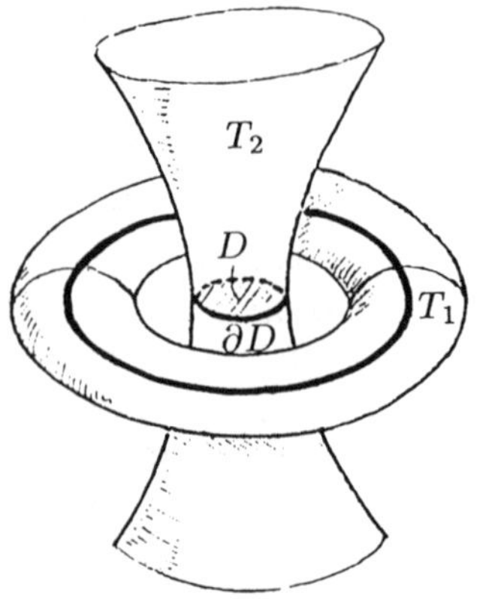

FIGURE 14.6. The parallel of T_1 is the meridian of T_2

framing $p/(q \pm np) = 1/(\pm n + 1/r)$. Here the solid torus T_1 is an ε-neighborhood of the circle J, and the restriction of f to $T_2 = M^3 - T_1$ is a twist along the parallel of the solid torus T_1 (Fig.14.6).

The twist of the solid torus T_2 shown in Fig.14.7,a adds an extra revolution in the direction of α, while the twist shown in Fig.14.7,b adds a revolution in the opposite direction. Hence a twist of the first type corresponds to a plus sign in front of n, while a twist of the second type corresponds to a minus sign. This completes our description of the homeomorphism f.

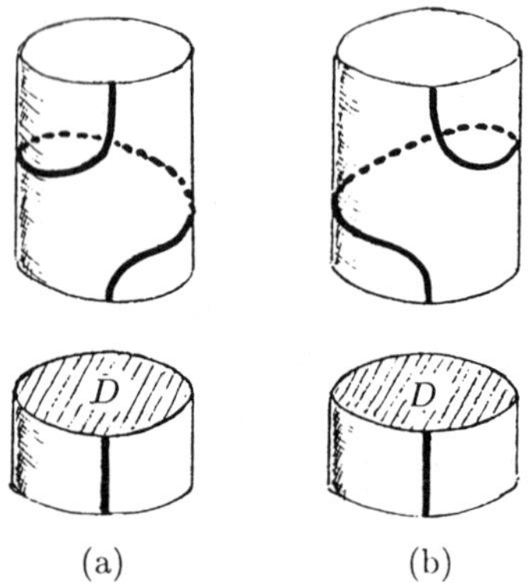

FIGURE 14.7. One-revolution twist in different directions

§15. Linking numbers

15.1. In the previous section we showed that surgery of the sphere S^3 along the trivial knot J is determined by a rational number r, the framing of J. To determine the framing of the knot, we had to choose a parallel and a meridian of its tubular ε-neighborhood, and this choice was unique up to isotopy.

In the more general case of a nontrivial knot, there are no problems in choosing the meridian: we can simply take a section of the tubular ε-neighborhood by a plane perpendicular to the knot at some point (Fig.15.1,a). However, the choice of the parallel presents difficulties. In the case of the unknot, it was sufficient to take a diagram without self-intersections and for the parallel choose the curve

K shown by the dotted line in Fig.15.1,b. But if we take the diagram with one crossing to represent the unknot, the curve K' that we get similarly (Fig.15.1,c) is not equivalent to K in the sense that the two-component links $\{J, K\}$ and $\{J, K'\}$ are not isotopic. In the case of the trefoil, the two parallels chosen in Fig.15.2 are also nonequivalent.

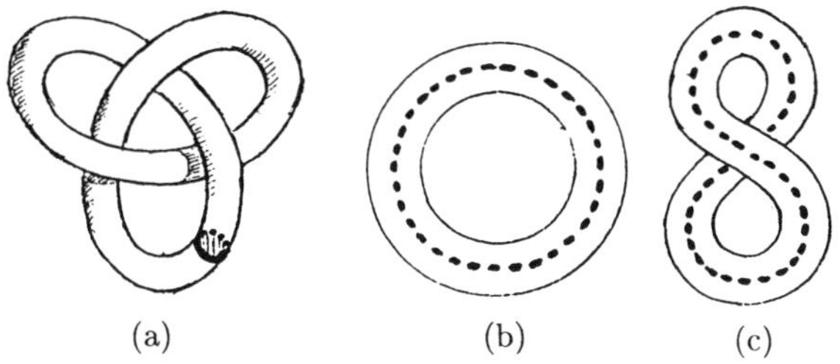

FIGURE 15.1. Choosing the meridian and parallel

FIGURE 15.2. Nonequivalent parallels for the trefoil

15.2. In order to obtain a definition of the parallel that does not depend on the choice of knot diagram, we shall introduce one of the oldest invariants of topology, the Gauss linking number of two curves J and K in S^3. For our purpose, the following definition is most convenient. Consider the diagram of a link consisting of two oriented curves J and K. Let us pay attention only to those crossing points where the curve K passes over J. These crossing points can be of two types (Fig.15.3). For each of the crossing points considered, take the corresponding value $\varepsilon_i = \pm 1$ and sum all these ε_i. The integer thus obtained is called the *linking number* of the curves J and K in S^3, and is denoted by $\text{lk}(J, K)$.

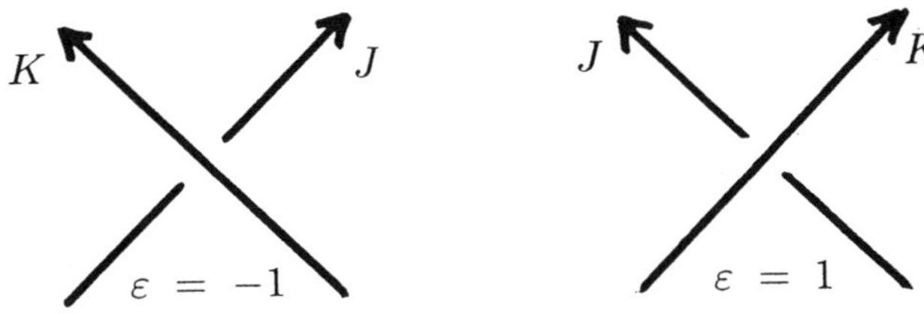

FIGURE 15.3. Two types of crossing points

15.3. THEOREM. *The linking number is an invariant of the link* $\{J, K\}$, *i.e., it does not depend on the choice of the diagram of the link.*

Proof. According to the Reidemeister Theorem 1.7, it suffices to prove that $\text{lk}(J, K)$ does not change under Reidemeister moves. The moves Ω_1 and Ω_3 never

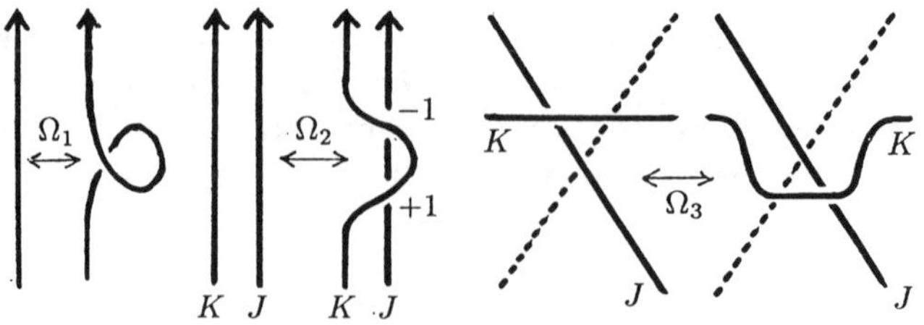

FIGURE 15.4. Invariance of linking numbers

change the set of values of ε_i, while the move Ω_2 either does nothing, or adds (or destroys) two numbers ε_i of opposite signs (Fig.15.4). □

Problem 15.1. Suppose $-J$ is the curve obtained from J by changing the orientation. Prove that $\text{lk}(-J, K) = -\text{lk}(J, K)$ and $\text{lk}(J, -K) = -\text{lk}(J, K)$.

Problem 15.2. Prove that $\text{lk}(J, K) = \text{lk}(K, J)$

15.4. Two closed curves J and K in S^3 are called *unlinked* if there exists an isotopy that takes them to two curves J' and K' lying in two nonintersecting 3-disks. In the converse case, they are called *linked.* The linking number is a very simple invariant that allows one to prove (in many cases) that two curves are linked. Indeed, we have the following statement.

15.5. THEOREM. *If* $\text{lk}(J, K) \neq 0$, *then* J *and* K *are linked.*

Proof. Suppose the curves J and K are unlinked. Consider a diagram on which the projections of the 3-disks containing the curves J' and K' do not intersect. On this diagram the curves J' and K' have no common crossing points, and so $\text{lk}(J', K') = 0$. Therefore by the previous theorem $\text{lk}(J, K) = 0$. This contradicts the assumption of the theorem; hence J and K are linked. □

Problem 15.3. Prove that the pairs of curves shown in Fig.15.5,a-d are linked.

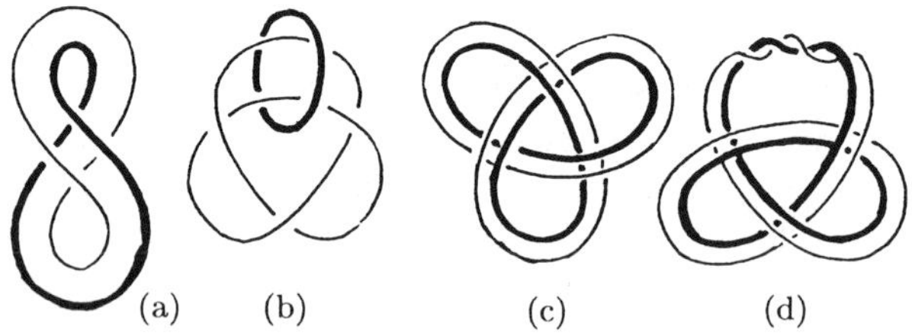

FIGURE 15.5. Examples of linked curves

Note that the relation $\text{lk}(J, K) = 0$ does not imply that the curves J and K are unlinked. For example, the two pairs of curves shown in Fig.15.6. are linked (as can be shown by computing the Jones polynomials of the two links that they constitute) although their linking numbers are zero.

15.6. We are now ready to give an invariant definition of the parallel on the tubular ε-neighborhood of an arbitrary oriented knot J. First we must exclude the situation shown in Fig.15.7,a. To that end, we require that the curves J and K be *codirected* (Fig.15.7.b). In other words, if u and v are the orientation vectors on J and K at points of the same meridional disk, their scalar product (u, v) must be positive. A curve K on the boundary of the tubular ε-neighborhood of J is said to be a *parallel* of J if K and J are codirected and $\text{lk}(J, K) = 0$.

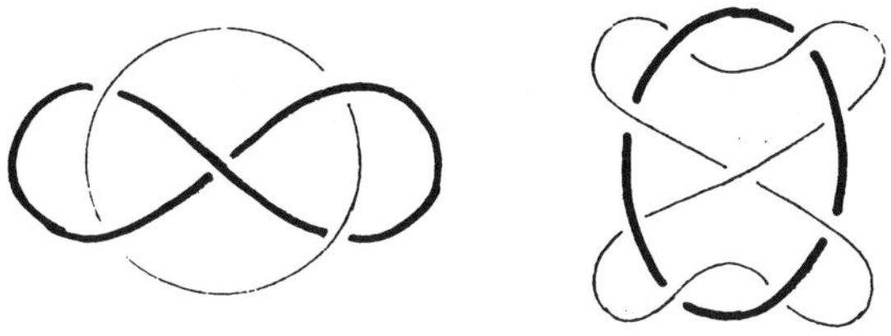

FIGURE 15.6. Linked curves with zero linking numbers

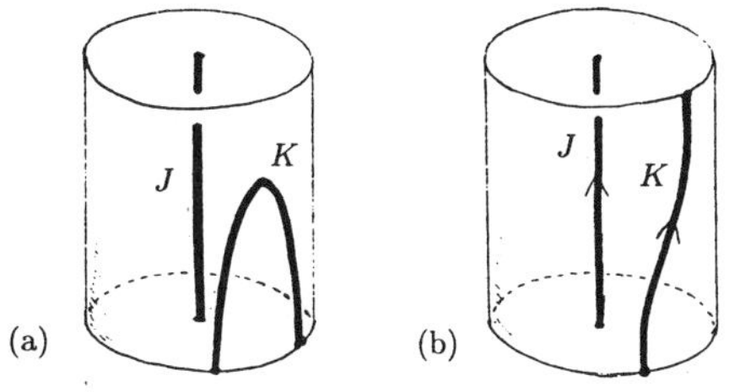

FIGURE 15.7. Not codirected and codirected curves

Problem 15.4. Let J be an oriented knot, and let n be an integer. Prove that there exists a curve K codirected with J such that $\mathrm{lk}(J, K) = n$.

Problem 15.5. Draw the parallels for the ε-neighborhoods of the curves shown in Fig.15.8.

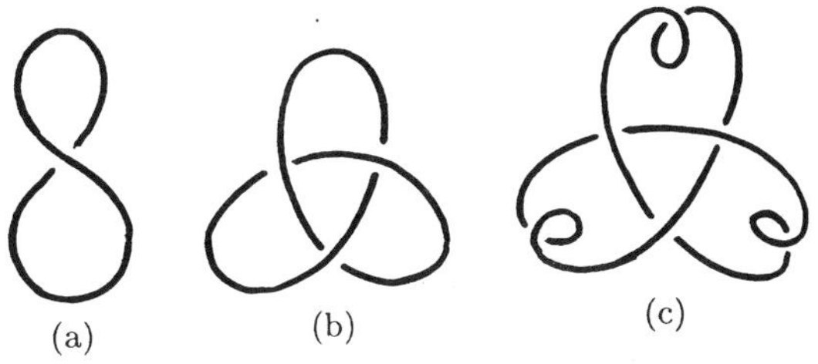

FIGURE 15.8. Knots requiring parallels

§16. Integer surgery

16.1. Now that we have agreed on the choice of the meridian and parallel, surgery of the sphere S^3 is uniquely determined by a knot J and its framing, a rational number r. Indeed, given the data (J, r), we begin by arbitrarily orienting the knot J, choose the meridian α (as explained in 15.1) and the parallel β (as explained in 15.6), and orient them so that J and β are codirected and we have $\mathrm{lk}(\alpha, J) = +1$ (Fig.16.1).

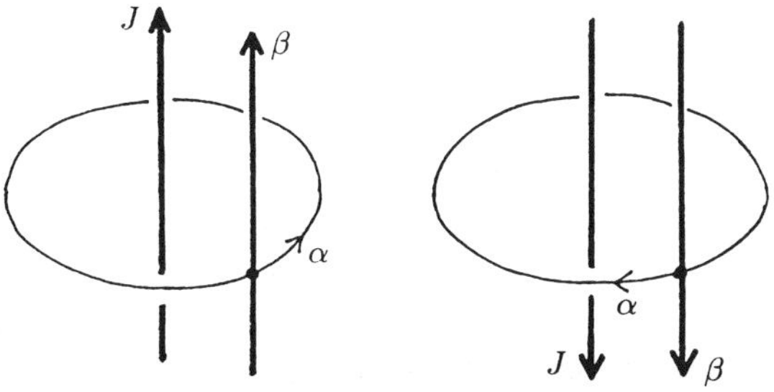

FIGURE 16.1. Orienting the parallel and meridian

Under the homeomorphism that determines the surgery, the meridian α is taken to a curve of the form $p\alpha + q\beta$, where $p, q \in \mathbb{Z}$ are coprime (recall Proposition 14.1). The integers p and q do not depend on the choice of the orientation of J, because if the orientation of J is reversed, then so is the orientation of α, so that we get the same curve (see Fig.16.2, where the curve $\alpha + \beta$ is shown for both orientations). Thus *to specify a surgery of the sphere* S^3 *it suffices to indicate a (nonoriented) knot diagram* J *and its framing* $r = p/q$.

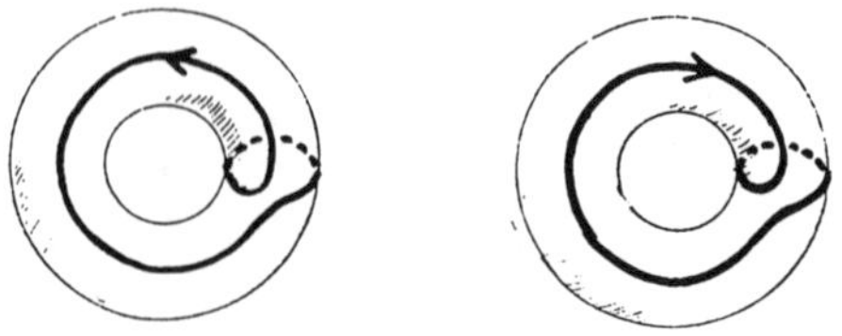

FIGURE 16.2. Result of orientation reversal

In the present section we are primarily interested in surgery along links whose components have integers for framing indices. We call this *integer surgery.*

16.2. THEOREM. *Any compact orientable* 3*-manifold without boundary can be obtained from the sphere* S^3 *by integer surgery.*

Proof. This theorem is simply a restatement of Corollary 12.4, because one torus switch is an integer surgery along a trivial knot with framing 0, while a sequence of torus switches is an integer surgery along a link (consisting of unknotted components with framings equal to 0). □

16.3. Surgery along ribbons. Under integer surgery along a component J of the given link, the meridian α is mapped to the curve $K = p\alpha + \beta$. This curve effects exactly one revolution in the direction of the parallel β, so we can assume that the curve K and the knot J are codirected. In that situation they span a narrow ribbon (Fig.16.3).

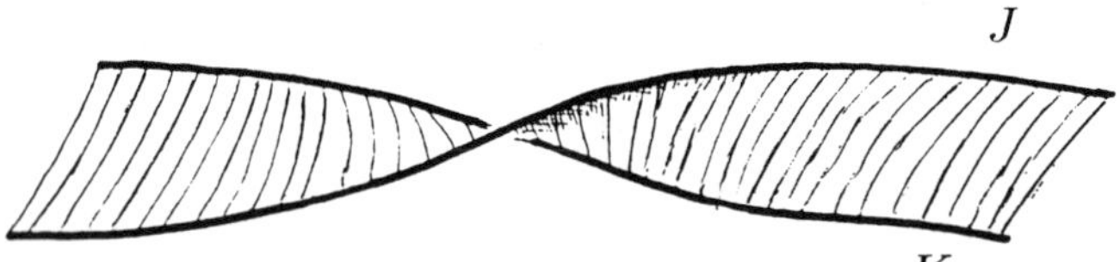

FIGURE 16.3. Ribbon spanning J and K

Therefore, an integer surgery along J with framing $r = p$ is entirely determined by a ribbon with boundary components J and K, where

$$\mathrm{lk}(J, K) = \mathrm{lk}(J, p\alpha + \beta) = p,$$

since $\mathrm{lk}(J, \beta) = 0$ by definition of the parallel and $\mathrm{lk}(J, \alpha) = 1$. Moreover, it is not necessary to distinguish the boundary components, because $\mathrm{lk}(J, K) = \mathrm{lk}(K, J)$. Thus our integer surgery presentation is determined by a set of twisted ribbons (which may be knotted and linked). We call this method of defining integer surgery a *ribbon surgery presentation.* Two examples of ribbon presentations determining

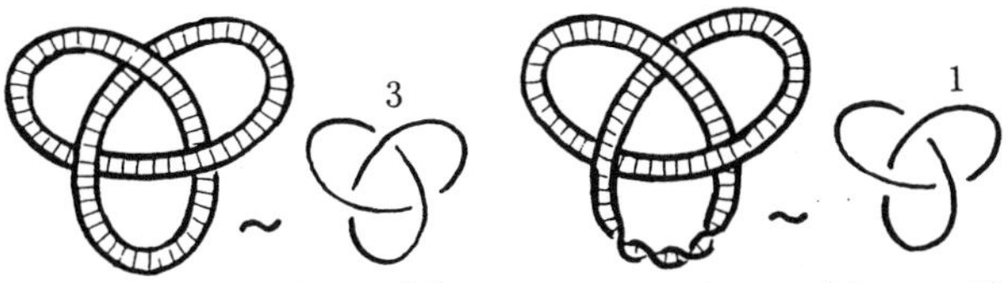

FIGURE 16.4. Two ribbon presentations of 3-manifolds

integer surgery on the sphere along the trefoil with framing $r = 1$ and $r = 3$ are shown in Fig.16.4.

16.4. Equivalent surgeries. Surgery on the sphere S^3 along different framed links can produce the same manifold (up to homeomorphism). Two such surgeries are called *equivalent*. For example, in Proposition 14.6 we saw that surgery along the circle with framing $\pm 1/n$, $n \in \mathbb{N}$, is the identity. We also showed (Proposition 14.7) that the manifolds obtained from S^3 by surgeries along the trivial knot J with framings r and $1/(\pm n + 1/r)$, where n is any positive integer, are homeomorphic and hence the two sugeries are equivalent.

The sphere S^3 can be obtained by gluing the solid torus T_1 (which is a tubular neighborhood of the circle J) and the solid torus T_2 by a homeomorphism interchanging parallel and meridian (Fig.16.5). Let us cut T_2 along the meridional disk D and perform a twist by n full turns. This changes the value of the framing r of J to $1/(\pm n + 1/r)$; the sign in front of n depends on the direction of the twist (Fig.16.6).

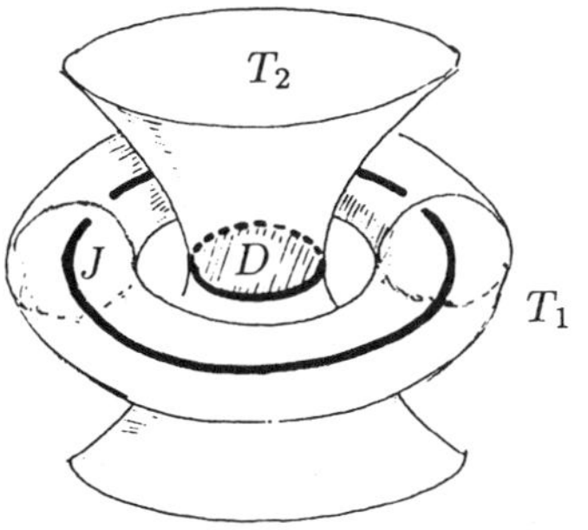

FIGURE 16.5. The sphere as two solid tori glued together

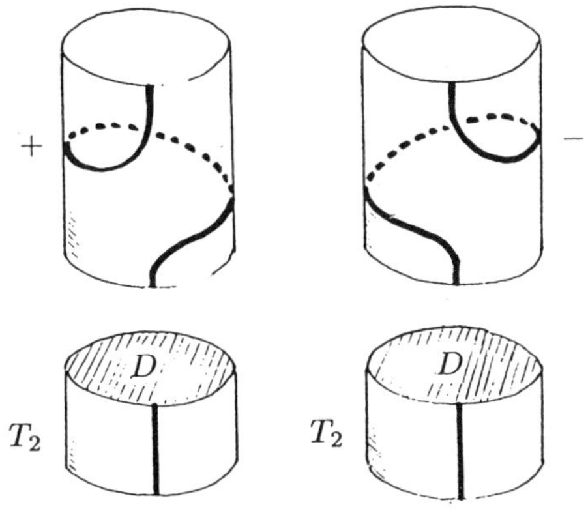

FIGURE 16.6. Twists of T_2 in opposite directions

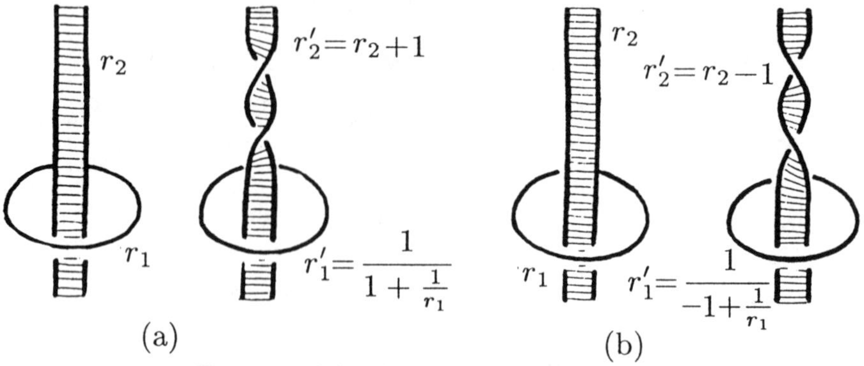

FIGURE 16.7. Two equivalent surgeries

In the case when the surgery is not only performed along the circle J, but along a framed link containing J as a component, the above homeomorphism establishes the equivalence of the surgeries represented in Fig.16.7. Here we assume that r_2 is an integer, while r_1 can be any rational number. The change in the framing r_2 is determined by the fact that the linking number of the boundary components of the ribbon, as can easily be seen, changes by ± 1 (recall that these components are codirected and reversal of their orientation does not affect their linking number). When we perform a twist by n turns, the framings r_1 and r_2 change to $r_1' = 1/(\pm n + 1/r_1)$ and $r_2' = \pm n + r_2$.

Problem 16.1. Prove that

i.e., surgery along the indicated framed links produces homeomorphic manifolds.

16.5. The twisting of our ribbon can be represented in a different way, which is sometimes more convenient (Fig.16.8). (The reader may recall that this transformation was called the belt trick in §2.)

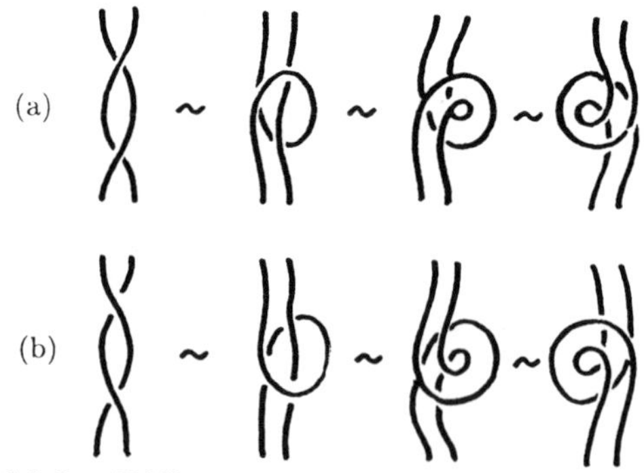

FIGURE 16.8. Different representations of twisted ribbons

In particular, this is very convenient when more than one ribbon pierces the meridional disk D (Fig.16.9). This picture can be used to find how the framing changes under twists of the disk D when the link pierces D in more than one point. We are interested in the linking number of one of the two boundary components of the ribbon with the other, so we can assume that Fig.16.9 only shows parts of the

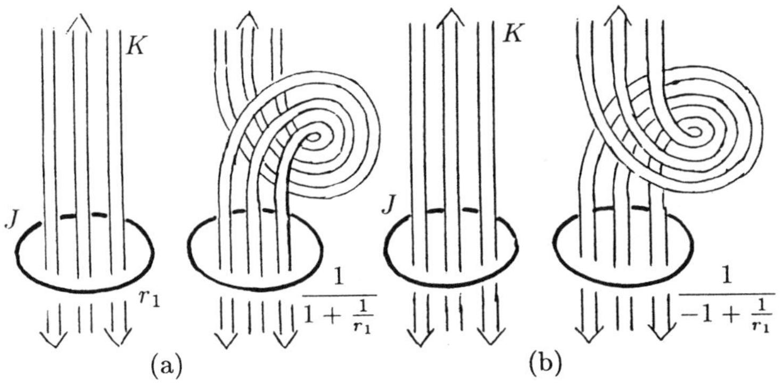

FIGURE 16.9. Twisting several parts of a ribbon

same ribbon. Suppose s parts are in one direction, t in the other. For the twist shown in Fig.16.9,a, the crossing where one part of the ribbon passes above another part directed the same way gives a contribution of $+1$ to the linking number of the components, while the crossing where it passes above an oppositely directed part contributes -1. For the twists shown in Fig.16.9,b the corresponding contributions are -1 and $+1$, respectively. Therefore in case a) we get

$$r_2' = r_2 + s^2 + t^2 - st - ts = r_2 + (s-t)^2 = r_2 + \mathrm{lk}^2(K, J),$$

where K is the knot under consideration and $J = \partial D$. In case b) we obtain $r_2' = r_2 - \mathrm{lk}^2(K, J)$. When $\pm n$ full revolutions are performed, we obtain

$$r_1' = \frac{1}{\pm n + 1/r_1}, \qquad r_2' = r_2 \pm n\mathrm{lk}^2(J, K).$$

Problem 16.2. a) Prove that the surgeries along the framed links shown in Fig.16.10 are equivalent.

b) Prove that the surgeries along the framed links shown in Fig.16.11 are equivalent.

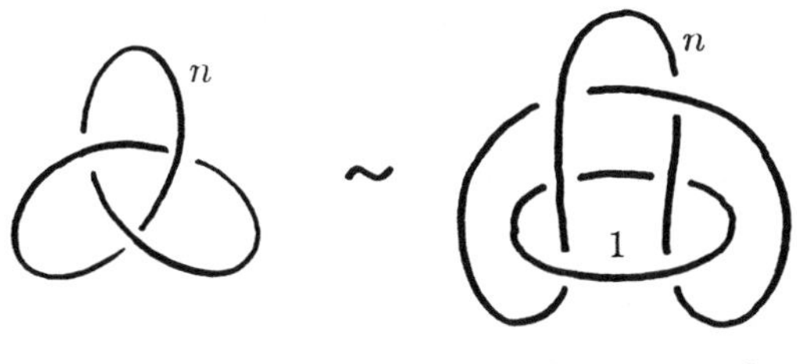

FIGURE 16.10. Two equivalent surgeries

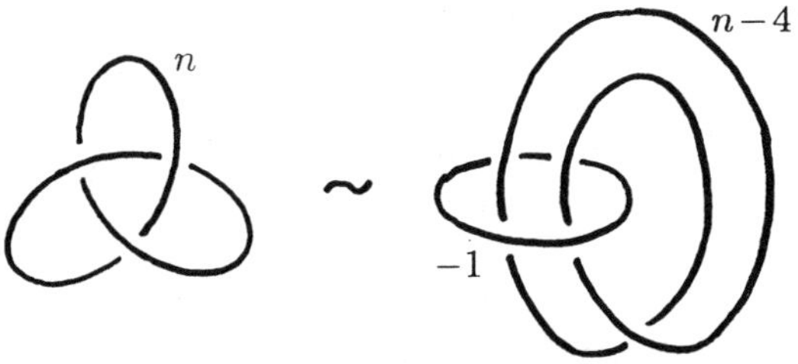

FIGURE 16.11. Two more equivalent surgeries

§17. Lens spaces revisited

17.1. In this section we consider several ways to represent lens spaces $L(p,q)$ via surgery on the sphere S^3 along framed links. Recall that the lens space $L(p,q)$ can be defined as the result of surgery along a circle with rational framing p/q.

17.2. PROPOSITION. *Suppose $m, n \in \mathbb{N}$. Then*

$$= L(mn-1, m) = L(mn-1, n).$$

Proof. First let us perform a twist by t revolutions along the circle with framing m, and then a twist by s revolutions along the circle with framing n. As the result, the framings m and n will first be replaced by $1/(t+1/m)$ and $t+n$, and then by

$$s + \frac{1}{t+1/m} \qquad \text{and} \qquad \frac{1}{s+1/(t+n)},$$

respectively. To destroy one of the circles, we must make its framing equal to ∞. This may be achieved by setting $s=-1$ and $t = 1-n$. As the result the remaining circle will acquire the framing

$$-1 + \frac{1}{1-n+1/m} = \frac{mn-1}{m-mn+1}.$$

Thus the surgery under consideration produces the manifold $L(mn-1, m-mn+1)$, which is the same as $L(mn-1,m)$, since m and $m-mn+1$ are the same modulo $mn-1$. □

17.3. PROPOSITION. *Suppose $a_1, \ldots, a_n \in \mathbb{Z}$ and $a_{n+1} \in \mathbb{Q}$. Then*

$$= L(p,q),$$

where p/q is given by the continued fraction

$$\frac{p}{q} = a_1 - \cfrac{1}{a_2 - \cfrac{1}{a_3 - \cdots - \cfrac{1}{a_n - \cfrac{1}{a_{n+1}}}}}. \tag{17.1}$$

Proof. First let us perform the twist by $t = -(a_n+1)$ revolutions along the circle with framing a_{n+1}. As the result, the nth and $(n+1)$st circles will acquire the framing indices

$$a_n' = -1 \quad \text{and} \quad a_{n+1}' = \frac{1}{-a_n - 1 + 1/a_{n+1}},$$

respectively, while the framing indices of the other circles will remain the same.

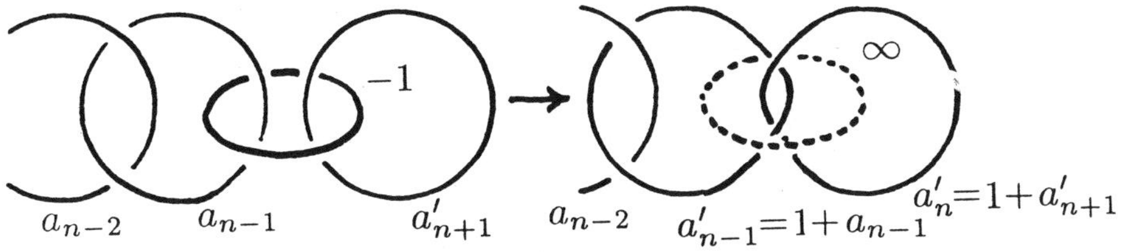

FIGURE 17.1. Destroying the nth circle

Now we can destroy the nth circle by performing a one-revolution twist along it (Fig.17.1).

The new framings after that will be

$$a'_{n-1}=1+a_{n-1},\quad a'_n=1+a'_{n+1}=1+\frac{1}{-a_n-1+1/a_{n+1}}=\cfrac{1}{1+\cfrac{1}{a_n-\cfrac{1}{a_{n+1}}}}.$$

(The new nth circle comes from the old $(n+1)$st.)

Next by similar twists along the last and next-to-last circles we can kill the circle with number $n-1$. As the result we get new framings of the two last circles

$$a''_{n-2}=1+a_{n-2},\quad a''_{n-1}=\cfrac{1}{1+\cfrac{1}{a_{n-1}-\cfrac{1}{a_n-\cfrac{1}{a_{n+1}}}}}.$$

Continuing in his way, we can destroy all the circles except one. The remaining circle will clearly have the framing

$$\cfrac{1}{1+\cfrac{1}{a_1-\cfrac{1}{a_2-\cfrac{1}{a_3-\dots}}}}.$$

But this continued fraction equals $1/(1+1/r)$, where r is the required framing (see (17.1)). To get the latter, it suffices to notice that the one-revolution twist establishes the equivalence of surgeries with framings r and $1/(1+1/r)$. □

§18. Homology spheres

18.1. The famous Poincaré sphere, which played a crucial role in the history of algebraic topology and was apparently the first concrete example of a 3-manifold obtained by means of what is now called a Heegaard presentation (see 18.3 below and the Comments to this chapter), will be described in this section, as well as more general objects – homology spheres. We are now in a position where our surgical techniques allow us to look at surgered manifolds from different points of view, and we shall discover that the manifolds under study can appear in many different and beautiful guises, involving several familiar objects: trefoils, the Whitehead link, the Borromeo rings, as well as the dodecahedral and icosahedral groups, and the Dynkin diagram for the Lie group E_8.

18.2. A compact 3-manifold (without boundary) M^3 is said to be a *homology sphere* if its fundamental group $\pi_1 = \pi_1(M^3)$ coincides with its commutant

$$\pi_1' = \{aba^{-1}b^{-1} | a, b \in \pi_1\}.$$

The equality $\pi_1' = \pi_1$ means that the quotient group π_1/π_1' consists only of the unit element.

It can be proved that the homology groups of any homology sphere are the same as those of S^3. Let us also mention without proof that surgery of S^3 along any knot with framing ± 1 produces a homology sphere. Hence it is not surprising that there are infinitely many nonhomeomorphic homology spheres, although surgery along different knots with framing ± 1 may produce the same homology sphere.

Among the various homology spheres, the most famous is the Poincaré homology sphere. Poincaré at first conjectured that any homology sphere is homeomorphic to S^3. But soon afterwards he himself constructed a counterexample, which we describe in the next subsection, and used the fundamental group (which he had conveniently invented before that) to prove that it was not the true sphere. He then conjectured that any compact oriented 3-manifold (without boundary) with trivial fundamental group is homeomorphic to S^3 (the *Poincaré conjecture*). This conjecture is the most famous unsolved problem in three-dimensional topology.

18.3. The manifold obtained by surgery on the sphere S^3 along the right trefoil with framing 1 is called the *Poincaré homology sphere*, or briefly the *Poincaré sphere* (Fig.18.1,a).

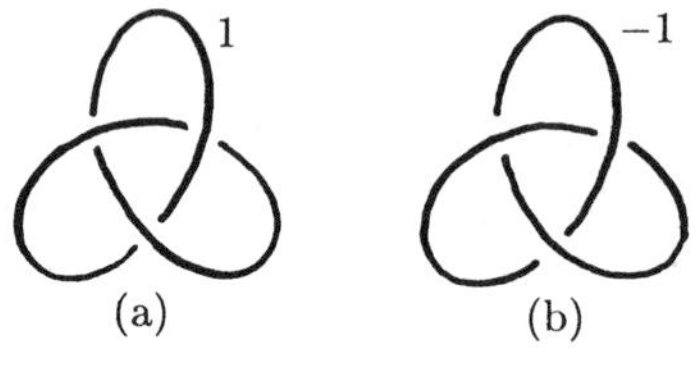

FIGURE 18.1. Two surgery presentations of the Poincaré sphere

It is easy to verify that under symmetry in any plane the framing of a knot changes its sign (Fig.18.2). Therefore the Poincaré sphere may also be obtained by surgering along the left trefoil with framing -1 (Fig.18.1,b).

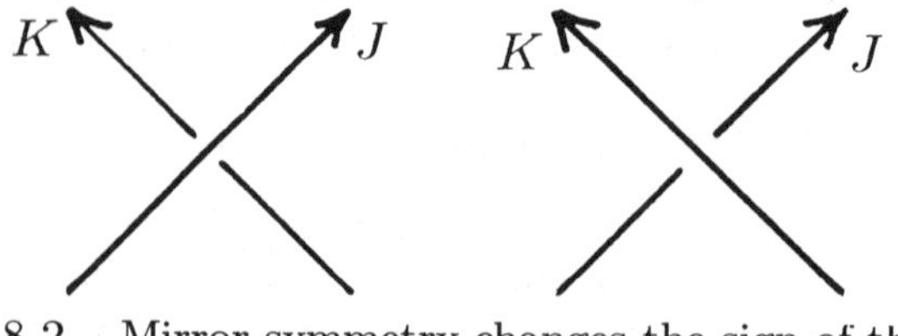

FIGURE 18.2. Mirror symmetry changes the sign of the framing

To prove that the Poincaré sphere is not S^3, we shall compute its fundamental group. We begin this computation by finding a presentation of $\pi_1(S^3 - K)$, where K is the right trefoil. We assume that the base point O is at infinity. Any loop from

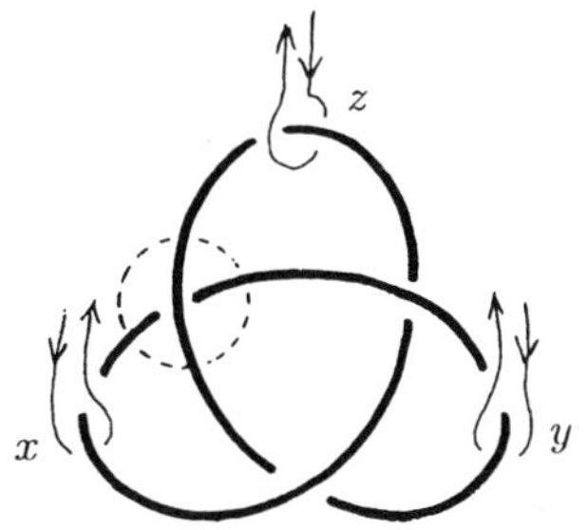

FIGURE 18.3. Generators of $\pi_1(S^3 - K)$

O can clearly be represented as the composition of the loops x, y, and z (Fig.18.3) and their inverses.

In other words, these loops generate the group $\pi_1(S^3 - K)$. The crossing points yield the defining relations. For example, the crossing shown by the dotted circle in Fig.18.3 and represented separately in Fig.18.4,a gives us $z = xyx^{-1}$, i.e., $zx = xy$ (see Fig.18.4,b).

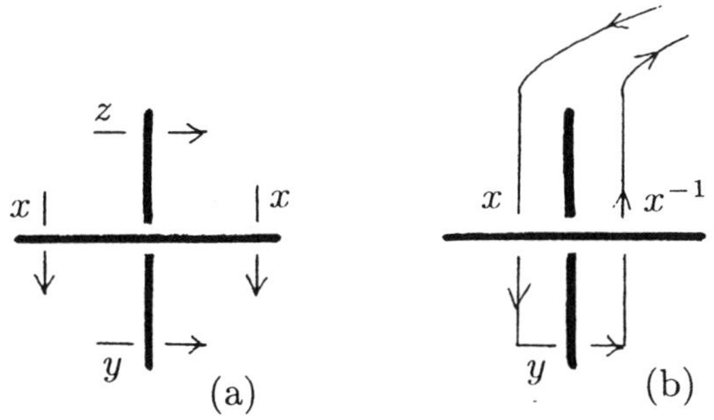

FIGURE 18.4. Relations in $\pi_1(S^3 - K)$

Similarly, the two other crossings yield $xy = yz$ and $yz = zx$. Therefore

$$\pi_1(S^3 - K) = \{x, y, z\,;\, xy = yz = zx\}.$$

Since $z = xyx^{-1}$, we can get rid of the generator z, obtaining the group

$$\pi_1(S^3 - K) = \{x, y\,;\, xyx = yxy\}.$$

When we perform surgery on S^3 along the right trefoil with framing 1, to the curve J (Fig,18.5,a) we attach the spanning meridional disk of the solid torus.

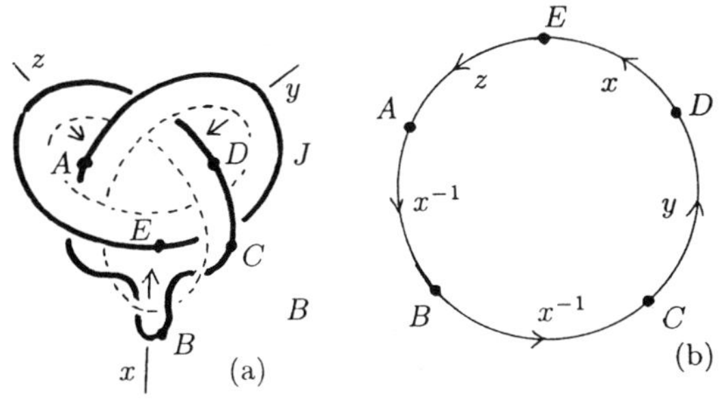

FIGURE 18.5. Relation in π_1 of the Poincaré sphere

Hence the fundamental group of the Poincaré sphere is obtained from the group $\pi_1(S^3 - K)$ by adding the relation $J = 1$. It is easy to see (Fig.18.5,b) that $J = x^{-2}yxz = x^{-2}yx^2yx^{-1}$. Thus the fundamental group of the Poincaré sphere is isomorphic to the group

$$I^* = \{x, y\,;\, xyx = yxy,\, yx^2y = x^3\}.$$

Problem 18.1. Prove that the group I^* coincides with its commutant $I^{*\prime}$.

Problem 18.2. Put $a = x$, $b = xy$. Prove that

$$I^* \cong \{a, b\,;\, a^5 = b^3 = (ba)^2\},$$

and use this fact to prove that the Poincaré sphere is not S^3.

Remark. Suppose $I \subset SO(3)$ is the group of self-isometries of the icosahedron. It can be proved that the group I^* is isomorphic to the inverse image of I under the double covering $SU(2) \to SO(3)$. For this reason the group I^* is known as the *binary icosahedral group.*

18.4. The Poincaré sphere can be obtained not only by surgery along trefoils, but by surgery along many other links as well. Here we describe some of the more interesting examples of such surgery presentations.

First we note that to our trefoil we can add a circle with framing ∞, since the corresponding surgery is obviously identical (see 14.3). We are interested in two ways of adding this circle (Fig.18.6). They differ in that in one case the linking number of the circle with the trefoil is equal to 0 and in the other it is equal to ± 2.

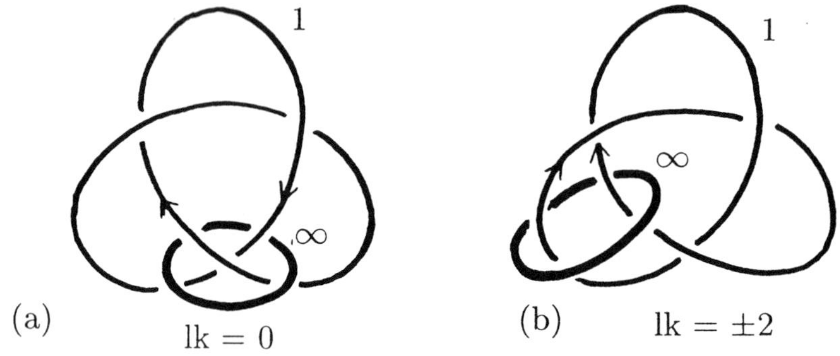

FIGURE 18.6. Two ways of adding a circle to the trefoil

A twist along the additional circle will unknot the trefoil. Here in case (a) we must perform $+1$ revolution, while in case (b), -1 revolution is needed. Using 16.4, we see that if the twist is by n revolutions, then the framing of the circle will become $1/n$, while that of the former trefoil will be $1+n\mathrm{lk}^2$, where lk stands for the linking number of the circle and the trefoil (Fig.18.6). Indeed, in our case $r_1 = \infty$ and $r_2 = 1$; therefore $r_1' = 1/n$ and $r_2' = 1 + n\mathrm{lk}^2$. As the result, after unknotting the trefoil we obtain the framed links shown in Fig.18.7.

The link shown in Fig.18.7,a is the Whitehead link (see 1.1 and Fig.1.15). Thus we have proved the following statement.

18.5. PROPOSITION. (a) *Surgery of the* 3*-sphere along the framed Whitehead links shown in* Fig.18.7,a *produces the Poincaré sphere.* (b) *Surgery of the* 3*-sphere along the framed link shown in* Fig.18.7,b *also produces the Poincaré sphere.*

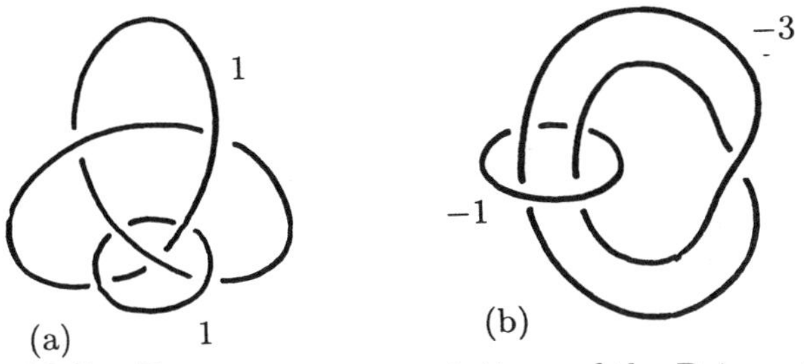

FIGURE 18.7. Two surgery presentations of the Poincaré sphere

18.6. The Whitehead link is a two-component link which is symmetric in the sense that there exists an isotopy interchanging its components. Fig.18.8 shows several diagrams of the Whitehead link (neighboring diagrams are easily obtained from each other by the appropriate Reidemeister moves; compare Fig.1.15). The link corresponding to the diagram drawn in thick lines has an axis of symmetry.

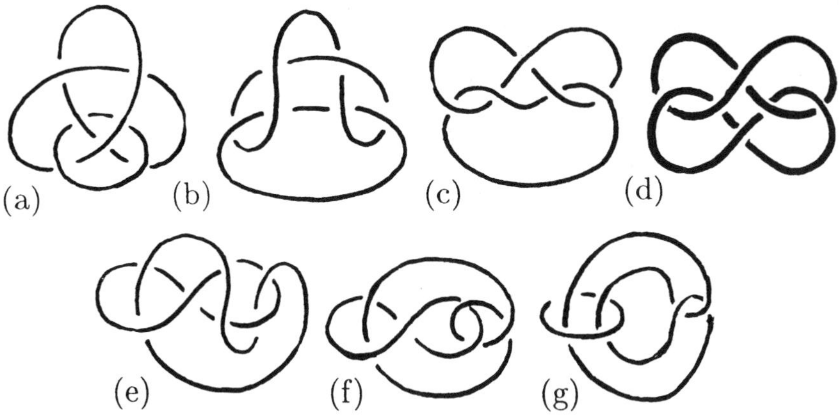

FIGURE 18.8. Seven versions of the Whitehead link

The Whitehead link leads to two interesting ways of presenting the Poincaré sphere by framed links possessing axes of symmetry of degree 3 and 5 repectively. This is related to the fact that the Poincaré sphere can also be obtained by an appropriate identification of the faces of the dodecahedron, which possesses symmetry axes of orders 2, 3, and 5 (Fig.18.9).

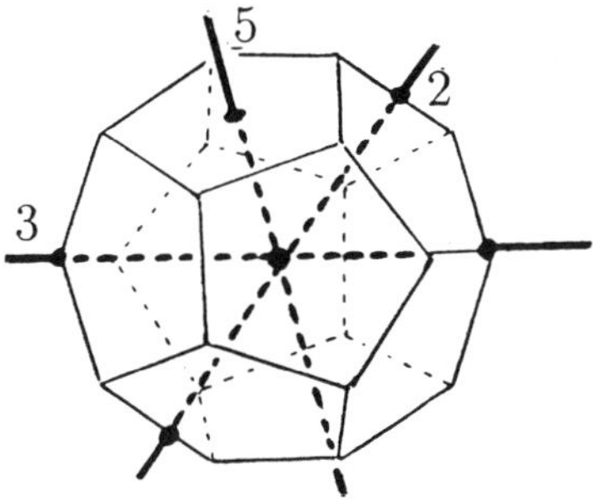

FIGURE 18.9. Symmetry axes of the dodecahedron

FIGURE 18.10. Framed Borromeo rings presenting the Poincaré sphere

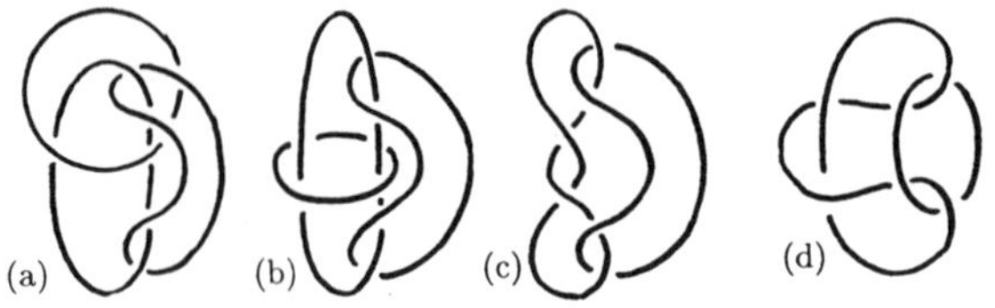

FIGURE 18.11. Four modifications of the Borromeo rings

18.7. PROPOSITION. *Surgery of the 3-sphere along the Borromeo rings with all three framings equal to* 1 (Fig.18.10) *produces the Poincaré sphere.*

Proof. Fig.18.11,a,b shows two isotopic defomations of the Borromeo rings. After a twist destroying one of the circles, we obtain the framed link shown in Fig.18.11,c. The framing indices do not change because the circles are pairwise unlinked. After the isotopy producing the framed link in Fig.18.11,d, we get the link in Fig.18.8,b. But this is the same link as the one in Fig.18.7,a, and it has the same framing indices. But as we saw earlier, surgery along that link yields the Poincaré sphere. □

18.8. PROPOSITION. *Surgery along the five linked circles with framing indices* 1 *shown in* Fig.18.12 *produces the Poincaré sphere.*

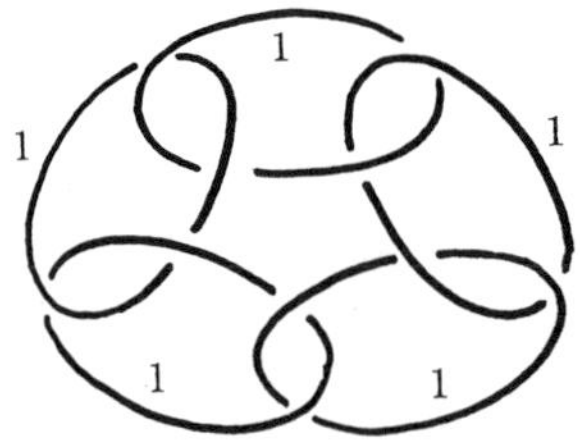

FIGURE 18.12. Five linked circles presenting the Poincaré sphere

Proof. First let us destroy the uppermost circle. To this end we shall perform a twist of the disk spanned by this circle by -1 revolution. In order to see what will happen to the circles directly linked to the circle that we are destroying, let us picture them as shown in Fig.18.13,a.

After the twist, these circles will be linked as shown in Fig.18.13,b. The linking number of these circles with the destroyed circle was equal to ± 1, hence their framings will become equal to 0. As the result, we obtain the framed link shown in Fig.18.13,c.

Now let us destroy one of the circles with framing 1, say the left one. The result will be the framed link in Fig.18.14,a. Finally, let us destroy the circle with framing

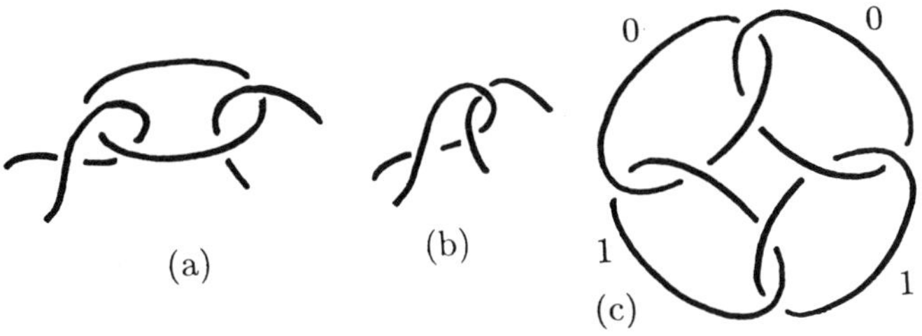

FIGURE 18.13. Destruction of the upper circle

-1 (Fig.18.14,b). The result is shown in Fig.18.14,c, but it can also be found among the modifications of the Whitehead link in Fig.18.8,c. Thus our statement follows from Proposition 18.5. □

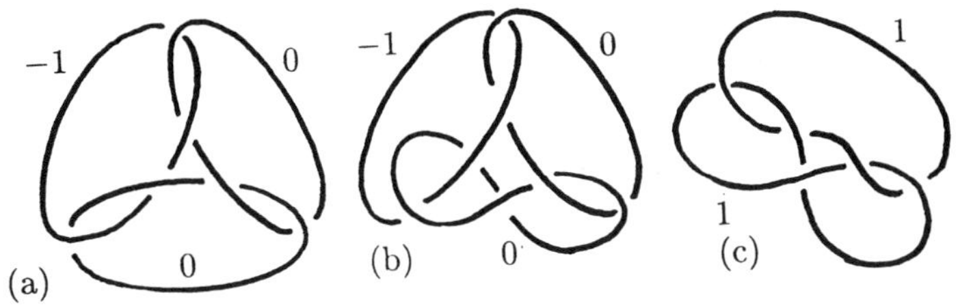

FIGURE 18.14. Reduction to the Whitehead link.

18.9. For the next proposition it will be convenient to determine certain links by means of graphs. This can be done when the link consists of n unknotted components, each two of which are either unlinked or linked in the simplest way (as the components of the Hopf link). To such a link we assign an n-vertex graph as follows: to each component of the link corresponds a vertex, two vertices are joined by an edge iff the corresponding components are linked. If the graph thus obtained is a tree (i.e., has no cycles), then it is easy to see that the link can be uniquely recovered from the graph. To specify the framing, we can write the framing index of each circle next to the corresponding vertex.

18.10. PROPOSITION. *Surgery along the framed link corresponding to the graph $\mathcal{E}_8$ with framing indices -2 shown in* Fig.18.15 *produces the Poincaré sphere. (The graph $\mathcal{E}_8$ is the Dynkin diagram of the root system of the simple Lie algebra E_8.)*

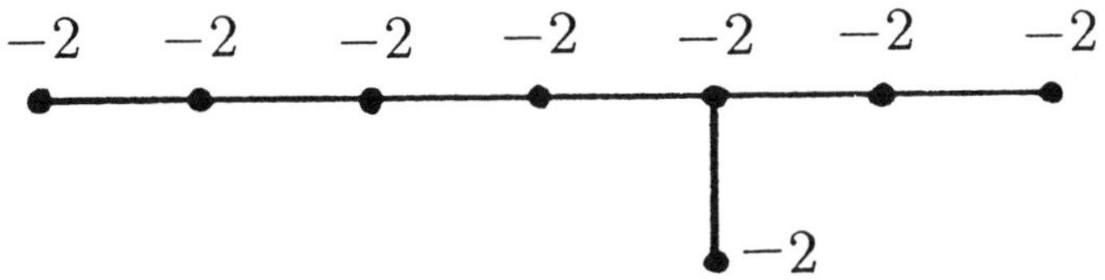

FIGURE 18.15. Weighted Dynkin diagram for E_8

Proof. Let us perform twists by $+1$ revolutions of three circles: the first circle from the left, the first from the right, and the lower circle. The result will be the framed link corresponding to the weighted graph shown in Fig.18.16,a. Successively

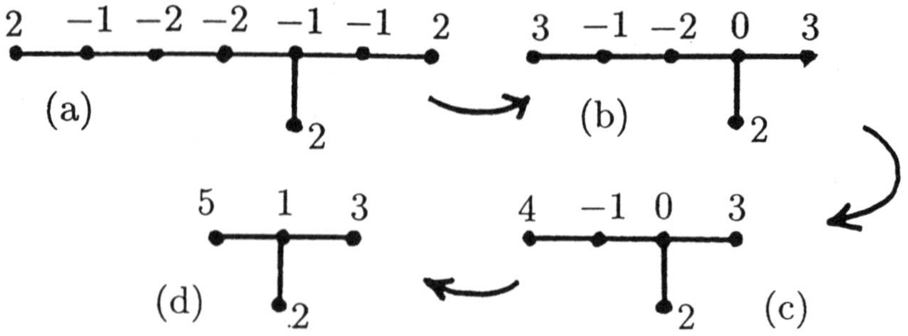

FIGURE 18.16. Modifications of the weighted Dynkin graph

destroying circles with framing −1, after three steps we get the weighted graph in Fig.18.16,d.

The corresponding framed link appears in Fig.18.17,a. Destroying the circle with framing index 1, we get the framed link in Fig.18.17,b, which is isotopic to the one in Fig.18.17,c. Again destroying the circle with framing 1, we get the framed link in Fig.18.17,d, which is isotopic to the ones in Fig.18.17,e,f. But the last one can be obtained from one of the previously visited framed links (namely the one in Fig.18.7,b) by mirror symmetry and sign changes of the framings, so that our statement follows from Proposition 18.5. □

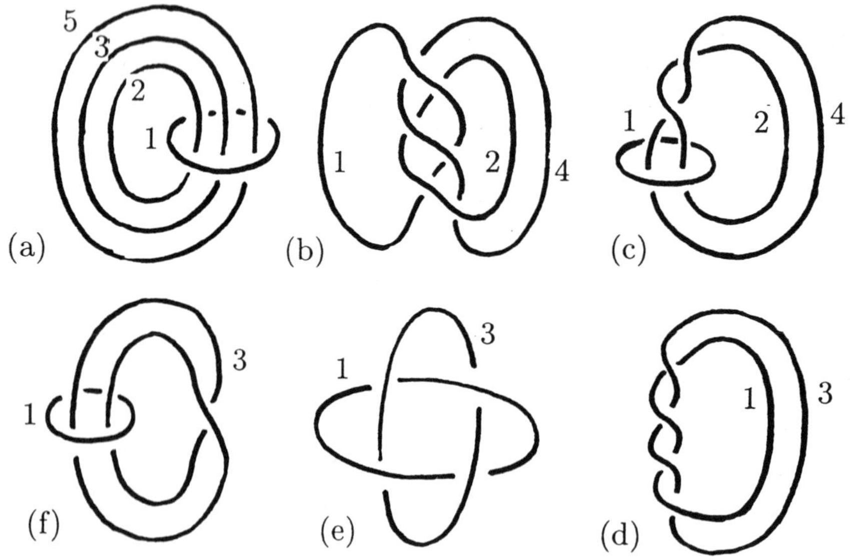

FIGURE 18.17. Modifications of framed links

18.11. Another example of a homology sphere. As we saw at the beginning of this section, the Poincaré sphere is obtained by surgery along the right trefoil with framing +1. If we do surgery along the same knot but with framing −1, we get a different manifold (which we denote M^3_-), which is also a homology sphere.

Problem 18.3. Prove that

$$\pi_1(M^3_-) = \{x, y\,;\, xyx = yxy,\ yx^2y = x^5\}$$
$$= \{a, b\,;\, a^7 = b^3 = (ba)^2\}.$$

Problem 18.4. Prove that M^3_- is a homology sphere.

Remark 1. It can be proved that the group $\pi_1(M^3_-)$ is infinite, while the group I^* has exactly 120 elements. Therefore M^3_- is not the Poincaré sphere.

Remark 2. The manifold M^3_- may also be produced by surgery along a left trefoil with framing $+1$. Further, it turns out that M^3_- may be produced as the result of surgery along the figure eight knot with framing ± 1 as well (this knot is isotopic to its mirror image, hence surgery along it with framing $+1$ and -1 gives the same manifold). To prove that M^3_- can be obtained in these two ways, let us note that the link shown in Fig.18.8 is symmetric in the sense that there is an isotopy interchanging its components. But now let us endow the components with different framings (Fig.18.18). Destroying the component with framing $+1$, we get the (left) trefoil; destroying the other component, we get the eight. □

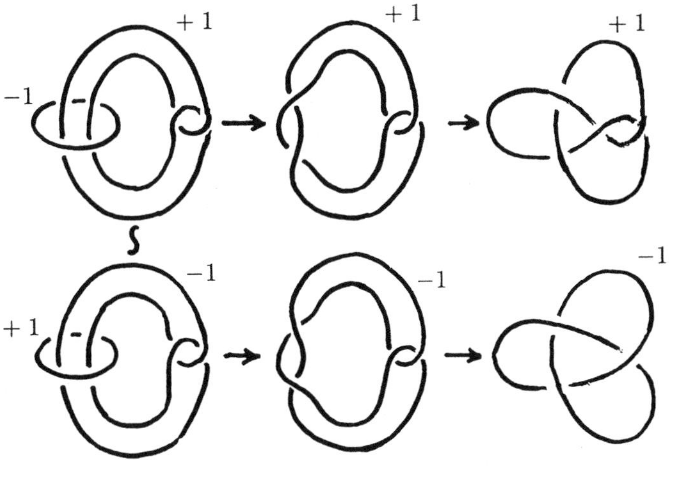

FIGURE 18.18. Two modifications of the Whitehead link

§19. The Kirby calculus

The Kirby calculus is what we have been practicing in the last two sections: modifying framed links so that by the surgery presentations they determine the same manifold. We saw that two important types of manifolds – lens spaces and homology spheres – have extremely varied presentations by framed links. Our goal now is a more systematic description of the modifications of framed links that do not change the resulting manifold.

19.1. The first Kirby move. The *Kirby move of the first kind* consists in adding to (or deleting from) the given framed link $L \subset S^3$ an unknotted circle with framing ± 1 provided that it is unlinked with the other components of L, i.e., there exists a sphere S^2 embedded in S^3 enclosing the circle and bounding a 3-disk that does not intersect any other components of L. This transformation does not change the resulting 3-manifold, because when we perform surgery on S^3 along a circle with framing $\pm 1/n$, we get S^3 again (by Proposition 14.6).

In terms of ribbons, the first Kirby move consists in adding (deleting) a ribbon with one full twist (in either direction), provided the ribbon is not linked with the other ribbons of the ribbon presentation. For the ribbon presentation, as well as the integer surgery presentation, the first Kirby move may be described as *adding a trivial component.* This is shown graphically in Fig.19.1.

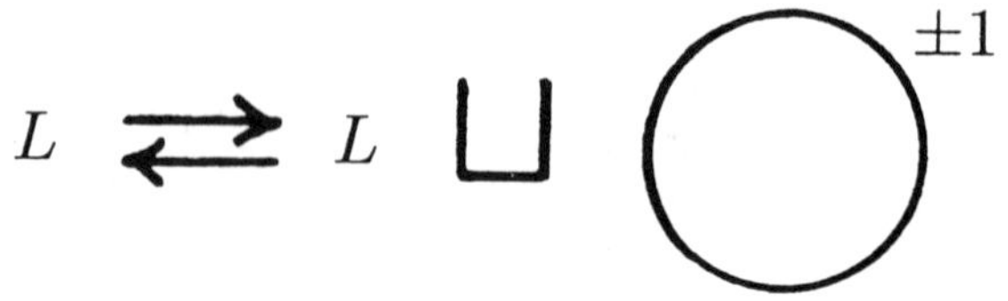

FIGURE 19.1. The first Kirby move

19.2. The second Kirby move. This transformation is more complicated than the previous one, but also arises naturally. We begin by explaining where it comes from.

Let us recall how integer surgery (see 16.1) is carried out. The framing of the component K of the link L determines (and is determined by) a closed curve J lying on the boundary of a tubular neighborhood (homeomorphic to the solid torus) of the curve K. (The curves K and J constitute the two boundary components of a ribbon that also defines the surgery.) The surgery itself consists in cutting out the ε-neighborhood of K and pasting back in a solid torus along a homeomorphism that identifies the meridian of the solid torus with the curve J (Fig.19.2,a). When this is occurs, the meridional disk D^2 will span the curve J (Fig.19.2,b).

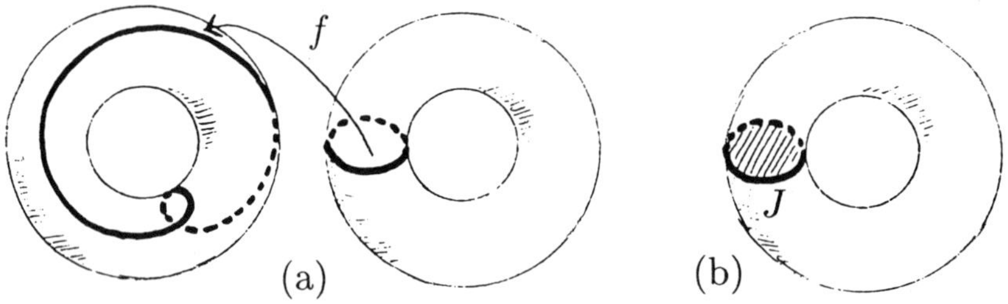

FIGURE 19.2. Integer surgery

This means that in the manifold M_1^3 obtained by surgery along K, any component C of our framed link running parallel to J (Fig.19.3,a) can be isotoped to the position C' by sliding it off J along the spanning disk D^2 (Fig.19.3,b). However, the corresponding curves are in general not isotopic in S^3, because J does not necessarily have an embedded spanning disk in $S^3 - K$. Nevertheless, it is possible to modify the framed link (without changing the end result of surgery) so that after the surgery along K the curve C will occupy the position C' directly. This modification is precisely the second Kirby move. It is easier to describe in the language of ribbons.

We first consider the case when the ribbons R_1 and R_2 in S^3 are unlinked (Fig.19.4,a). Suppose R_1 and R_2 have n and k twists, respectively. Let R_2' be the ribbon that coincides with R_2 except in the part P that *encircles* R_1, i.e., P

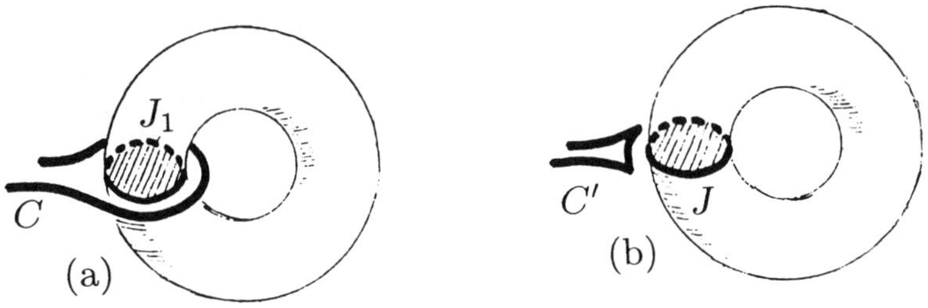

FIGURE 19.3. Isotopy after integer surgery

runs around R_1, remaining parallel to it and therefore performing n extra twists as compared with R_2 (Fig.19.4,b). Then we claim that *the surgery presented by the two ribbons* $\{R_1, R_2\}$ *produces the same manifold as that presented by the two ribbons* $\{R_1, R_2'\}$ *shown in* Fig.19.4,a and b.

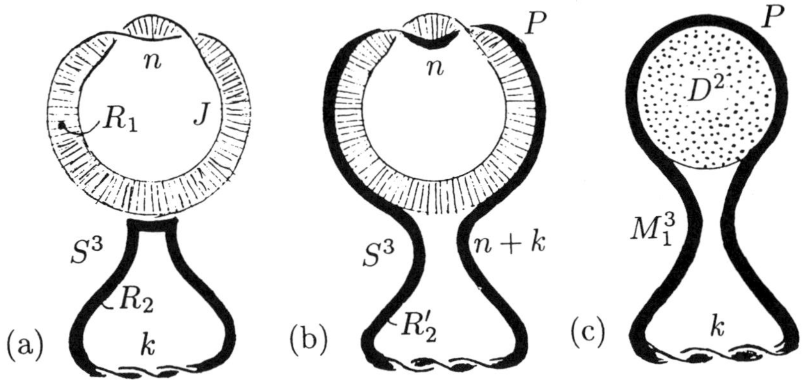

FIGURE 19.4. Equivalent ribbon surgery presentations

To prove that, first let us perform surgery along $R_1 \subset S^3$ (Fig.19.4,b). Then in M_1^3 the spanning disk D^2 is attached to the curve J in the boundary of the ribbon R_1 (Fig.19.4,c). We can therefore slide the encircling part $P \subset R_2'$ through R_1 until it becomes the collar of D^2 and then slide it off D^2; in the process, the n extra twists of P disappear. Now the surgery presented by R_2' can be done in M_1^3 before or after the slide (which is an isotopy in M_1^3) with the same result. But the result of surgery of M_1^3 along R_2' after the isotopy is clearly the same as that of surgery of S^3 along R_1 and R_2. □

In the construction described above, the assumption that the ribbons R_1 and R_2 are unlinked was only used to *explicitly* determine the framing of link component corresponding to R_2' from those corresponding to R_1 and R_2. The ribbon R_2' may be constructed (just as above) in the case when the ribbons R_1 and R_2 are arbitrarily linked; of course, the topology (i.e., the number of twists) of the ribbon thus constructed defines the framing of the link component corresponding to R_2'.

The replacement of $\{R_1, R_2\}$ by $\{R_1, R_2'\}$ or vice versa is called a *Kirby move of the second kind.* A concrete example of the second Kirby move (in which the ribbons are linked) is shown in Fig.19.5. Notice that in the case considered, the encircling process produces two extra twists in the ribbon R_2 (because of the little kink in the encircling ribbon due to linking); in the general case, as we shall see in §29, the number of extra twists due to linking will be equal to the doubled linking number of the ribbons.

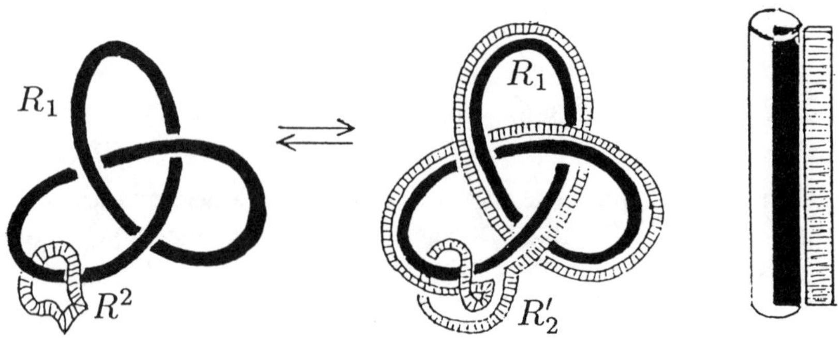

FIGURE 19.5. Second Kirby move (for linked ribbons)

The second Kirby move can also be described in terms of framed links: in a given framed link with two distinguished *unlinked* components C and K with framing indices n and k (Fig.19.6,a), respectively, the first of these components is modified, the curve C being replaced by the curve C' that differs from C only in that it encircles K (Fig.19.6,b) and in that its framing index is changed to $n+k$, the other components remaining the same. The general case (not necessarily unlinked components) will only be needed in §29 and is described there.

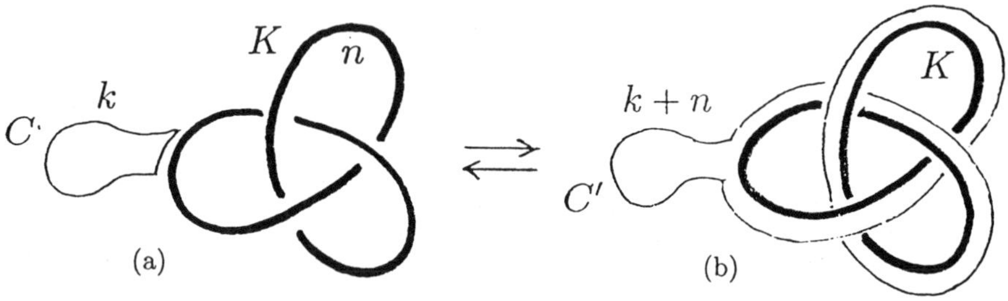

FIGURE 19.6. Example of the second Kirby move (for framed links).

We have shown that the second Kirby move, like the first one, does not change the result of the surgery. Its turns out that the converse is also true.

19.3. KIRBY'S THEOREM. *Two links in S^3 with integer framings produce the same 3-manifold if and only if they can be obtained from each other by a finite sequence of Kirby moves of the first and second kinds and isotopies.*

We have already established the easy part of this theorem (the "if" part) in the particular case of unlinked ribbons. For arbitrary ribbons, the proof is similar, except that the extra twists due to linking must be taken into account. The hard part ("only if") is hard indeed. For a proof, see [**Kir**], [**FR**], or [**Lu**]. □

19.4. Instead of the two Kirby moves, it is possible to use only one modification of framed links, which will be discussed in this subsection. Recall that besides the addition (deletion) to the given framed link of an unlinked circle with framing ± 1, we have made use (e.g. in 16.4) of the following modification (which also does not affect the resulting manifold). It consists in deleting a circle with framing ± 1 through which k ($k \geqslant 0$) parallel strands of the given link pass, while adding a kink to these strands, or vice versa (removing the kink and adding the circle). In Fig.19.7,a and b this move is shown for framing indices $+1$ and -1 respectively (note that the kinks are different).

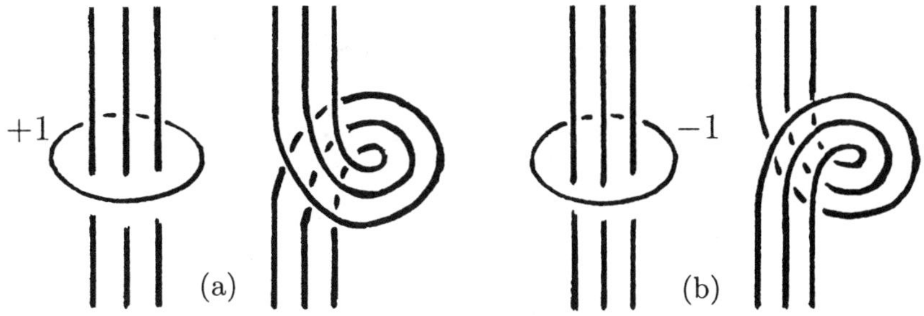

FIGURE 19.7. The Fenn-Rourke move

We call this transformation of framed links the *Fenn-Rourke move*, although G. Fenn and C. Rourke were not its inventors (see the Comments to this chapter).

We do not exclude the case $k = 0$ (when the circle is unlinked with the rest of the link), so that the first Kirby move is a particular case of the Fenn-Rourke move. The next statement says that the Fenn-Rourke move suffices.

19.5. FENN-ROURKE THEOREM. *A framed link L_1 can be transformed by Kirby moves (of the first and second kind) into the framed link L_2 if and only if this can be done by Fenn-Rourke moves.*

Proof. First let us check that a Fenn-Rourke move can be represented as the composition of Kirby moves. Fig.19.8,a shows a circle with framing $+1$ through which passes a strand from one of the other components of L_1.

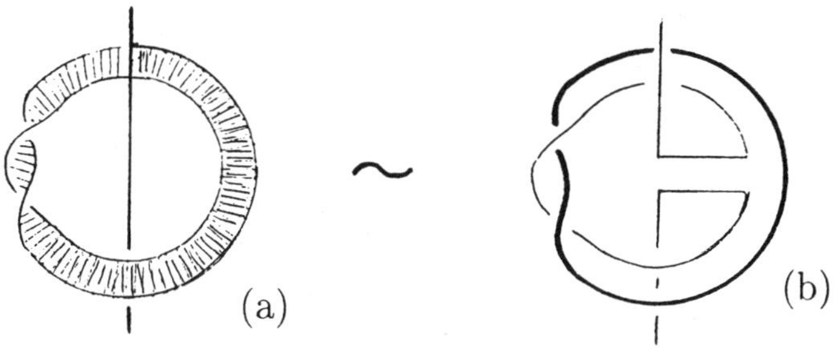

FIGURE 19.8. Unlinking via the second Kirby move

The second Kirby move changes the picture to the one in Fig.19.8,b. It can be clearly seen that the circle with framing $+1$ is now unlinked from our strand. Hence we can destroy the circle by using the first Kirby move. When several strands pass through our circle, the previous construction can be iterated. This proves the "only if" part of the theorem.

Now let us verify that the first and second Kirby moves can be presented as compositions of the Fenn-Rourke move. The first, as we have noted above, is a particular case of the Fenn-Rourke move, so it suffices to consider the second. Suppose that to the ribbon K_2 we want to add a ribbon parallel to K_1 (Fig.19.9). We begin by performing several Fenn-Rourke moves to modify K_1 to a more manageable form.

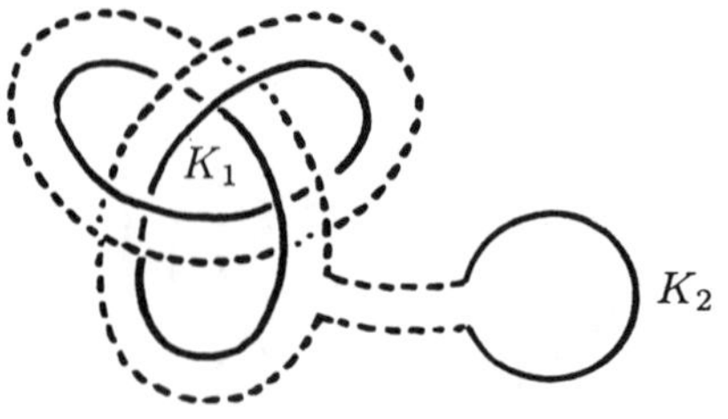

FIGURE 19.9. Adding a parallel ribbon

By means of moves such as the one in Fig.19.10,a (and similar moves with framing -1), we can untie the knot K_1 (Theorem 3.8). Then by means of moves such as the one in Fig.19.10,b, we can make the framing of the unknotted curve obtained from K_1 equal to 1. We can assume that all these moves do not touch a certain neighborhood of the ribbon K_2. So we can now assume that K_1 was an (unknotted) circle with framing $+1$.

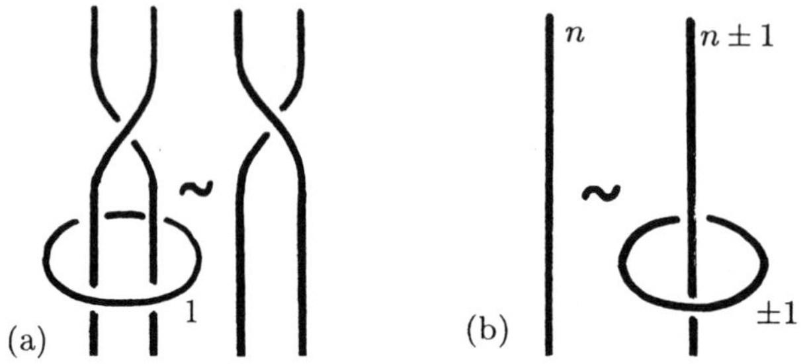

FIGURE 19.10. Unknotting and changing the framing of K_1

Recall that the addition (deletion) of an unknotted curve with framing 1 produces a twist (Fig.19.11). Now consider the second Kirby move in which a ribbon parallel to the (unknotted) curve K_1 is to be added to the ribbon K_2 (Fig.19.12,a). This move can be represented as the composition of the following Fenn-Rourke moves. First we add the circle K_1' (with framing $+1$), which is parallel to K_1 and is linked to K_2 (Fig.19.12,b). Then we remove the circle K_1 (Fig.19.12,c). When we added the circle K_1', the strands passing through K_1 were twisted, but when we deleted K_1, they were untwisted back to their initial position. Thus we have obtained the second Kirby move as required. □

19.6. Framed link diagrams. In conclusion of this section, we consider one more way of looking at integer surgery presentation. It is based on the fact that the ribbon specifying an integer surgery can be taken by an isotopy to a ribbon lying flat on the plane of the diagram or just above it (the belt trick again, cf. §2). Such a ribbon can then be replaced by one of its boundary components (the framing need not be specified by an integer; it is determined by the curve itself, in

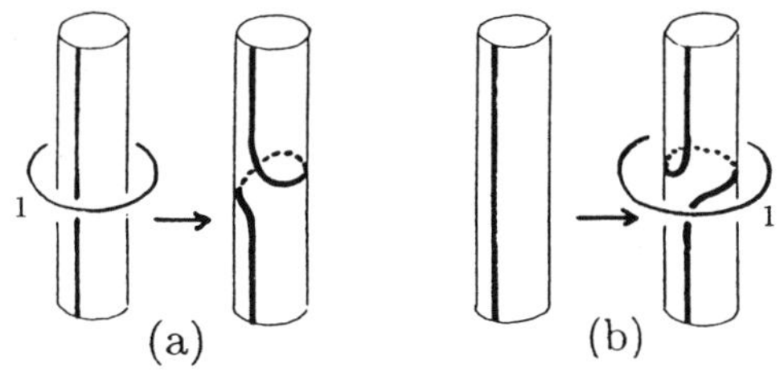

FIGURE 19.11. Twist produced by deleting (adding) a framed curve

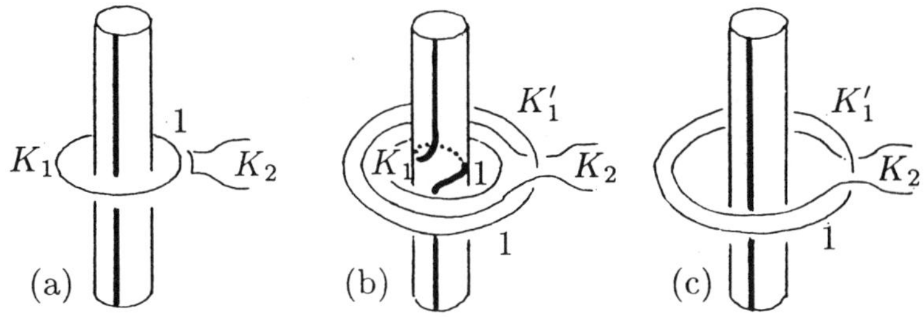

FIGURE 19.12. Second Kirby move via Fenn-Rourke moves

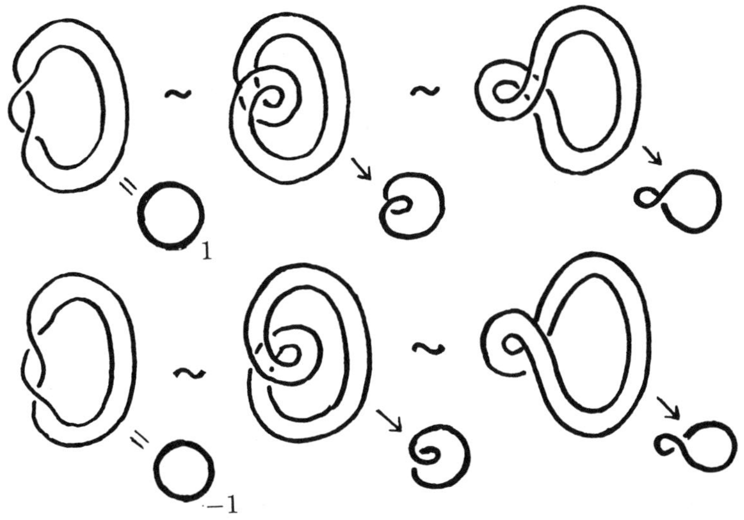

FIGURE 19.13. Adding kinks to circles with framing index ±1

particular by the number of kinks that it contains). For the cases of framings equal to ±1, this is shown in Fig.19.13.

To framed link diagrams obtained in this way, we can apply the second and third Reidemeister moves (without changing the framings): this does not affect the resulting 3-manifold. The first Reidemeister move cannot be applied, because it changes the framing by ±1, so that by using it we would be able to change any framing $m \in \mathbb{Z}$ to any other framing $n \in \mathbb{Z}$. To preserve the framing information, we can only use the composition of two "opposite" Reidemeister moves, i.e., carry out the move Ω_1' shown in Fig.19.14. The belt trick and Fig.19.13 convince us that the move Ω_1' transforms our ribbon into an isotopic ribbon and therefore does not change the resulting 3-manifold.

FIGURE 19.14. The double Reidemeister move $\Omega_1{}'$

Note that by using the move Ω_1', we can always position two kinks of "opposite signs" as shown in Fig.19.15,a. After that we can cancel them with each other by using the second and third Reidemeister moves (Fig.19.15,b,c).

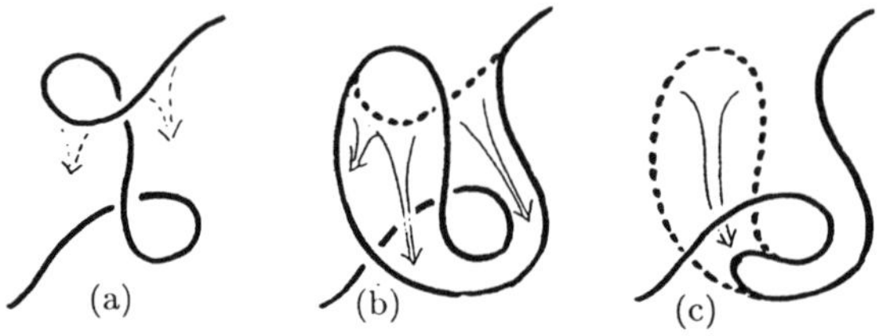

FIGURE 19.15. Cancellation of two opposite kinks

The arguments presented in this subsection prove the following statement, which is the analog of the Reidemeister theorem for framed links.

19.7. THEOREM. *Two framed link diagrams correspond to isotopic links with the same framings (and therefore produce the same 3-manifold) if and only if these diagrams can be obtained from each other by the moves Ω_1', Ω_2, and Ω_3 and plane isotopies.* □

19.8. Finally let us describe the Fenn-Rourke move in the language of framed link diagrams.

For the case of the circle with framing +1 through which three strands pass, the corresponding Fenn-Rourke move is shown in Fig.19.16,a. In the case when the circle is not removed, but added, it is more convenient to picture the corresponding move as in Fig.19.16,b.

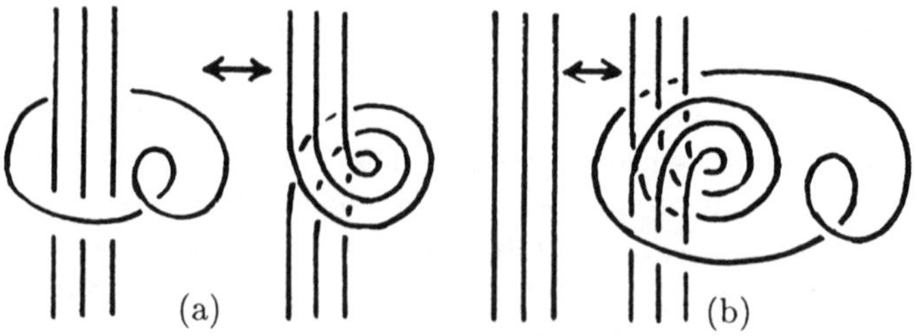

FIGURE 19.16. The Fenn-Rourke move for framed link diagrams

FIGURE 19.17. Two framed link diagrams giving the same manifold

Problem 19.1. Prove that the two framed link diagrams shown in Fig.19.17 produce the same 3-manifold by joining them via Fenn-Rourke moves.

Comments

The main geometrical construction discussed in this chapter – cutting out a solid torus and pasting it back in differently – is due, as far as we know, to M. Dehn ([**De1**], 1910)

The fundamental importance of this construction in the study of manifolds of dimension three became clear to topologists much later, in particular after the beautiful work of V. Rokhlin in the 1950's and that – no less striking – of R. Kirby in the 1970's. By that time the main tool in the topology of manifolds, good in all dimensions, was handle decomposition. This tool works only in the smooth category, being based on Morse theory. The basic operation of adding handles of various indices corresponding to singular points of Morse functions is closely related to surgery in the corresponding dimensions. The Kirby moves are clearly inspired by this theory. In particular, the second Kirby move comes from a classical move transforming handle decompositions and called "handle slide". In dimension three, however, one surgery operation (cutting out a solid torus and pasting it back in in a different way) is sufficient for the presentation of all 3-manifolds, so it is not necessary to develop the powerful and fairly complicated machinery of differential topology and Morse theory all the way back to Sard's theorem to learn most of what is known about the topology of 3-manifolds.

Note that while the terminology that we use is quite standard for "Kirby moves", the term "Fenn–Rourke move" is less standard, probably because this move was *not* invented by G. Fenn and C. Rourke, although those two authors were the first to prove its universality (Theorem 19.5). As a formal modification of framed links, the move was used earlier by Rolfsen, in particular in his book [**Rol**] (1976), but as a change of surgery presentation, it appeared even earlier than that, e.g. in the work of J. Hempel [**Hemp**] (1962).

Surgery is by no means the only elementary topological tool for studying this subject. Let us mention W. Haken's theory of normal surfaces, which allowed him to solve the unknotting problem and led to the fundamental work of F. Waldhausen on the classification of "sufficiently large" manifolds, such notions as incompressible surfaces, or S. Matveev's special spine presentations of 3-manifolds, on which the first serious computer analysis of these objects was based and which led Matveev to define a natural notion of complexity for them. Recent achievements along these lines include the solution of the algorithmic recognition problem for the 3-sphere due to J. H. Rubinstein [**Rub**] and to A. Thompson [**Tho**]. Although in our book we do not discuss the work of the authors mentioned in this paragraph, nor that of their followers and competitors, its subject matter and methods are extremely important in geometric topology.

The first homology sphere appeared in the work of H. Poincaré (and the whole series of homology spheres in Dehn's work). As often happened in the work of Poincaré, it originated in one of his mistakes: after having invented homology,[1] he erroneously asserted that the homology invariants of the 3-sphere distinguish it among all 3-manifolds. But he soon realized his mistake, and using a Heegaard presentation succeeded in constructing a counterexample (which he was able to justify by using his recent invention, the fundamental group). Although it is widely believed (see, for example, the survey [**KS**]) that Poincaré knew the dodecahedral construction of his homology sphere (see §18), we have not found confirmation of this in his works. The dodecahedral construction is attributed by the authors of the famous book [**ST**] to H. Kneser ([**Kne**]).

Our main sources in this chapter are D. Rolfsen's book [**Rol**], the original papers by R. Kirby [**Kir**], R. Fenn and C. Rourke [**FR**], the survey by R. Kirby and M. Scharlemann [**KS**], and numerous lecture notes from several seminars in three-dimensional topology that one of us conducted jointly with L. V. Keldysh and A. V. Chernavsky, and more recently with Yu. P. Soloviev. None of these sources is followed closely. Unfortunately, our efforts to make the proof of the "hard part" of Kirby's theorem accessible enough for an introductory course did not meet with success, and we regretfully omit it (the simplest proof we know is in [**Lu**]).

The subject matter of the chapter is open-ended and very much alive; there are a good many people working on it at present (see the book [**MF**]). Among the initial problems of the theory that are still unsolved, let us note the following two:

- to find an efficient (computer implementable) geometric algorithm for comparing 3-manifolds;
- to find an efficiently computable complete system of algebraic invariants for 3-manifolds.

[1] Poincaré attributes this great invention to the Italian mathematician and physicist Enrico Betti. It would seem more plausible that it was, at least in part, due to B. Riemann. Somewhat forgotten in Germany and no longer publishing, in the 1860s he was dying of tuberculosis in the small town of Verbana in northern Italy, where he was mainly preoccupied with religious and philosophical questions, but would gladly talk mathematics with his rare visitors, mainly Italian colleagues, one of whom was Betti.

CHAPTER VII

Branched Coverings

This chapter is also about the presentation of 3-manifolds. We shall see that any 3-manifold can be presented as a finite covering of the sphere S^3 branching along a knot or a link (§23). In fact there exist certain universal knots and links in S^3 (e.g. the Borromeo rings) from which any 3-manifold can be obtained as the total space of a covering whose branching set is that specific link (§25). These results are due to J. M. Montesinos and H. M. Hilden, and their collaborators (see the Comments at the end of the chapter).

We begin, however, with branched coverings of two-dimensional manifolds (§20), which will be the foundation of the three-dimensional theory, just as the study of homeomorphisms of 2-manifolds was the foundation of surgery in dimension 3. We then stay in dimension 2 to digress about a classical formula concerning branched coverings of surfaces (§21), and only then go on to the main topic of this chapter: the study of various branched covering presentations of 3-manifolds (§§22–25).

§20. Branched coverings of surfaces

20.1. Definitions. Suppose M^2 and N^2 are two-dimensional manifolds. Recall that a continuous map $p : M^2 \to N^2$ is said to be a *covering* (with *fiber* Γ, where Γ is a fixed discrete space) if for every point $x \in N^2$ there exist a neighborhood U and a homeomorphism $\varphi : p^{-1}(U) \to U \times \Gamma$ such that the restriction of p to $p^{-1}(U)$ coincides with $\pi \circ \varphi$, where $\pi : U \times \Gamma \to U$ is the projection on the first factor. Then M^2 is called the *covering manifold* or *cover*, while N^2 is the *base manifold* or *base*. If the fiber Γ consists of n points, then the covering p is said to be *n-fold*.

A continuous map $p : M^2 \to N^2$ is said to be a *branched covering* if there exists a finite set of points $x_1, \dots, x_n \in N^2$ such that the set $p^{-1}(\{x_1, \dots, x_n\})$ is discrete and the restriction of p to the set $M^2 - p^{-1}(\{x_1, \dots, x_n\})$ is a covering. In other words, after we delete a finite set of points, we get a covering. The points $x_1, \dots, x_n \in N^2$ that must be deleted are called the *branch points* of p. The following obvious statement not only provides an example of a branched covering, but shows how branched coverings behave near branch points.

20.2. PROPOSITION. *Let $D^2 = \{z \in \mathbb{C} : |z| \leqslant 1\}$ and let $p : D^2 \to D^2$ be the map given by the formula $p(z) = z^n$. Then p is an n-fold branched covering with unique branch point $z = 0$.* □

The example in the proposition for different n describes the structure of an arbitrary branched covering near its branch points. Indeed, it turns out that if p is a k-fold branched covering and U is a sufficiently small disk neighborhood of a branch point, then $p^{-1}(U)$ consists of one or several disks on which p has the same structure as the map in the proposition. We shall not prove this fact in the general case, but in all the examples considered it will be easy to verify that this is indeed the case. If in a small neighborhood of a point x of the covering manifold the covering map is equivalent to the map $z \mapsto z^m$, we shall say that x has *branching index* m. It is easy to see that for an n-fold branched covering, the sum of branching indices of all the preimages of any branch point is equal to n.

20.3. PROPOSITION. *Consider the map $f : \mathbb{C} - \{0\} \to \mathbb{C}$ given by the formula $f(z) = 2(z + 1/z)$. This map is a 2-fold branched covering with branch points ± 4. The preimages of these points are the points ± 1, and the branching index of each is 2.*

Proof. The equation $2(z + 1/z) = c$ is quadratic. Its discriminant $c^2/4 - 4$ vanishes iff $c = \pm 4$. This value is assumed by the function f when $z = \pm 1$. $\square$

20.4. PROPOSITION. *Let p be the restriction of the map f from Proposition* 20.3 *to the annulus $C = \{z \in \mathbb{C} : 1/2 < |z| < 2\}$. If $z = \rho e^{i\varphi}$, then*

$$p(z) = 2((\rho + 1/\rho)\cos\varphi + i(\rho - 1/\rho)\sin\varphi),$$

so that the image of the annulus C is the set of points located inside the ellipse (*see* Fig.20.1) $\{z = 5\cos\varphi + 3i\sin\varphi,\ 0 \leqslant \varphi < 2\pi\}$. $\square$

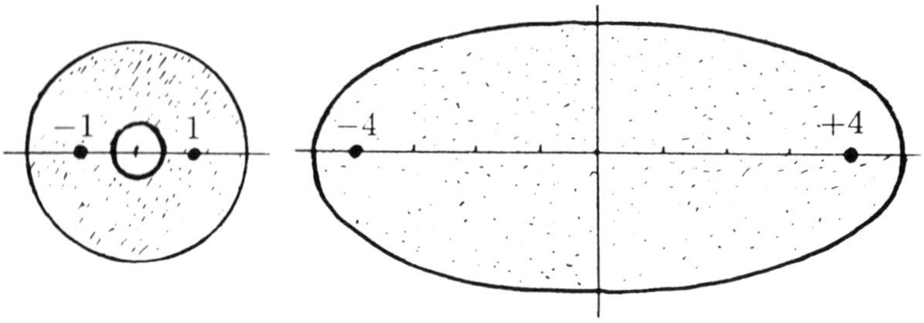

FIGURE 20.1. Branched covering of the ellipse by the annulus

A more geometric description of p is the following. Imagine the annulus C as a sphere with two holes and an axis of symmetry l (Fig.20.2). Let us identify points of the sphere symmetric with respect to l. It is easy to see that the resulting space is homeomorphic to the open disk D^2. The quotient map $p : C \to D^2$ thus constructed is a 2-fold branched covering with two branch points.

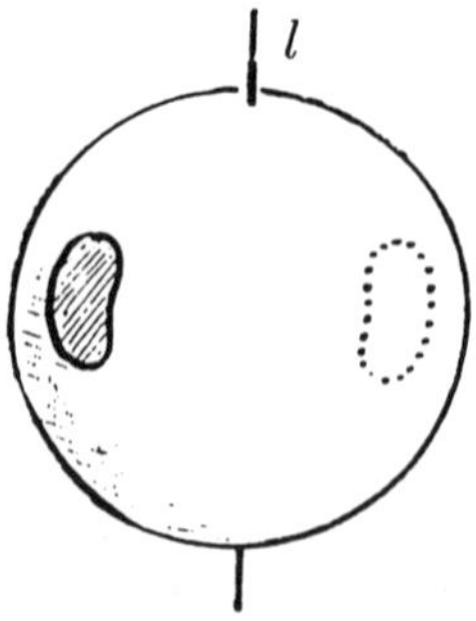

FIGURE 20.2. Sphere with two holes and symmetry axis

Problem 20.1. Prove that if the base manifold M^2 of a branched covering $p : N^2 \to M^2$ is orientable, then so is the covering manifold N^2.

20.5. THEOREM. *Let M_g^2 be the sphere with g handles. Then there exists a branched covering $p : M_g^2 \to S^2$.*

First Proof. Consider a copy of the sphere with g handles with an axis of symmetry l (Fig.20.3). Identify all pairs of points symmetric with respect to l. The resulting quotient space is (homeomorphic to) the ordinary sphere S^2. The natural projection $p : M_g^2 \to S^2$ is a 2-fold branched covering which has $2g + 2$ branch points. □

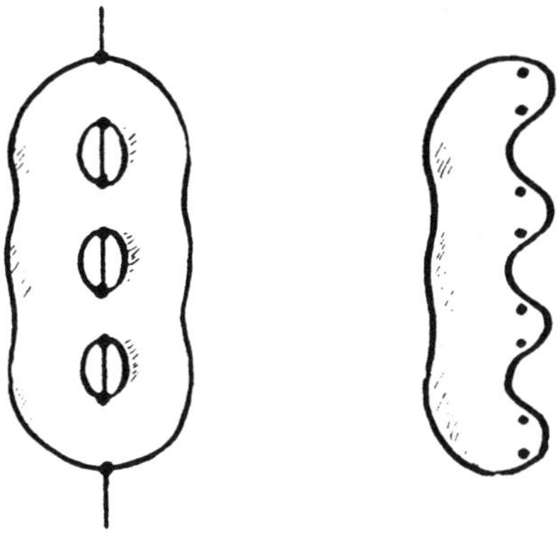

FIGURE 20.3. Branched covering of the 2-sphere

Second Proof. Consider a *triangulation* of the manifold M_g^2. (This means that M_g^2 is cut up into (curvilinear) triangles, any two of which either intersect along a common side, or intersect in a common vertex, or have no common points.) Let $A_1, \dots, A_n$ be the vertices of the triangulation. On the sphere S^2 choose n points $B_1, \dots, B_n$ situated in general position in the following sense: no three of them lie on one and the same great circle and no two are antipodes. Then any three points B_i, B_j, B_k uniquely determine a spherical triangle Δ_1. Suppose Δ_2 is the closure of its complement $S^2 - \Delta_1$; then Δ_2 is also homeomorphic to the triangle. Therefore there exist homeomorphisms

$$f_1 : A_iA_jA_k \to \Delta_1, \qquad f_2 : A_iA_jA_k \to \Delta_2$$

that are linear in the following sense. We can assume that the length of a curve is defined both on the manifold M_g^2 and on the sphere S^2; we require that an arbitrary point X divide the arc A_pA_q in the same ratio as the point $f_r(X)$, $r = 1, 2$, divides the arc B_pB_q.

Let us fix orientations of M_g^2 and S^2. The orientations of the triangles $A_iA_jA_k$ and $B_iB_jB_k$ induced by their vertex order may agree with or be opposite to that of M_g^2 and S^2. If both orientations agree, or both are opposite, then we map $A_iA_jA_k$ onto $\Delta_1 = B_iB_jB_k$ by the homeomorphism f_1. If one orientation agrees and the other doesn't, we map $A_iA_jA_k$ onto Δ_2 (the complement to $B_iB_jB_k$) via f_2. Defining such maps on all the triangles of the triangulation of M_g^2, we obtain a map $f : M_g^2 \to S^2$. We claim that this map is a branched covering.

For each interior point x_0 of triangle $A_iA_jA_k$ there obviously exists a neighborhood $U(x_0)$ mapped homeomorphically onto its image. Let us prove that this is the case not only for interior points of the triangles, but also for inner points of their sides. Indeed, let the side A_iA_j belong to the two triangles $A_iA_jA_k$ and $A_iA_jA_l$. On the sphere S^2, the great circle passing through the points B_i and B_j may either separate the points B_k and B_l or not separate them. In the first case we must have used the maps f_1 and f_1 (or f_2 and f_2), in the second one f_1 and f_2 (or f_2 and f_1). In all cases a sufficiently small neighborhood of a point chosen inside A_iA_j will be mapped bijectively (and hence homeomorphically) onto its own image (Fig.20.4). So at all points except possibly the vertices $A_1, \ldots, A_n$ we have a covering, so f is a branched covering. □

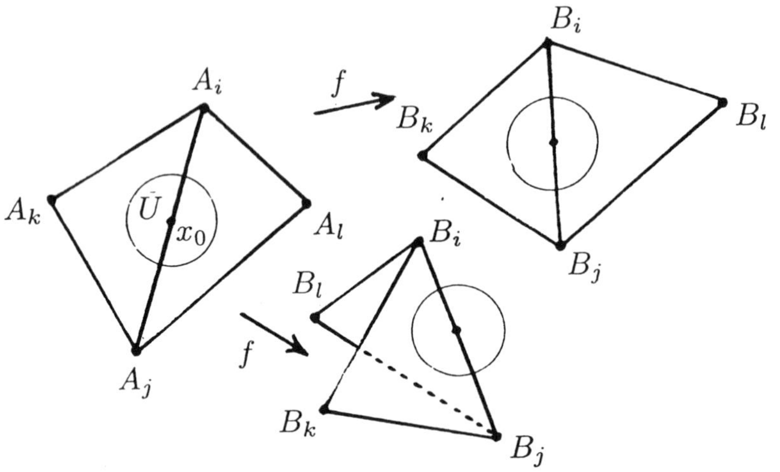

FIGURE 20.4. Structure of f at inner points of the sides

In fact not all the points $B_1, \ldots, B_n$ are necessarily branch points, although some must be when $g > 0$. How few can there be? The answer is contained in the next statement, and surprisingly does not depend on g.

20.6. THEOREM. *Let M_g^2 be the sphere with g handles, where $g \geqslant 1$. Then there exists a branched covering $p : M_g^2 \to S^2$ with exactly three branch points.*

Proof. Choose an arbitrary triangulation of the manifold M_g^2 and take its barycentric subdivision, i.e., subdivide each triangle into 6 triangles by its three medians. To the vertices of the barycentric subdivision assign the numbers $0, 1, 2$ as follows:

0 to the vertices of the initial triangulation;

1 to the midpoints of the sides;

2 to the barycenters of the triangles.

If the orientation of some triangle 012 induced by its vertex order agrees with that of the manifold M_g^2, then we paint the triangle black; otherwise we leave it white (Fig.20.5).

We can consider the sphere S^2 as the union of two triangles glued along their sides; denote their vertices by $0', 1', 2'$ and paint one of these triangles black, leaving the other white. Now we map each white triangle 012 from the barycentric subdivision in M_g^2 to the white triangle $0'1'2'$ in S^2, and each black one to the black $0'1'2'$. In more detail this map was described in the second proof of Theorem 20.5,

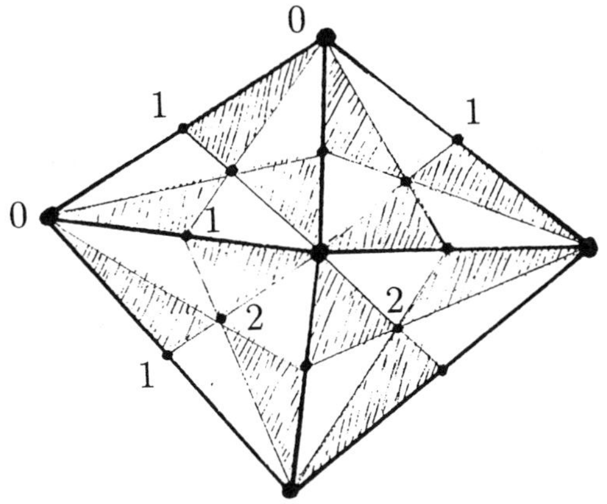

FIGURE 20.5. Black and white coloring of the barycentric subdivision

where it was established that this map is a branched covering. The three branch points are of course $0', 1', 2'$. □

§21. Riemann–Hurwitz formula

21.1. In this section, we begin by recalling the definition and some properties of the Euler characteristic, which will then be used to express and prove the classical Riemann–Hurwitz formula relating the topological properties of the base and covering manifolds in a branched covering with the branching indices.

21.2. The Euler characteristic. Let M^n be an n-dimensional compact triangulated manifold. Its *Euler characteristic* $\chi(M^n)$ is defined as the alternating sum

$$\chi(M^n) = a_0 - a_1 + a_2 - \cdots + (-1)^n a_n,$$

where a_k is the number of simplices of dimension k. The fact that the number $\chi(M^n)$ for arbitrary n does not depend on the choice of the triangulation follows from homology theory and will not be proved here for $n > 2$.

For $n = 2$, however, we shall need this fact. To establish it, recall that a triangulation K' is said to be a *subdivision* of the triangulation K if any simplex of K is the union of simplices from K'. Two triangulations K_1 and K_2 of a two-dimensional manifold are called *transversal* if their edges intersect transversally at a finite number of points. It is not difficult to show that any two triangulations of a 2-manifold can be made transversal by a small move.

Problem 21.1. (a) Verify that the Euler characteristic of a 2-manifold does not change when we pass to a subdivision of its triangulation.

(b) Prove that any two transversal triangulations of a 2-manifold have a common subdivision.

(c) Using (a) and (b), show that the Euler characteristic for 2-manifolds does not depend on the choice of the triangulation.

It is easy to verify that for any (compact) manifold $A \cup B$ obtained by gluing the manifolds A and B along the manifold $A \cap B$ we have

$$\chi(A \cup B) + \chi(A \cap B) = \chi(A) + \chi(B). \tag{21.1}$$

In dimension 2, when $A \cap B$ consists of one or several circles, we have the simpler formula

$$\chi(A \cup B) = \chi(A) + \chi(B). \tag{21.2}$$

Indeed, in this case $\chi(A \cap B) = 0$, because the Euler characteristic of the circle is zero.

21.3. PROPOSITION. *If M_g^2 is the sphere with g handles, then $\chi(M_g^2) = 2-2g$.*

Proof. Let $F_g^2 = M_g^2 - (D^2 \cup D^2)$, i.e., F_g^2 is the sphere with g handles and two holes (Fig.21.1). Then

$$\chi(M_g^2) = \chi(F_g^2) + 2\chi(D^2), \quad \text{i.e.,} \quad \chi(F_g^2) = \chi(M_g^2) - 2.$$

But M_{g+1}^2 can be obtained from F_g^2 by adding a handle (see Fig.21.1), which has Euler characteristic 0, so we get

$$\chi(M_{g+1}^2) = \chi(F_g^2) = \chi(M_g^2) - 2.$$

It is easy to check that $\chi(S^2) = 2$. Hence by induction $\chi(M_g^2) = 2 - 2g$. □

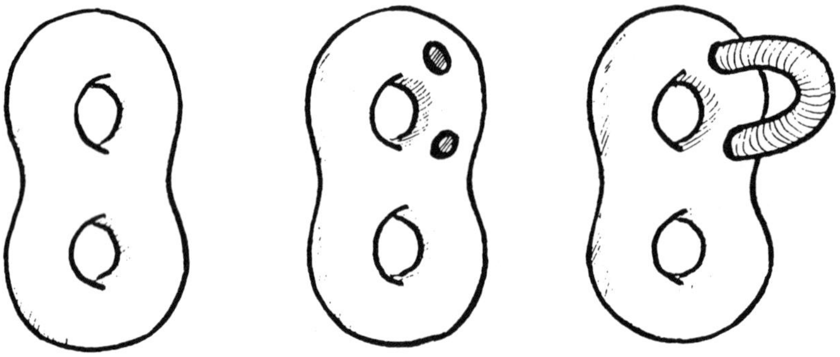

FIGURE 21.1. Adding a handle

Thus the topological type of any oriented compact 2-manifold without boundary is entirely determined by its Euler characteristic.

21.4. THEOREM (RIEMANN-HURWITZ FORMULA). *Suppose $p : M^2 \to N^2$ is an n-fold branched covering of compact 2-manifolds, $y_1, \dots, y_l$ are the preimages of the branch points, and $d_1, \dots, d_l$ are the corresponding branching indices. Then*

$$\chi(M^2) + \sum_{i=1}^{l}(d_i - 1) = n\chi(N^2). \tag{21.3}$$

The proof of Theorem 21.4 will be easier to understand if we first work out its particular case when p is an ordinary (nonbranched) covering.

21.5. THEOREM. *Let $p : M^2 \to N^2$ be an n-fold covering of compact 2-manifolds. Then*

$$\chi(M^2) = n\chi(N^2). \tag{21.4}$$

Proof. If the triangles in the triangulation of N^2 are sufficiently small, then the inverse image of each triangle consists of n nonintersecting triangles. Moreover, to each vertex, to each side, to each triangle in N^2 correspond n vertices, n sides, and n triangles of the triangulation of M^2. Therefore the formula $\chi(N^2) = V - E + F$, where V is the number of vertices, E the number of edges (sides), F the number of faces (triangles), implies that $\chi(M^2) = n\chi(N^2)$. □

Problem 21.2. a) Let $p : M_g^2 \to N_h^2$ be a covering of the sphere with h handles by the sphere with g handles. Prove that $g - 1$ is divisible by $h - 1$.

b) Suppose that $g, h \geqslant 2$ and $g - 1$ is divisible by $h - 1$. Prove that there exists a covering $p : M_g^2 \to N_h^2$.

21.6. Proof of the Riemann–Hurwitz formula. First let us rewrite formula (21.3) in a more convenient form. We can assume that $d_1, \dots, d_{a_1}$ are the branching indices of all the preimages of one branch point, $d_{a_1+1}, \dots, d_{a_1+a_2}$ are the branching indices of all the preimages of another branch point, and so on. Since (see 20.2)

$$d_1 + \cdots + d_{a_1} = d_{a_1+1} + \cdots + d_{a_1+a_2} = \cdots = n,$$

we obtain

$$\sum_{i=1}^{l}(d_i - 1) = (n - a_1) + (n - a_2) + \cdots = kn - a_1 - a_2 - \cdots - a_k,$$

where k is the number of branch points and a_i is the number of preimage points of the ith branch point. Hence formula (21.3) can be rewritten as

$$\chi(M^2) = n(\chi(N^2) - k) + a_1 + \cdots + a_k. \tag{21.5}$$

It will be more convenient for us to prove the Riemann–Hurwitz formula in this form.

The manifolds M^2 and N^2 may be presented as follows:

$$M^2 = A_M \cup B_M \quad \text{and} \quad N^2 = A_N \cup B_N,$$

where A_N is the union of the closures of small disk neighborhoods of all the branch points, A_M is the inverse image of A_N, while B_M and B_N are the closures of the complements $M^2 - A_M$ and $N^2 - A_N$. The sets $A_M \cap B_M$ and $A_N \cap B_N$ consist of nonintersecting circles, and so we can use formula (21.2). As the result we get

$$\chi(M^2) = \chi(A_M) + \chi(B_M), \quad \chi(N^2) = \chi(A_N) + \chi(B_N).$$

The restriction of the map p to the set B_M is a (nonbranched) covering, so by Theorem 21.5 we have

$$\chi(B_M) = n\chi(B_N).$$

The set A_N consists of k disks, while A_M consists of $a_1 + \cdots + a_k$ disks. Therefore

$$\chi(A_M) = a_1 + \cdots + a_k, \quad \chi(A_N) = k.$$

Combining the displayed formulas, we get

$$\chi(M^2) = a_1 + \cdots + a_k + n\chi(B_N) = a_1 + \cdots + a_k + n(\chi(N^2) - k),$$

which is the required formula (21.5). This completes the proof of the Riemann–Hurwitz formula. □

The Riemann–Hurwitz formula has numerous applications. The first one that we discuss has to do with Theorem 20.6, which asserts the existence of a branched covering $p : M_g^2 \to S^2$ with exactly 3 branch points (when $g \geqslant 1$). Can this number be decreased? The answer is "no", as the next statement shows.

21.7. THEOREM. *If $g \geqslant 1$, there exists no branched covering $p : M_g^2 \to S^2$ of the sphere by the sphere with g handles having less than 3 branch points.*

Proof. By formula (21.5), we have

$$2 - 2g = \chi(M_g^2) = n(\chi(S^2) - k) + a_1 + \dots + a_k = n(2 - k) + a_1 + \dots + a_k.$$

If $k \leqslant 2$, then $n(2-k) \geqslant 0$ and hence $n(2-k)+a_1+\dots+a_k > 0$, because the relations $n(2-k) = 0$ and $a_1 + \dots + a_k = 0$, i.e., $k = 0$, cannot hold simultaneously. Thus $2 - 2g > 0$, hence $g = 0$, contradicting the assumption of the theorem. Therefore $k > 2$, as asserted. □

Problem 21.3. Prove that if $p : D^2 \to D^2$ is a branched covering of the disk by the disk with exactly one branch point, then the preimage of the branch point consists of one point.

21.8. Genus of complex algebraic curves. The Riemann–Hurwitz formula can be applied to the computation of the genus of algebraic curves in $\mathbb{C}P^2$. We begin with the necessary background material. An *algebraic curve* of degree n in $\mathbb{C}P^2$ is the set of points satisfying the homogeneous equation $F(x, y, z) = 0$, where

$$F(x, y, z) = \sum_{i+j+k=n} a_{ij} x^i y^j z^{n-i-j}, \quad x, y, z \in \mathbb{C}P^2.$$

When $z = 1$ and $x, y \in \mathbb{R}$, we get a *plane algebraic curve*:

$$\sum_{i+j \leqslant n} a_{ij} x^i y^j = 0.$$

If the gradient

$$\operatorname{grad} F = \left(\frac{\partial F}{\partial x}, \frac{\partial F}{\partial y}, \frac{\partial F}{\partial z} \right)$$

does not vanish for every point of an algebraic curve in $\mathbb{C}P^2$, then the curve is called *nonsingular*. If the polynomial F is irreducible, i.e., cannot be represented as the product of two homogeneous polynomials of lesser degree, then the curve is said to be *irreducible*. It can be proved that any nonsingular irreducible algebraic curve in $\mathbb{C}P^2$ is homeomorphic to the sphere with g handles for some $g \geqslant 0$; the nonnegative integer g is called the *genus* of the curve.

21.9. PROPOSITION. *Fermat's curve $x^n + y^n + z^n = 0$ is of genus*

$$(n-1)(n-2)/2$$

Proof. First note that Fermat's curve $\Gamma \subset \mathbb{C}P^2$ is nonsingular, because

$$\operatorname{grad} F = n\left(x^{n-1}, y^{n-1}, z^{n-1}\right) \neq 0 \quad \forall (x : y : z) \in \mathbb{C}P^2.$$

Now consider the map $p : \mathbb{C}P^2 - \{(0:0:1)\} \to \mathbb{C}P^1$ that takes the point $(x : y : z) \in \mathbb{C}P^2$ to $(x : y) \in \mathbb{C}P^1$. The point $(0 : 0 : 1)$ does not lie on the curve Γ; therefore the map p induces the map $p' : \Gamma \to \mathbb{C}P^1 = S^2$. The inverse image of the point $(x_0 : y_0) \in \mathbb{C}P^1$ consists of all points $(x_0 : y_0 : z) \in \mathbb{C}P^2$ such that

$z^n = -(x_0^n + y_0^n)$. When $x_0^n + y_0^n \neq 0$, the inverse image consists of n points; when $x_0^n + y_0^n = 0$, of only one. Hence p' is an n-fold branched covering with branch points $(1 : \varepsilon_n) \in \mathbb{C}P^1$, where ε_n is a root of unity of degree n. So there are n branch points, and the inverse image of each consists of one point. According to formula (21.5), we get

$$\chi(\Gamma) = n(\chi(S^2) - n) + n = n(2-n) + n = -n^2 + 3n.$$

Therefore,

$$g = \frac{2-\chi(\Gamma)}{2} = \frac{n^2 - 3n + 2}{2} = \frac{(n-1)(n-2)}{2}. \quad \square$$

21.10. PROPOSITION. *The hyperelliptic curve $y^2 = P_n(x)$, where P_n is a polynomial of degree $n \geqslant 5$ without multiple roots, is of genus $[(n+1)/2] - 1$.*

Remark. This statement is also true for $n < 5$. When $n = 3, 4$ the curve $y^2 = P_n(x)$ is called *elliptic.*

Proof. Let $P_n(x) = a_0 + \dots + a_n x^n$. Then the hyperelliptic curve Γ in $\mathbb{C}P^2$ is given by the equation

$$y^2 z^{n-2} = \sum_{k=0}^{n} a_k x^k z^{n-k}. \tag{21.6}$$

For the curve Γ we have $\operatorname{grad} F = 0$ at the point $(0:1:0) \in \Gamma$, so the hyperelliptic curve is singular.

Consider the map

$$p : \mathbb{C}P^2 - \{(0:1:0)\} \to \mathbb{C}P^1, \quad \mathbb{C}P^2 \ni (x:y:z) \mapsto (x:z) \in \mathbb{C}P^1.$$

Let $p' : \Gamma - \{(0:1:0)\} \to \mathbb{C}P^1$ be the restriction of p. We claim that for $x = 1$ the preimage of the point $(x : z)$ tends to the singular point $(0:1:0) \in \Gamma$ as $z \to 0$. Indeed, by (21.6) we have $y^2 \approx a_n z^{2-n} \to \infty$, so that

$$(1 : y : z) = (1/y : 1 : z/y) \to (0:1:0).$$

Therefore the map p' can be extended to a map of the whole curve, $p' : \Gamma \to \mathbb{C}P^1$, by putting $p'((0:1:0)) = (1:0)$.

In order to find the inverse image under p' of the point $(x_0 : z_0) \in \mathbb{C}P^1$ when $z_0 \neq 0$, we must solve the equation

$$y^2 = z_0^2 \sum_k a_k \cdot \left(\frac{x_0}{z_0}\right)^k = z_0^2 \, P_n\left(\frac{x_0}{z_0}\right).$$

If x_0/z_0 is not a root of the polynomial P_n, then this equation has exactly two roots. Therefore the map $p' : \Gamma \to \mathbb{C}P^1$ is a double branched covering, the branch points being $(x_0 : z_0) \in \mathbb{C}P^1$, where x_0/z_0 is a root of P_n, and, possibly, the point $(1:0)$. We claim that $(1:0)$ is a branch point iff n is odd. To prove this claim, note that for small z the preimage of the point $(1 : z)$ consists of points of the form $(1 : y : z)$, where $y^2 \approx a_n z^{2-n}$. Let $z = \rho e^{i\phi}$. When ϕ varies from 0 to 2π, i.e., when we go around the point $(1:0) \in \mathbb{C}P^1$, the argument of the point $y \in \mathbb{C}$ changes by $(2-n)2\pi/2 = (2-n)\pi$. Therefore for odd n the number y changes sign, i.e., we switch to a different branch, while for n even y does not change, i.e., we stay on the same branch.

Thus the number of branch points is $2[(n+1)/2]$. Let g be the genus of the curve Γ. Then according to (21.5),

$$2-2g = 2\left(2-2\left[\frac{n+1}{2}\right]\right)+2\left[\frac{n+1}{2}\right], \quad \text{i.e.,} \quad g=\left[\frac{n+1}{2}\right]-1. \quad \square$$

§22. Branched coverings of 3-manifolds

22.1. Many definitions and theorems from §20 can be carried over to the case of dimension three without significant modifications. The definition of covering does not require any changes at all. A *branched covering* of 3-manifolds is defined as a continuous map $p: M^3 \to N^3$ such that there exists a one-dimensional subcomplex L^1 in N^3 whose inverse image $p^{-1}(L^1)$ is a one-dimensional subcomplex on the complement to which, $M^3 - p^{-1}(L^1)$, the restriction of p is a covering. In this situation M^3 is called the *covering manifold*, N^3 is the *base*, and L^1 is the *branching set*.

Theorems 20.5 and 20.6 together with their proofs are easily carried over to the three-dimensional case. For Theorem 20.5 we have in mind the second proof. Thus we have the following statements.

22.2. ALEXANDER BRANCHED COVER THEOREM. *Suppose M^3 is a compact oriented three-dimensional manifold without boundary. Then there exists a branched covering $p: M^3 \to S^3$.* $\square$

22.3. THEOREM. *Suppose M^3 is a compact oriented three-dimensional manifold without boundary. Then there exists a branched covering $p: M^3 \to S^3$ whose branching set is the 1-skeleton of a tetrahedron (i.e., its 6 edges).* $\square$

Remark. Theorems 22.2 and 22.3 can be generalized to any dimension n. In the generalization of Theorem 22.3 the branching set will be the $(n-2)$-skeleton of an n-simplex.

Problem 22.1. Let M^3 be a compact oriented manifold without boundary. Prove that there exists a branched covering $p: M^3 \to S^3$ whose branching set is the set shown in Fig.22.1.

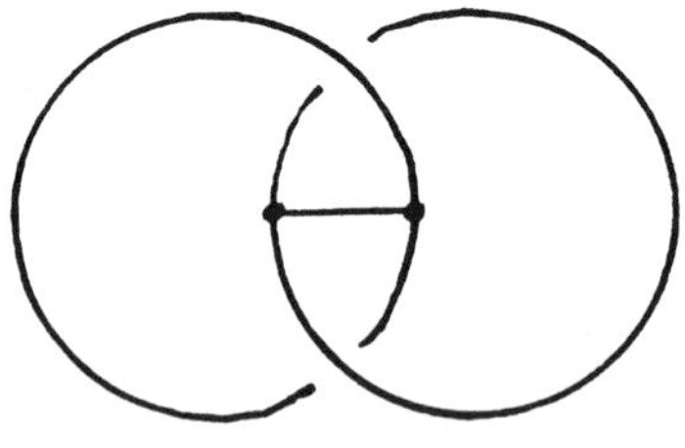

FIGURE 22.1. Universal branching set

22.4. In the sequel we shall be interested in branched coverings $p: M^3 \to N^3$ whose branching set is not just a subcomplex, but a submanifold (and thus, in dimension three, a link) whose inverse image is also a manifold. The analogs of Theorems 22.2 and 22.3 for such branched coverings are much more difficult to prove. This will be done in §§23–25.

22.5 Cyclic branched coverings. An important example of such branched coverings is the cyclic branched covering $p: S^3 \to S^3$. It is defined as follows. Let us represent S^3 as $\mathbb{R}^3 \cup \infty$. In $\mathbb{R}^3$ choose a straight line l and identify the points of $\mathbb{R}^3$ obtained from each other by a rotation by the angle $2\pi/n$ about the axis l. As the result, we get the map $p_1 : \mathbb{R}^3 \to \mathbb{R}^3/\mathbb{Z}_n$, where $\mathbb{Z}_n$ denotes the rotation group by angles $2\pi k/n$ about the axis l. The restriction of the map p_1 to a plane perpendicular to the line l is shown in Fig.22.2; the sides of the angles marked by arrows must be identified.

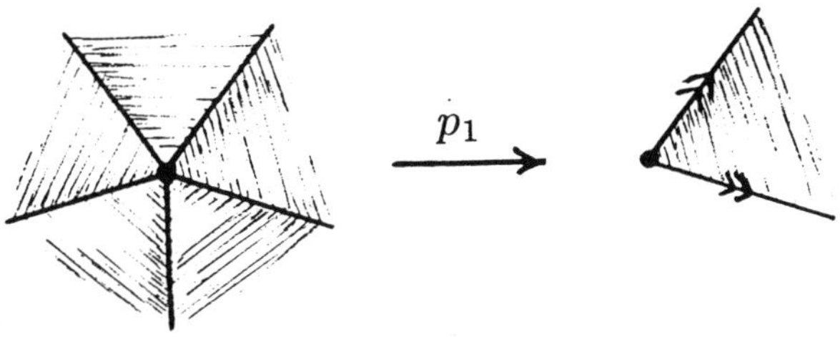

FIGURE 22.2. Plane section of a branched cyclic covering

Clearly, $\mathbb{R}^3/\mathbb{Z}_n \approx \mathbb{R}^3$, and the map $p_1 : \mathbb{R}^3 \to \mathbb{R}^3$ is a branched covering with branching set l. Adding the point at infinity to each of the spaces $\mathbb{R}^3$, we get a branched covering $p : S^3 \to S^3$, which is called *cyclic*. The branching set of the cyclic covering is the circle $l \cup \infty$. The inverse image of the branching set is also the circle $l \cup \infty$.

22.6. Using cyclic branched coverings $p : S^3 \to S^3$ as the starting point, it is possible to construct other examples of branched coverings by performing surgery along framed links (see §16). In the base of the branched covering p, let us choose a framed link L and do surgery along L, producing a manifold N^3. Consider the inverse image of L under the map p and perform surgery along it, obtaining another manifold M^3. The branched covering $p : S^3 \to S^3$ induces the branched covering $p' : M^3 \to N^3$; we shall also call such a branched covering *cyclic*.

As we did in §16, we can define the framing on L by specifying a curve on the boundary torus of the ε-neighborhood of each component of L. Then we get a curve on each boundary torus for the components of $p^{-1}(L)$. The next statement is an example illustrating this situation.

22.7. PROPOSITION. *The inverse image of the curve with framing* 1 *shown in* Fig.22.3,a *under the corresponding cyclic n-fold branched covering is the link depicted in* Fig.22.3,b.

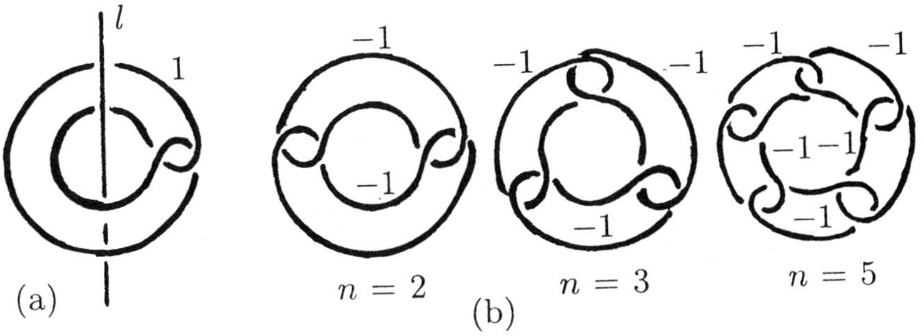

FIGURE 22.3. Example of a cyclic branched covering

Proof. The inverse image is obtained by identifying n copies of the linked curves shown in Fig.22.4. Therefore the inverse image does look as drawn in Fig.22.3,b. It remains to compute the framings.

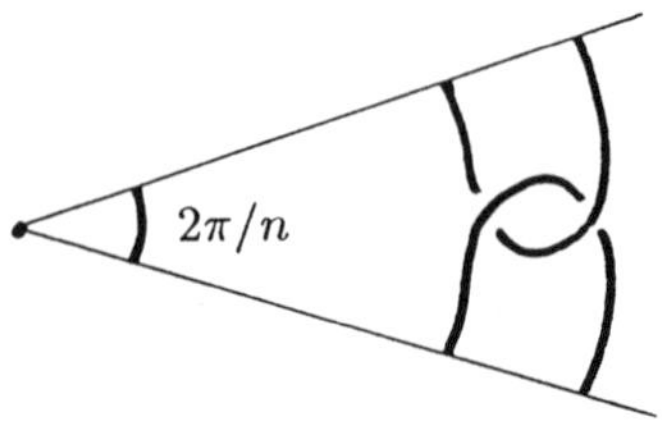

FIGURE 22.4. Building blocks for framed links

The framed link diagram of L (see 19.6) can be depicted as in Fig.22.5,a; the negative (-1) kink in the diagram must be added to compensate for the two $(+1)$ framings corresponding to the two crossing points of the original diagram. For $n = 2$ the corresponding framed link diagram in the covering manifold is shown in Fig.22.5,b. The framing indices are determined by the two kinks in its components; they are both equal to -1. □

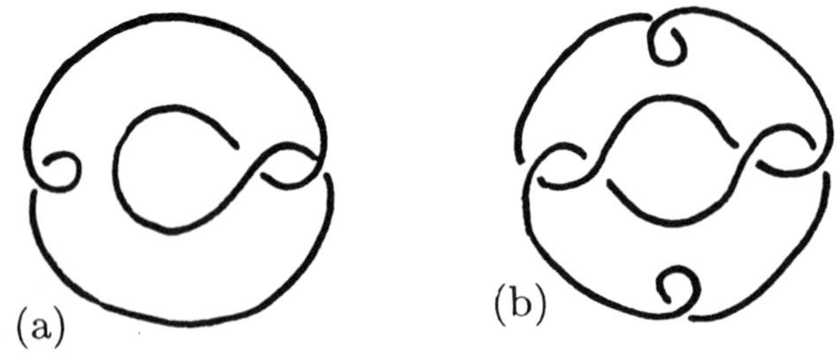

FIGURE 22.5. Framed link diagram and its inverse image

22.8. PROPOSITION. *For the manifold M^3 obtained from S^3 by surgery along the framed link shown in* Fig.22.3,b, *there exists an n-fold covering $p : M^3 \to S^3$ branching along the trefoil.*

Proof. According to Proposition 22.7, the inverse image of the curve J with framing $+1$ under the n-fold cyclic branched covering $p_1 : S^3 \to S^3$ consists of n curves $J_1, \ldots, J_n$ with framings equal to -1. It is not hard to verify that the link $\{J, K\}$, where $K = l \cup \infty$, from Fig.22.6,b, is isotopic to the link shown in Fig.22.7 (recall Fig.18.8).

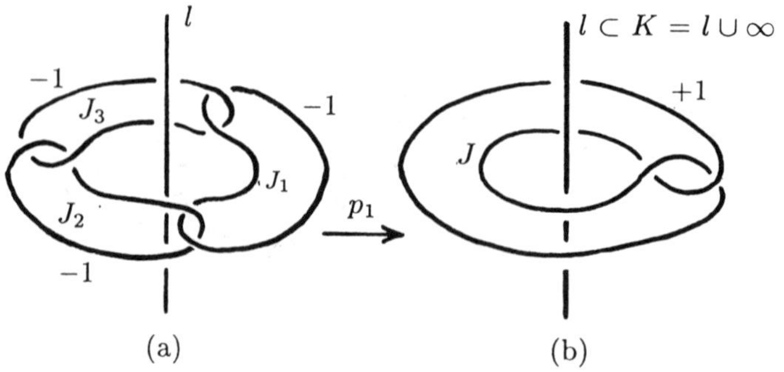

FIGURE 22.6. Three-fold cyclic branched covering

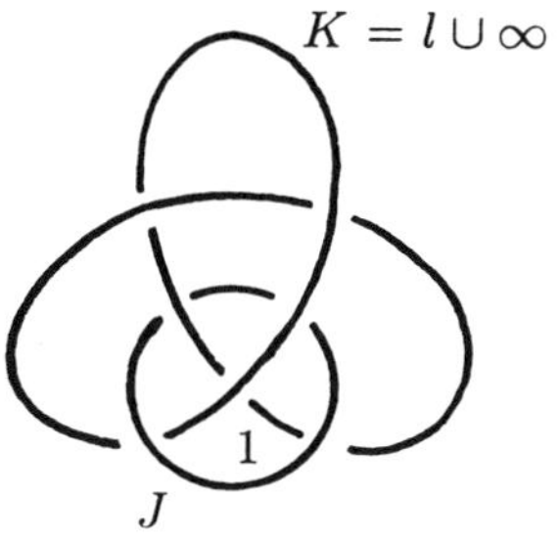

FIGURE 22.7. A curve K that becomes the trefoil after surgery

Let us perform surgery on S^3 along the curve J with framing $+1$. As the result, the branching set K will become the trefoil, while the manifold S^3 will not change (because J is the unknot). Therefore after surgery in the base and cover along the links J and $\{J_1, \dots, J_n\}$, respectively, the map $p_1 : S^3 \to S^3$ will become the required covering $p : M^3 \to S^3$ branching along the trefoil. □

Problem 22.2. Prove that there exists a 2-fold covering of the lens space $p : L(3,1) \to S^3$ branching along the trefoil.

22.9. PROPOSITION. *The Poincaré homology sphere is a* 5*-fold cover of* S^3 *branching along the trefoil.*

Proof. In the case $n = 5$, Proposition 22.8 yields the 5-fold covering $p : M^3 \to S^3$ branching along the trefoil; here the manifold M^3 is obtained from S^3 by surgery along the framed link shown in Fig.22.8,a. Performing a mirror symmetry, we get the framed link depicted in Fig.22.8,b. But the fact that the Poincaré sphere is presented by surgery along this link was established previously (see Fig.18.12). □

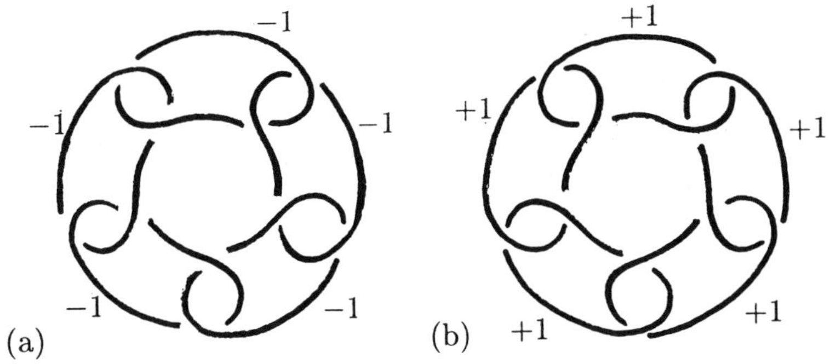

FIGURE 22.8. Mirror symmetric framed links

22.10. Suppose α and β are the meridian and parallel of the standardly embedded torus, and p and q are coprime integers. Then the curve $p\alpha + q\beta$ is called a *torus knot* of type (p, q).

Problem 22.3. Prove that the torus knot of type (p, q) is isotopic to the torus knot of type (q, p).

22.11. PROPOSITION. *There is a* 3*-fold covering ot the sphere* S^3 *by the Poincaré homology sphere branching along the torus knot of type* $(2, 5)$.

Proof. Consider the 3-fold cyclic covering $p_1 : S^3 \to S^3$ (Fig.22.9).

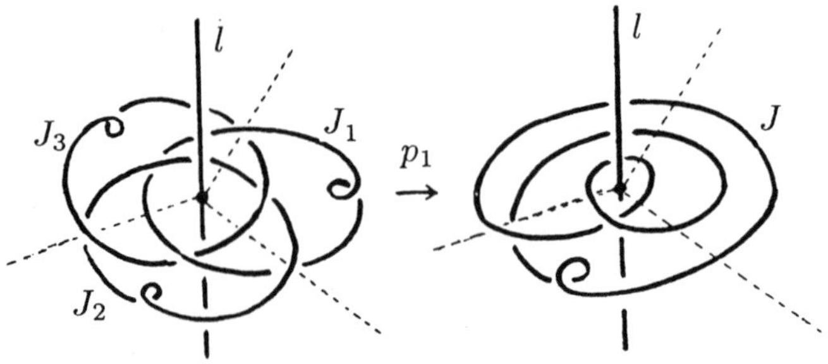

FIGURE 22.9. A 3-fold branched covering of the 3-sphere

The curve J is unknotted and its framing is equal to 1. Hence after surgery of the base manifold S^3 along J we still get S^3. As the result of this surgery, the curve $K = l \cup \infty$ becomes the torus knot of type $(2, 5)$—see Fig.22.10. In the covering manifold S^3 we must perform surgery along the inverse image of J, i.e., along the link $\{J_1, J_2, J_3\}$. This surgery produces the Poincaré sphere M^3. After these modifications, the covering p_1 becomes the required branched covering $p : M^3 \to S^3$. □

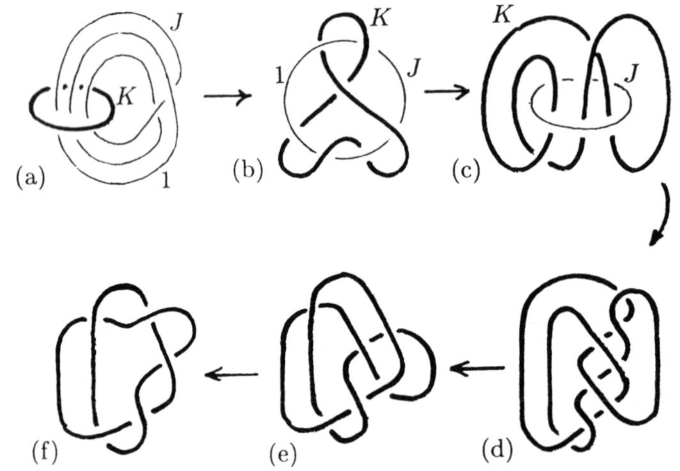

FIGURE 22.10. Transformations yielding the torus knot (2,5)

22.12. PROPOSITION. *There is a 2-fold covering ot the sphere S^3 by the Poincaré homology sphere branching along the torus knot of type* $(3, 5)$.

Proof. Consider the 2-fold cyclic branched covering $p_1 : S^3 \to S^3$ (Fig.22.11).

The curve J is unknotted and its framing is 1. Under the surgery of S^3 along J, the curve $K = l \cup \infty$ becomes the torus knot of type $(3, 5)$—see Fig.22.12. Surgery along the link $\{J_1, J_2\}$ transforms the covering manifold S^3 into the Poincaré sphere. □

Propositions 22.9, 22.11, and 22.12 show that the Poincaré sphere covers the standard 3-sphere as a p-fold covering branching along a torus knot of type (q, r) for any triple (p, q, r) such that $\{p, q, r\} = \{2, 3, 5\}$.

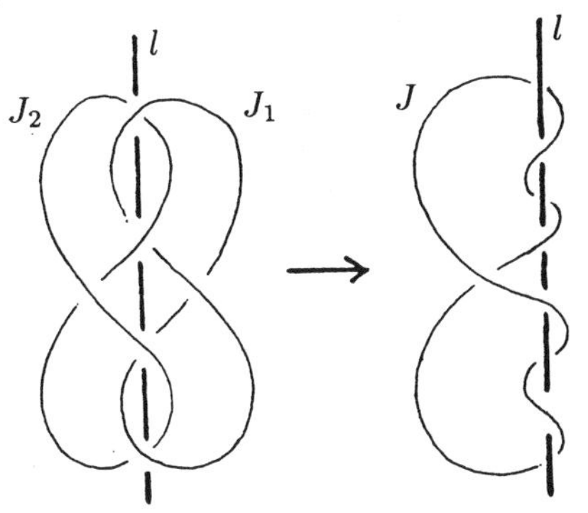

FIGURE 22.11. A 2-fold cyclic branched covering of S^3

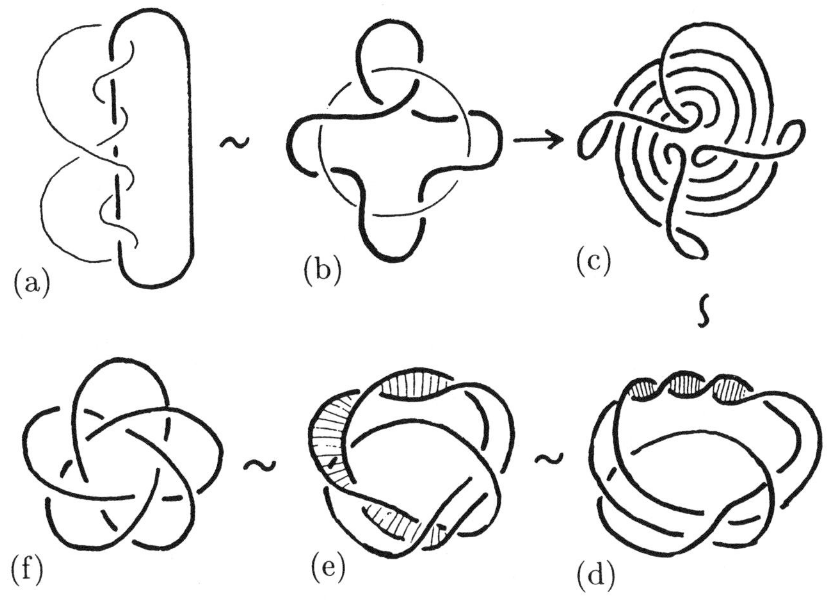

FIGURE 22.12. Obtaining the (3,5) torus knot

§23. Three-manifolds as branched covers of S^3

The aim of this section is to prove the following statement, which strengthens the Alexander branched covering theorem.

23.1. HILDEN–MONTESINOS THEOREM. *For any compact oriented* 3*-manifold* M^3 *(without boundary) there exists a* 3*-fold covering* $p : M^3 \to S^3$ *of the* 3*-sphere by this manifold branching along a knot.*

Note that this statement strengthens the Alexander theorem in three directions:
1) it specifies the number of folds of p;
2) it shows that for the branch set one can take not just some one-dimensional subcomplex, but a one-dimensional submanifold;
3) this submanifold may be assumed connected.

Thus the Hilden–Montesinos theorem provides a bridge between knot theory and the theory of 3-manifolds: the latter are none other than 3-fold coverings of the 3-sphere branching along different knots.

The proof of Theorem 23.1 is rather long; the general strategy is to start from a 3-fold branched covering $p : S^3 \to S^3$ and transform the covering manifold S^3 by surgery into the given manifold M^3. But first we must find out what kind of surgery of the base manifold and the covering manifold will still produce a branched covering.

Suppose the base of a branched covering $p : M^3 \to N^3$ is glued together from the manifolds N_1^3 and N_2^3, while the cover is glued from the manifolds M_1^3 and M_2^3 (not necessarily connected), and $M_i^3 = p^{-1}(N_i^3)$. We can assume that $\partial N_1^3 = \partial N_2^3 = N^2$ and $\partial M_1^3 = \partial M_2^3 = M^2$, where the gluings are performed along the identical homeomorphisms of the manifolds N^2 and M^2.

23.2. LEMMA. *Suppose p_1 is the restriction of the map p to the manifold M^2. Assume that the manifold N^2 intersects the branching set transversally. Then after we attach N_1^3 to N_2^3 and M_1^3 to M_2^3 along the homeomorphisms $f : \partial N_1^3 \to \partial N_2^3$ and $g : \partial M_1^3 \to \partial M_2^3$ respectively, the map p remains a branched covering if and only if $p_1 g = f p_1$.*

Proof. The condition $p_1 g = f p_1$ is necessary and sufficient for the map p to remain well-defined and continuous after the identifications described in the lemma are performed (Fig.23.1). The fact that the manifold N^2 intersects the branching set transversally implies that the map $p_1 : M^2 \to N^2$ is a branched covering (of surfaces). Since f and g are homeomorphisms satisfying $p_1 g = f p_1$, it follows that g maps $p^{-1}(f^{-1}(x))$ bijectively onto $p^{-1}(x)$. Hence f takes branch points to branch points, while g takes the preimage of each branch point to the preimage of a branch point. Thus in our identifications we actually perform identifications of the branch set and its preimage. Therefore after all these identifications, the restriction of p to the complement of the branch set will still be a covering. □

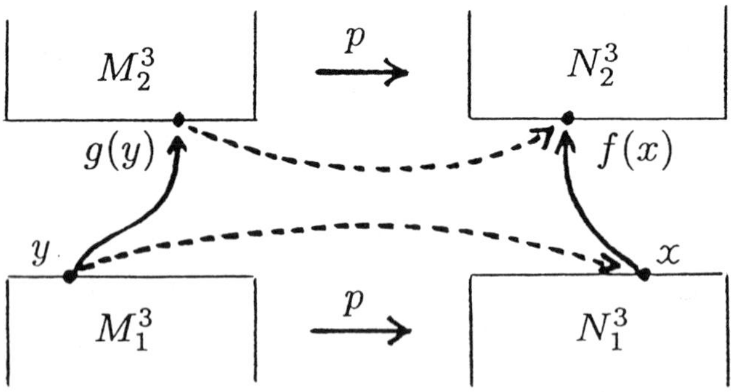

FIGURE 23.1. Homeomorphism g that covers f

23.3. Suppose $p_1 : M^2 \to N^2$ is a branched covering. We shall say that a homeomorphism $g : M^2 \to M^2$ *covers* a given homeomorphism $f : N^2 \to N^2$ with respect to p_1 if $p_1 g = f p_1$.

Lemma 23.2 shows that the description of simultaneous modifications of the base and covering manifolds of the branched covering $p : M^3 \to N^3$ compatible with the map p reduces to the description of the homeomorphism f and of the homeomorphism g that covers f. Let us present two examples of specific instances of this construction that will be useful in the sequel.

23.4. Example. Consider the 2-fold branched covering shown in Fig.23.2,c. It is easy to see that the covering manifold is (homeomorphic to) the annulus; see Fig.23.2,a,b.

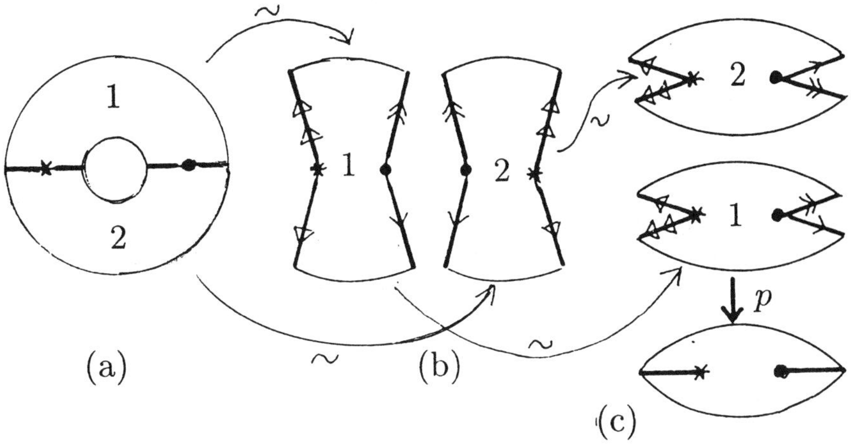

FIGURE 23.2. A 2-fold branched covering of the disk D^2

Suppose that f is a homeomorphism of the disk, identical on the boundary, interchanging the branch points by a 180° twist; Fig.23.3 shows the image of the line segment I under the homeomorphism f. For the homeomorphism f, *there exists a homeomorphism* g, *which is the identity on the boundary, that covers* f; this homeomorphism is shown in Fig.23.3.

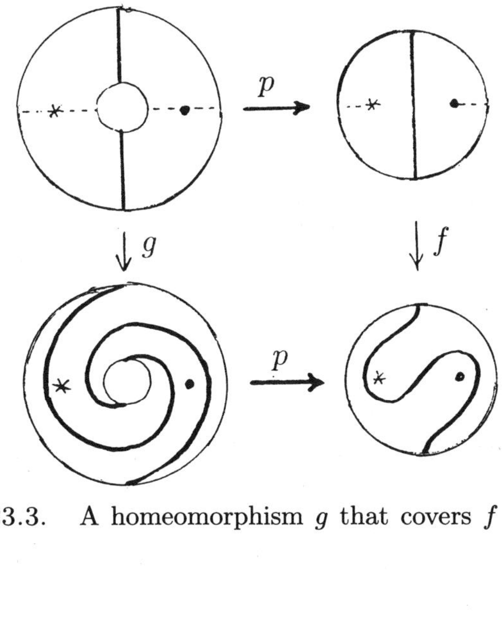

FIGURE 23.3. A homeomorphism g that covers f with respect to p

To construct g, let us cut two copies of the disk with the image of the segment I in it (Fig.23.4,a) and glue them together so as to obtain the covering manifold of the covering p (Fig.23.4,b).

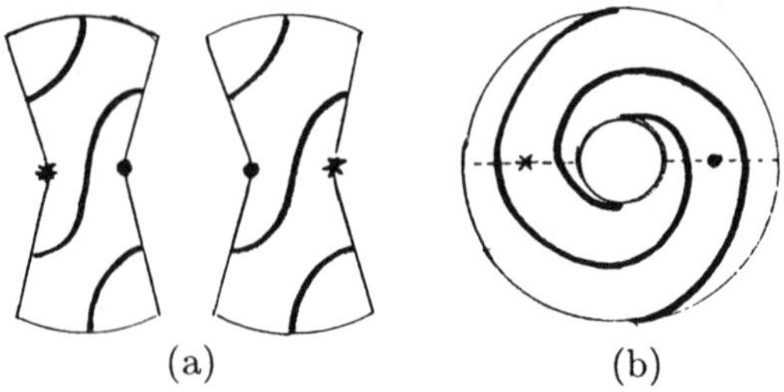

FIGURE 23.4. Construction of g

23.5. Example. Consider the 3-fold branched covering shown in Fig.23.5,a. It is easy to see that the covering manifold is homeomorphic to the disk (Fig.23.5,b-c).

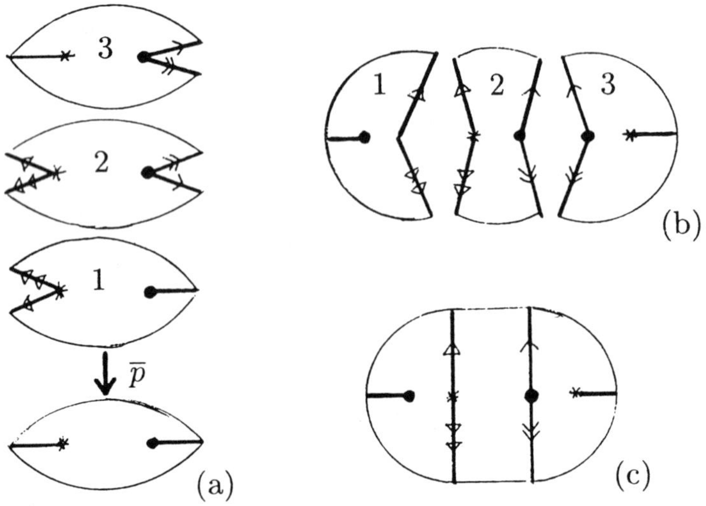

FIGURE 23.5. A 3-fold branched covering of the disk D^2

The structure of the branched covering $\bar{p}$ is shown in more detail in Fig.23.6. Note that the hatched regions are in fact taken to each other by parallel translations, possibly followed by symmetries in a point.

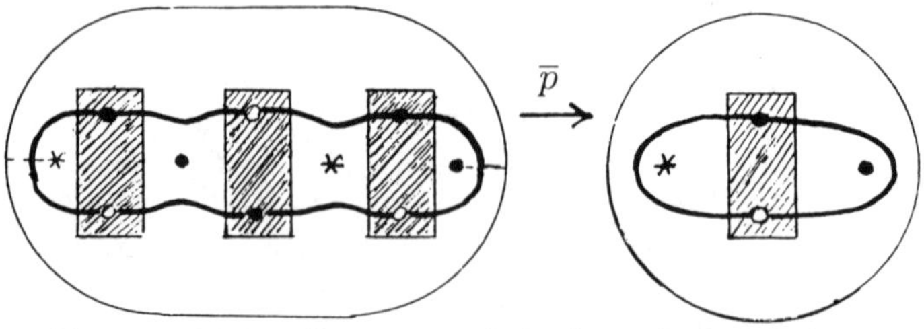

FIGURE 23.6. Structure of the branched covering $\bar{p}$

Suppose that f is a homeomorphism of the disk, identical on the boundary, interchanging the branch points by a 3π twist (Fig.23.7). Then *there exists a homeomorphism g, which is the identity on the boundary, that covers f*; this homeomorphism is shown in Fig.23.7.

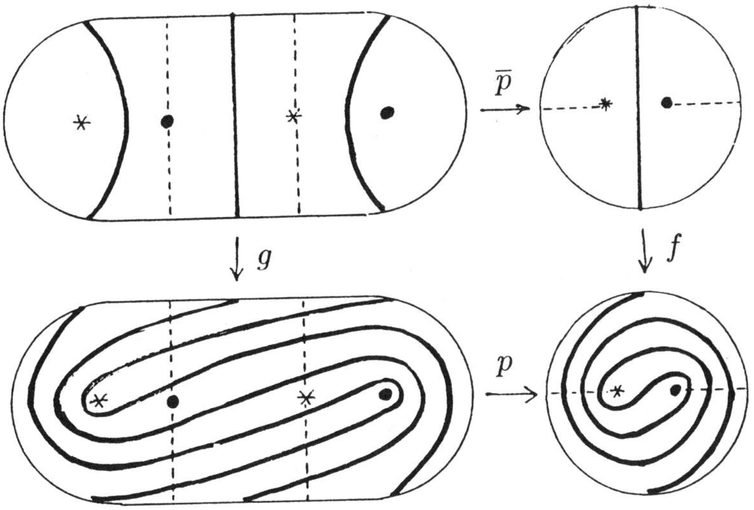

FIGURE 23.7. A homeomorphism g that covers f with respect to $\bar{p}$

To construct g, let us cut three copies of the base manifold and glue them together so as to obtain the covering manifold (Fig.23.8). It is clear that the homeomorphism g permutes the branch points via a rotation by π, so we have obtained the required map.

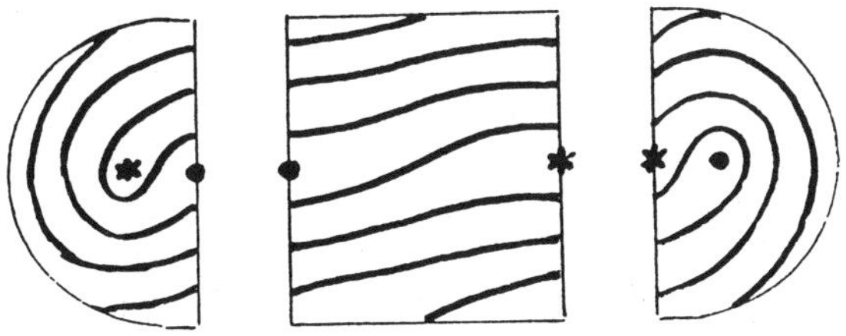

FIGURE 23.8. Constructing g

Problem 23.1. Prove that if in Example 23.5 we had taken twists f by π or 2π, it would not have been possible to find a homeomorphism g, identical on the boundary, that covers f.

23.6. Returning to the proof of the Hilden-Montesinos theorem, let us define the branched covering $p : S^3 \to S^3$ which will be modified to become the required map $p : M^3 \to S^3$; the construction of p will be based on Example 23.5. Take the Cartesian product of the covering map from that example by the segment $[0, 1]$

(Fig.23.9,a, where the distinguished segments in the cover correspond to points of branching index 2). Both the base and the cover of this branched covering are 3-disks. Therefore, attaching two copies of the base disk to each other by the identity map of their boundary 2-spheres, and doing the same with two copies of the covering disk, we obtain a 3-fold branched covering $p : S^3 \to S^3$ (Fig.23.9,b).

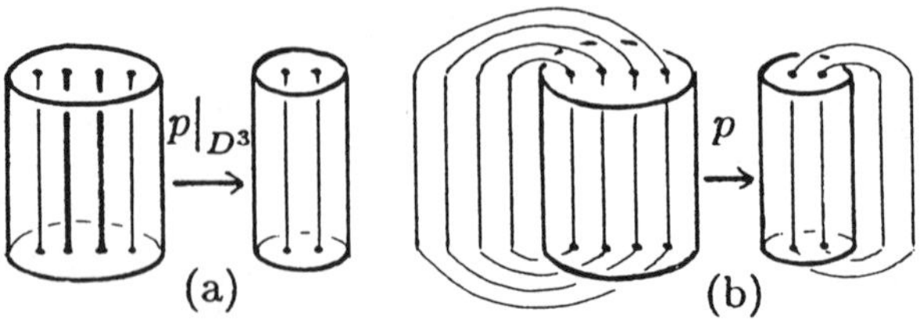

FIGURE 23.9. The branched covering $p : S^3 \to S^3$

In the base of the branched covering p consider a submanifold D_1^3 intersecting the branch set as shown in Fig.23.10.

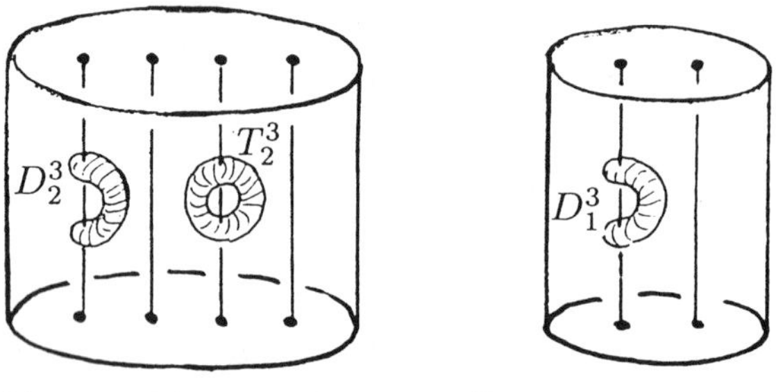

FIGURE 23.10. The disk D_1^3 and its preimage

The structure of the covering p is determined by the structure of the auxilliary covering $\bar{p}$ depicted in Fig.23.6; a careful study of this picture reveals that the preimage $p^{-1}(D_1^3)$ of the little crooked disk D_1^3 consists of the little disk D_2^3 and the little solid torus T_2^3 (Fig.23.10). Note that the restriction of the map p to D_2^3 is a homeomorphism. Further, let us show that the restricton of p to T_2^3 is the same as the cartesian product of the 2-fold branched covering considered in Example 23.4. To that end, it is convenient to present the map in the example as follows. Let us picture the annulus so that it has an axis of symmetry l and intersects l in two points (Fig.23.11,a). Identifying points of the annulus symmetric with respect to l, we get our branched covering. It is now clear that the product of this map by $[0, 1]$ has the same structure as the restriction of the map p to T_2^3 (Fig.23.11,b).

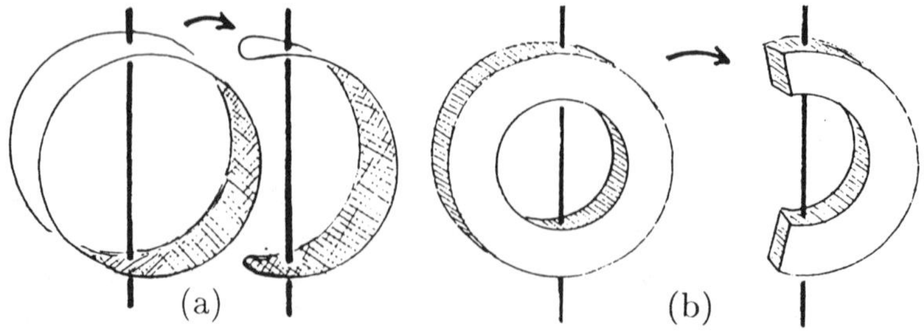

FIGURE 23.11. Restriction of p to the solid torus T_2^3

We shall make the appropriate modifications to the map $p : S^3 \to S^3$ by cutting out and pasting back in the little disks and the little solid torus so as to be able to apply Lemma 23.2.

23.7. Let $p_1 : M^2 \to N^2$ be the 2-fold branched covering from Example 23.4. Here M^2 is the annulus, N^2 is the disk. We must construct a homeomorphism $f_1 : \partial(N^2 \times I) \to \partial(N^2 \times I)$ and a second homeomorphism $g_1 : \partial(M^2 \times I) \to \partial(M^2 \times I)$ that covers the first one. Put $f_1(x, 1) = f(x)$ and $g_1(x, 1) = g(x)$, where f and g were defined in Example 23.4, and assume that the maps f_1 and g_1 do not move any other points (Fig.23.12). Such homeomorphisms f_1 and g_1 exist because the homeomorphisms f and g are identical on the boundary.

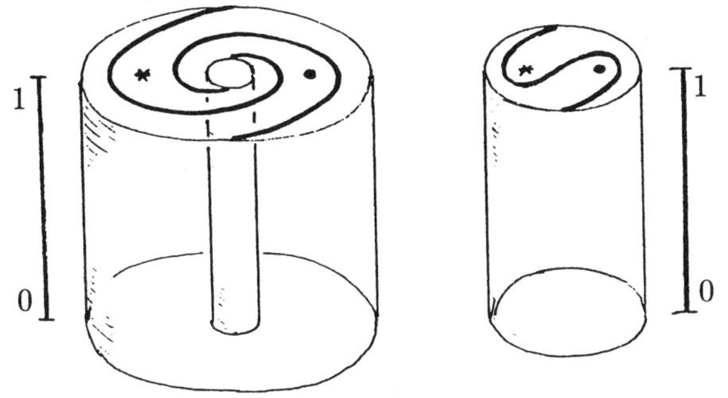

FIGURE 23.12. Auxilliary map of the torus

Now in the base of the branched covering $p : S^3 \to S^3$ let us cut out the submanifold D_1^3 and then paste it back in along the homeomorphism $f_1 : \partial(D_1^3) \to \partial(D_1^3)$. At the same time, we also cut out the manifolds D_2^3 and T_2^3 and then paste them back in along the maps $f_1 : \partial(D_2^3) \to \partial(D_2^3)$ and $g_1 : \partial(T_2^3) \to \partial(T_2^3)$. By Lemma 23.2 we still get a branched covering. The region where the homeomorphism f_1 is not the identity is displayed by hatching in Fig.23.13,a. This homeomorphism interchanges two branch points lying in this region, so that the branching set is modified as shown in Fig.23.13,b or c. Together with f_1, we can consider the homeomorphism f_1' that corresponds to the opposite twist. One of these homeomorphisms corresponds to the modification (b), the other to (c). Note that the number of components of the branch set does not change.

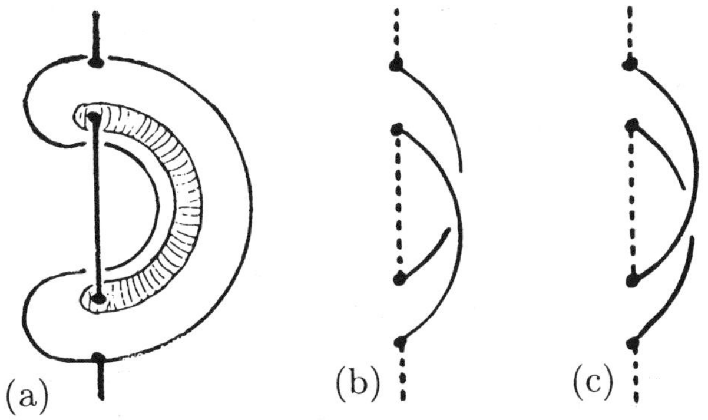

FIGURE 23.13. Modifications in the branch set

When we cut out and then paste back (along f_1) in the disks D_1^3 or D_2^3, we do not change the corresponding manifolds. However, when we cut out a solid torus like T_2^3 and then paste it back in along a homeomorphism like g_1, the covering manifold can change. Let us see in more detail what happens when we do surgery along the homeomorphism g_1 and the homeomorphism g_1' that covers f_1'. Under the homeomorphisms g_1 and g_1', the meridian α is transformed into the curve $\alpha \pm \beta$ (Fig.23.14). Thus our modification is a surgery along an unknotted curve with framing ± 1.

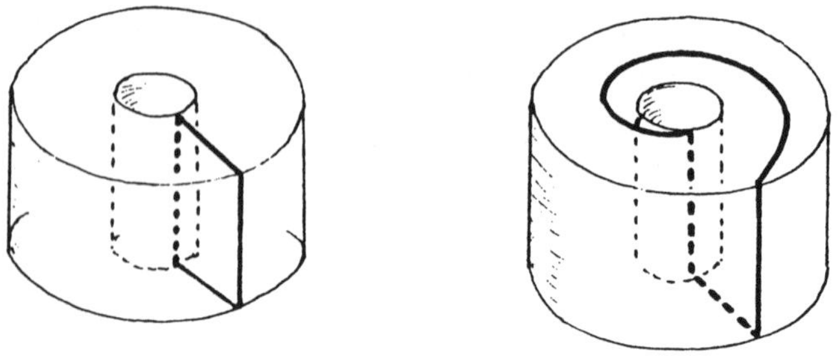

FIGURE 23.14. Modification of the meridian

The disk D_1^3 need not be positioned in S^3 as shown in Fig.23.10. By performing an isotopy that does not move the branching set, we can reposition the disk, say, as shown in Fig.23.15,b; then the corresponding torus T_2^3 will look as depicted in Fig.23.15,a. The new solid torus T_2^3 that we have thus obtained is unknotted and symmetric in the line l corresponding to one of the two components of the set of points with branching index 2. Thus we still obtain a 3-fold branched covering from $p : S^3 \to S^3$ after surgering the covering manifold S^3 along a system of unknotted curves with framing ± 1, each of which is symmetric in one of the two parallel branching lines l_1 and l_2, intersecting it in exactly two points, not intersecting the other branching line, and not linked with it. Let us call such a system of framed curves *special* for p.

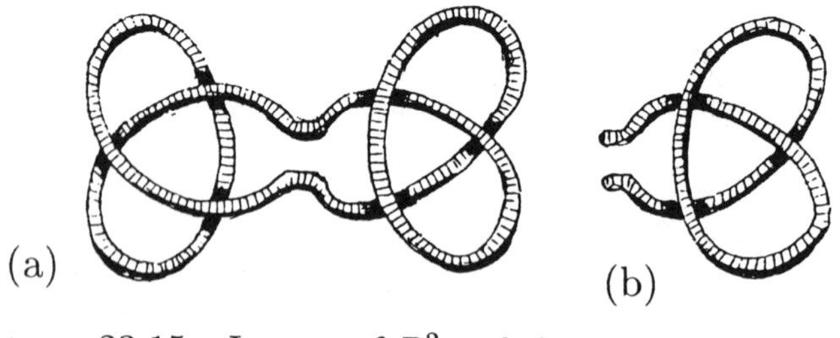

FIGURE 23.15. Isotopy of D^2 and the corresponding torus

23.8. LEMMA. *Any compact oriented manifold M^3 without boundary can be obtained as the 3-fold branched covering of S^3 by performing surgery along a framed link which is a special system of curves for p.*

Proof. According to 12.4 and 12.7, the manifold M^3 may be obtained from the sphere S^3 by integer surgery along a link L whose components are all unknotted.

The process of transforming such a link into a special system of curves for p is in three steps, depicted in Fig.23.16 and described below.

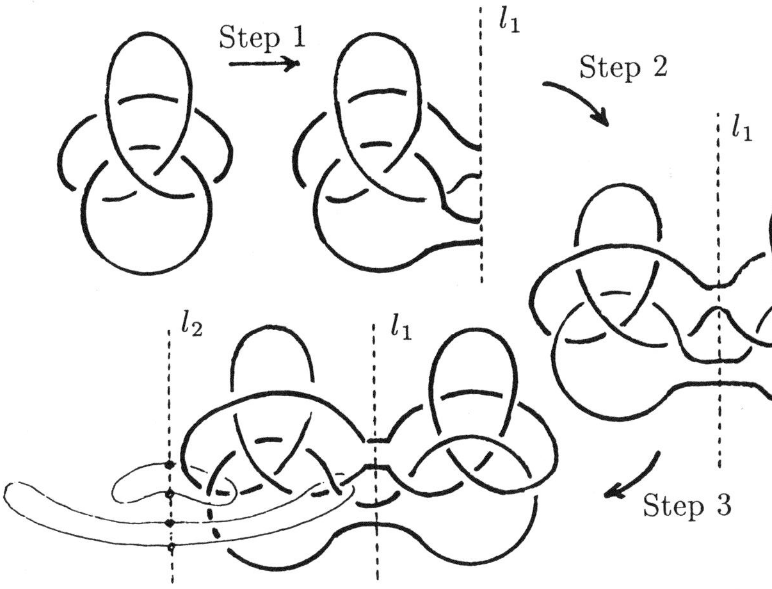

FIGURE 23.16. Making L special

Step 1. Choose a plane such that L lies entirely to one side of it. Pick a line l_1 in this plane. Move a section of each curve of the link L right up to l_1. This is all done by isotopy.

Step 2. Consider the link $S(L)$ symmetric to L in the line l_1 and construct the composition $L \# S(L)$ of the links L and $S(L)$. It may happen that $L \# S(L)$ is not isotopic to L. Moreover, we have not indicated the framings of its components. To get the required special link, one more step is needed.

Step 3. Let us add unknotted curves with framing ± 1 that encircle certain of the crossings of the link L in such a way that if we were to perform surgery along these curves, the link would fall apart into unknotted unlinked components. Without loss of generality the added curves may be assumed unlinked and symmetric in the line l_2. Supply each component K_i of the link $L \# S(L)$ with a framing such that after the surgery along the added curves K_i will acquire the same framing as the corresponding component of the link L. Thus we have obtained a framed link, consisting of all the curves of $L \# S(L)$ (with their chosen framings) and the added curves (with framings ± 1).

We claim that surgery along the framed link constructed above yields the same manifold as surgery along L. But this is quite obvious: when we do surgery along the added curves, the components of L unlink, the link $L \# S(L)$ becomes isotopic to $S(L)$ (which of course is isotopic to L), and each component has the same framing as the corresponding component of L.

The lemma is not proved yet, because our framed link need not be special: its components intersecting the line l_1 have framings that are not necessarily equal to ± 1 (as they should be if the link were special). To arrange that, first look at

Fig.23.17. It shows how the unknot with framing r may be modified to the unknot with framing $r+n\varepsilon$, where $\varepsilon = \pm 1$, $n \geqslant 0$. By an appropriate choice of the numbers n and ε, we can acheive the equality $r+n\varepsilon = \pm 1$. Thus the construction pictured in the figure allows to transform an unknot with any integer framing r to the required special form. □

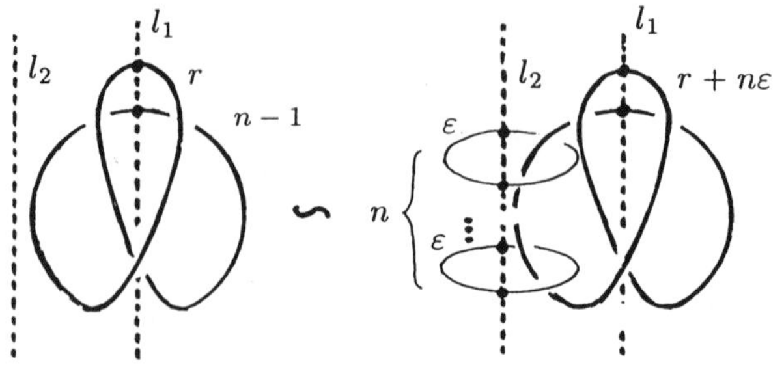

FIGURE 23.17. Changing the framing to ± 1

23.9. Lemma 23.8 gives us a 3-fold covering $p : M^3 \to S^3$ that branches along a two-component link. This is because we started with a two-component branch set in the base manifold S^3 (Fig.23.9) and subsequently did not change the number of components of this set. Thus to prove the Hilden-Montesinos theorem, it suffices to make the branch set connected.

Recall that our initial 3-fold map $p : S^3 \to S^3$ was glued from two copies of the product of the map from Example 23.5 by the segment $[0,1]$ (Fig.23.18).

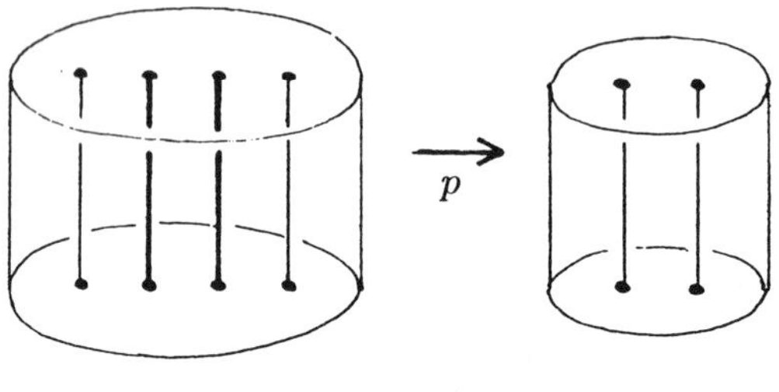

FIGURE 23.18. The map p (restricted to cylinders)

We can assume that all our surgery was performed inside one of the two copies; we shall use the other copy to make our branch set connected. To do that, take the homeomorphisms f and g constructed in Example 23.5. Using them, construct the homeomorphisms f_1 and g_1 of the boundaries of our cylinders that coincide with f and g on their upper bases and are identical elsewhere (Fig.23.19).

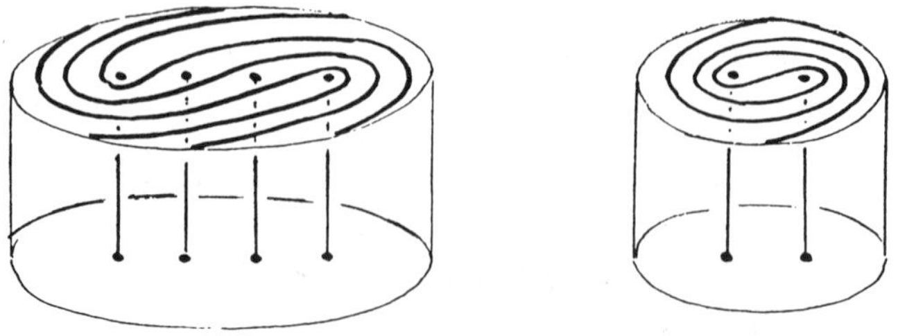

FIGURE 23.19. Constructing f_1 and g_1

If we cut out the cylinders and paste them back in along the maps f_1 and g_1, we will not change the base or the cover of our branched covering $p : M^3 \to S^3$. But since the maps f_1 and g_1 are actually twists by 3π and π, respectively, the branching set and its preimage will look as in Fig.23.20, and thus the branch set will become connected. This concludes the proof of the Hilden-Montesinos theorem. □

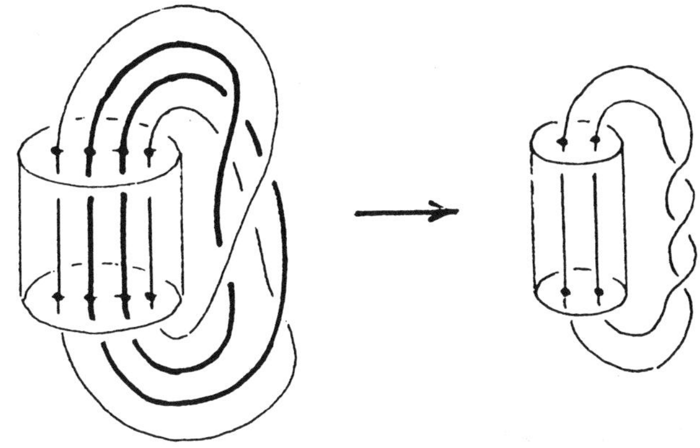

FIGURE 23.20. Modified branching set

Remark. A framed link representing a given manifold M^3 may be transformed to the special form obtained above in a different way, namely by using the following theorem.

23.10. THEOREM (Lickorish). *Any orientation-preserving homeomorphism of the sphere with g handles onto itself is isotopic to the composition of twists along the $3g-1$ curves shown in* Fig.23.21.

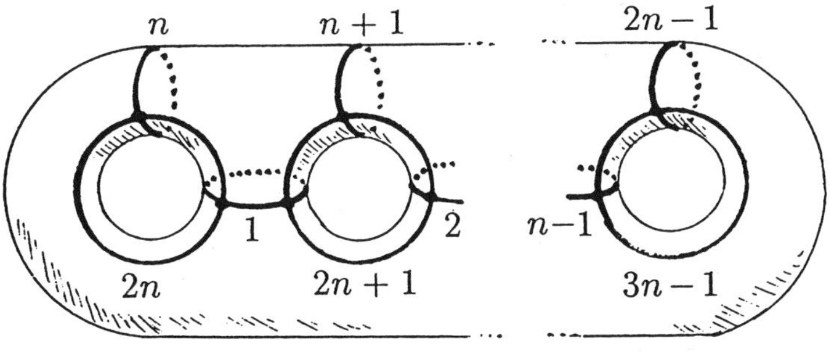

FIGURE 23.21. Canonical generators of the homeomorphism group

For the proof, see [**Lic2-3**], [**Bir1**], or [**MF**].

Theorem 23.10 implies that any compact oriented 3-manifold M^3 can be obtained from the sphere S^3 by surgery along a link whose components are circles parallel to those shown in Fig.23.21.

An example of such a link is presented in Fig.23.22. The framings must be chosen equal to ± 1. After an appropriate isotopy, one may assume that each of the circles intersects the line l_1 or l_2 at exactly two points and is symmetric with respect to that line.

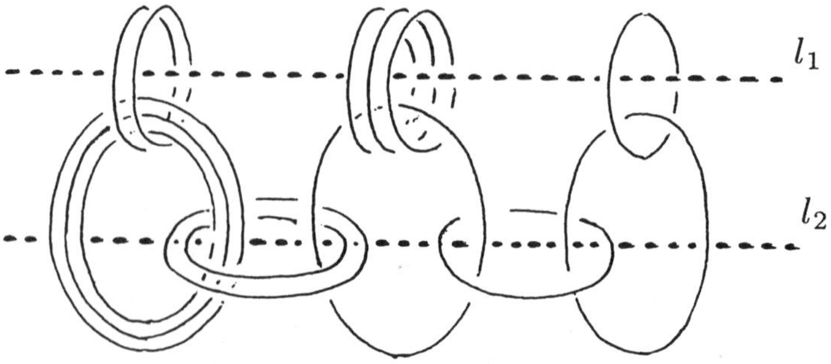

FIGURE 23.22. Special link corresponding to canonical generators

§24. Branched coverings and colored links

24.1. In this section we discuss some techniques for defining branched coverings of the sphere S^3 by means of links represented by diagrams whose arcs will be assigned "colors", i.e., will be marked in a certain way by elements of the permutation group. We are interested in these methods primarily because they will be needed in the next section, where a more refined version of the Hilden-Montesinos theorem (in which branching occurs along the Borromeo rings) is described. To attain this result, we shall need modifications of the branch set that do not change the covering manifold. Such modifications are provided by colored link diagram techniques, which therefore interest us from the viewpoint of finding the specific modifications that we shall need. We do not dwell on the nice existence and uniqueness theorems that are known in this area.

24.2. Correctly colored links. To any n-fold branched covering $p : M^3 \to S^3$ we can assign a homomorphism $\widetilde{p} : \pi_1(S^3 - L, x_0) \to S_n$, where L is the branch set, $x_0 \in S^3 - L$ is a fixed base point, and S_n is the permutation group. This is done in the following way. Let $y_1, \dots, y_n$ be the inverse images of the base point x_0. Consider a loop γ in $S^3 - L$ with extremities at x_0 and consider the path in M^3 originating at y_i and covering γ. Denote the endpoint of this path by $\sigma_\gamma(y_i)$. Thus we have assigned a permutation $\sigma_\gamma \in S_n$ to the loop γ. Clearly the permutation only depends on the homotopy class of γ, and the corresponding map $\widetilde{p} : \pi_1(S^3 - L, x_0) \to S_n$ is a homomorphism of groups.

In order to describe this homomorphism, it suffices to specify its values on the generators of $\pi_1(S^3 - L, x_0)$. Suppose that the link L is represented by a link diagram consisting of k arcs. Then the group $\pi_1(S^3 - L, x_0)$ is generated by k loops such as those shown in Fig.24.1,a; it is convenient to represent these loops symbolically as shown in Fig.24.1,b. Thus the arcs of our link diagram, on which perpendicular arrows have been drawn, can be "colored" by elements of the permutation group S_n in a well-defined way.

Reversing an arrow's direction corresponds to replacing a loop going around the arc in one direction by a loop going around it in the opposite direction; in

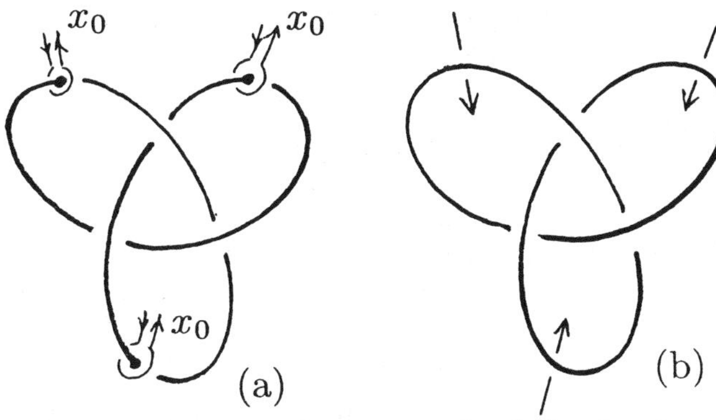

FIGURE 24.1. Generators of $\pi_1(S^3 - L, x_0)$

other words, to replacing a generator $a \in \pi_1(S^3 - L, x_0)$ by the inverse element a^{-1}. If $\widetilde{p}(a) = \widetilde{p}(a^{-1})$, then it is not necessary to specify the direction. The property $\sigma = \sigma^{-1}$ holds, in particular, for transpositions $\sigma = (ij)$. An arc of the link diagram L corresponds to the transposition (ij) provided that when a point X goes around this arc, the inverse image of X from the ith branch of the cover moves to the jth and back, while its inverse images from the other branches return to their initial positions. It is not difficult to check that *the branched* 3-*fold covering* $p : M^3 \to S^3$ *from Theorem* 23.1 *possesses this property.* Therefore the link diagram of the branching set of this covering is painted in the three colors $(12), (23), (31)$.

The coloring of the diagram of the branch set in our situation is not arbitrary, since the chosen generators of the group $\pi_1(S^3 - L, x_0)$ satisfy certain specific relations; namely, for each crossing we have $a = bcb^{-1}$ (Fig.24.2). For the link diagram painted in the colors $(12), (23), (31)$, this relation means that the three arcs appearing at a crossing point are all of the same color or all of different colors. Indeed, $(12)(13)(12)^{-1} = (23)$.

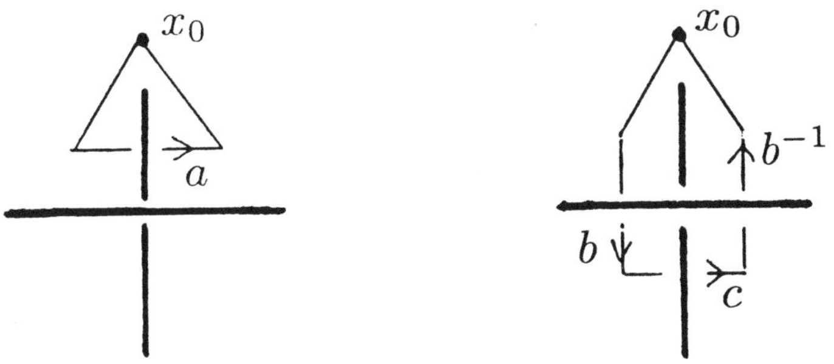

FIGURE 24.2. A relation in $\pi_1(S^3 - L, x_0)$

Let the arcs of a link diagram be painted in three colors; we shall say that the coloring is *correct* if at each crossing the three arcs are either all of the same color or all three of different colors. In the case when the colors are elements of the permutation group S_n and the arcs are marked by arrows (indicating the directions of the corresponding generators), such a coloring will be called *correct* provided we have the relation $a = bcb^{-1}$ at each crossing point.

Problem 24.1. Prove that the number of correct colorings of any link in three colors is invariant under Reidemeister moves. Use this invariant to prove that the trefoil cannot be unknotted.

24.3. Admissible transformations. Now we can return to our study of modifications of branching sets of our covering $p : M^3 \to S^3$. Suppose the branching set of p is correctly colored. Consider a transformation of p into a new covering with another correctly colored branching set such that the covering manifold M^3 does not change. Such a transformation of a colored diagram will be called *admissible.* Our goal in the remainder of this section is to establish a list of admissible transformations that will be sufficient for the purposes of §25.

Problem 24.2. Prove that plane isotopies and Reidemeister moves are admissible transformations.

24.4. The transformations in the previous problem are not the only admissible transformations that we have already used. In §23, in order to make the branch set of the 3-fold covering $p : M^3 \to S^3$ connected, we used the transformation shown in Fig.24.3. It is clearly admissible.

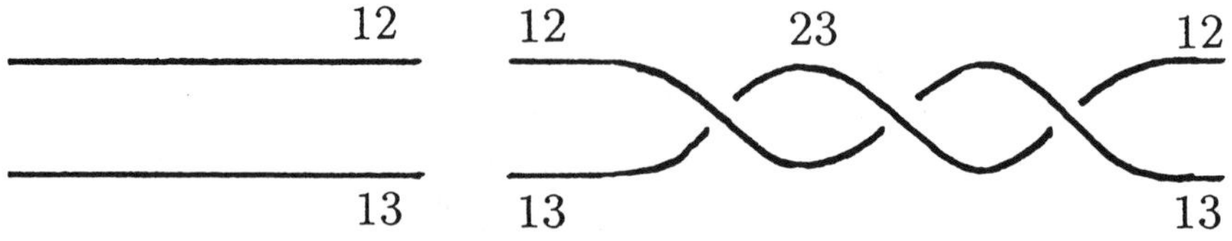

FIGURE 24.3. An admissible transformation from §23

Recall that this transformation was founded on the fact that the twist by 3π of the base of the cylinder in S^3 results in a twist by π of the base of the cylinder in the cover M^3.

Problem 24.3 Prove that all four transformations (and their inverses) of the branching set of a 3-fold branched covering $p : M^3 \to S^3$ that appear in Fig.24.4 are admisssible.

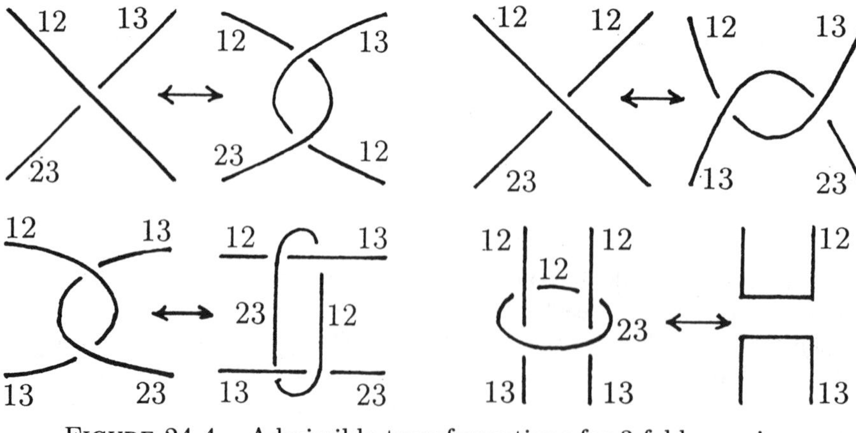

FIGURE 24.4. Admissible transformations for 3-fold coverings

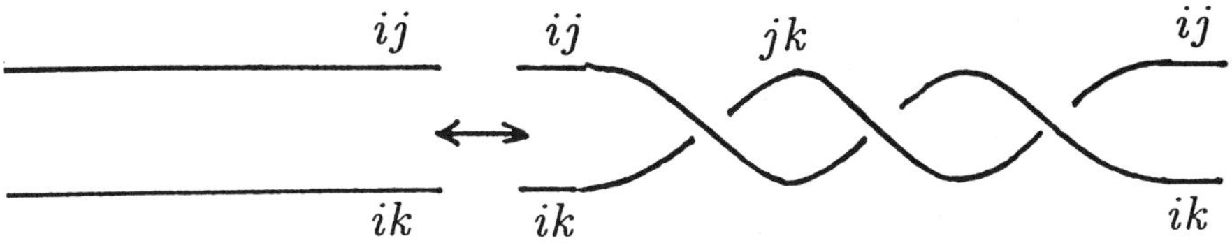

FIGURE 24.5. An admissible transformation for n-fold coverings

24.5. For n-fold branched coverings, *when the three numbers i, j, k are all different, the transformation depicted in* Fig.24.5 *is admissible.*

The proof is almost the same as for 3-fold coverings. Indeed, the inverse image under the n-fold covering of a 3-disk D^3 in the base consists of a 3-disk D_1^3 located in the branches with numbers i, j, k and of several other 3-disks, each of which is entirely contained in some other branch of the cover; the disk D_1^3 covers D^3 3-fold (just as in the case $n = 3$ considered in Fig.24.3); the other disks are mapped homeomorphically, so there is no problem in finding the covering homeomorphism (needed to apply Lemma 23.2), which guarantees that the covering manifold remains unchanged. □

24.6. Now let us consider similar transformations for parallel strands painted in the colors (i, j) and (k, l), where the four integers i, j, k, l are pairwise distinct. Let us prove that in this case *the two transformations shown in* Fig.24.6 *are admissible.* (It is easy to check that these transformations are equivalent in the sense that one can be taken to the other by Reidemeister moves.)

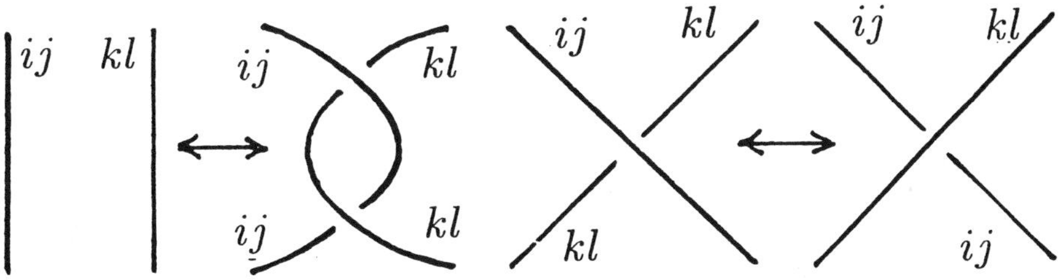

FIGURE 24.6. Two other admissible transformations

First of all let us note that the coloring in Fig.24.6 is correct, since we have $(ij)(kl) = (kl)(ij)$ when the four integers are all different. The covering of the 2-disk with two branch points painted in the colors (ij) and (kl) is represented in Fig.24.7.

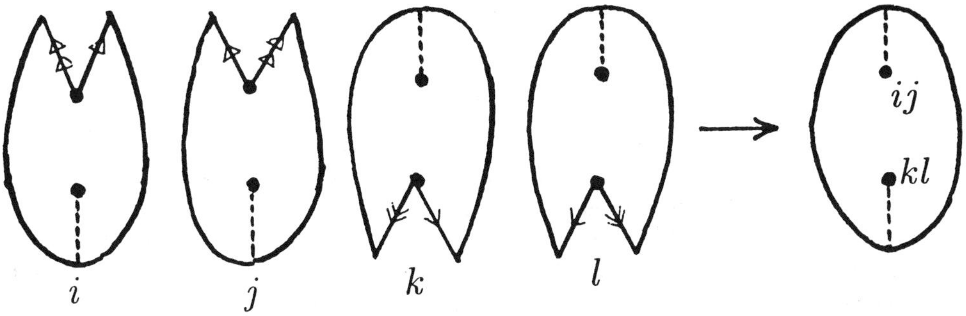

FIGURE 24.7. A 2-fold covering of the 2-disk

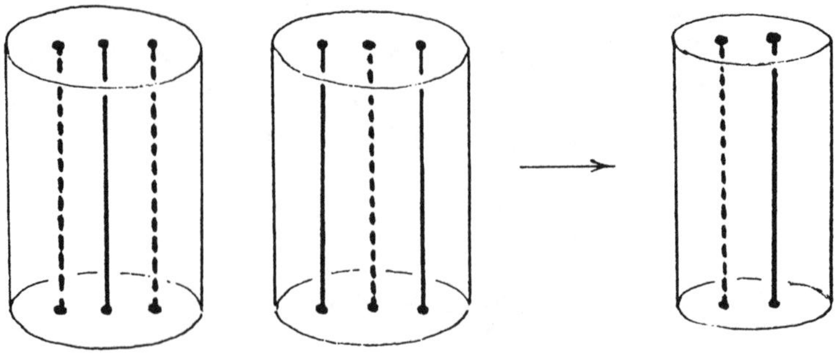

FIGURE 24.8. Part of a 2-fold branched covering

Hence after multiplication by $[0,1]$ the part of the covering map that interests us will appear as shown in Fig.24.8 (on the other components of the inverse image of the cylinder, p is a homeomorphism). For each of the two components shown in the figure, the covering homeomorphism may be constructed by using the homeomorphisms f and g defined for the 2-fold covering in Example 23.4. The modification in the manifold S^3 thus consists of a twist by 2π of the base of the cylinder. Lemma 23.2 now implies admissibility. □

24.7. The last admissible transformation that we consider is called the *addition of a trivial sheet*. We shall first describe it globally for manifolds, and then see how it affects the branch set. Consider an n-fold branched covering $p : M^3 \to S^3$. In Fig.24.9, a region containing the branch set is displayed by hatching, and so are the components of its inverse image.

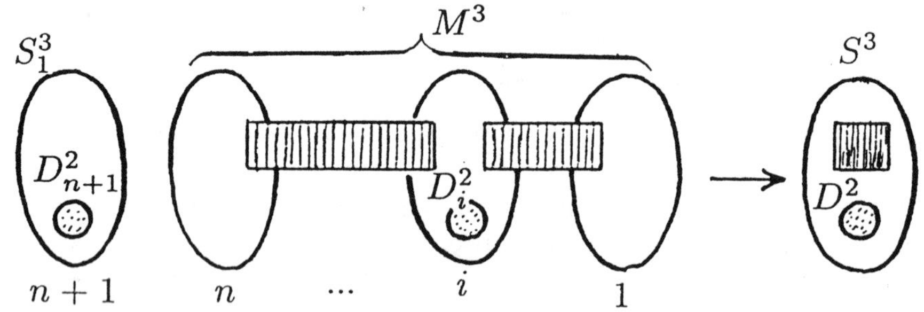

FIGURE 24.9. Addition of a trivial sheet

Let us add to M^3, as a disjoint summand, a new $(n+1)$st sheet, the 3-sphere S_1^3, mapped identically on the base sphere S^3. In order to obtain an $(n+1)$-fold branched cover with a connected covering manifold, we do the following. In the base S^3, we choose a 2-disk D^2 outside the hatched region, and consider the components D_i^2 and D_{n+1}^2 of its inverse image located in the ith and $(n+1)$st sheet of the cover, respectively. Cut the ith and $(n+1)$st sheets of the cover along the disks D_i^2 and

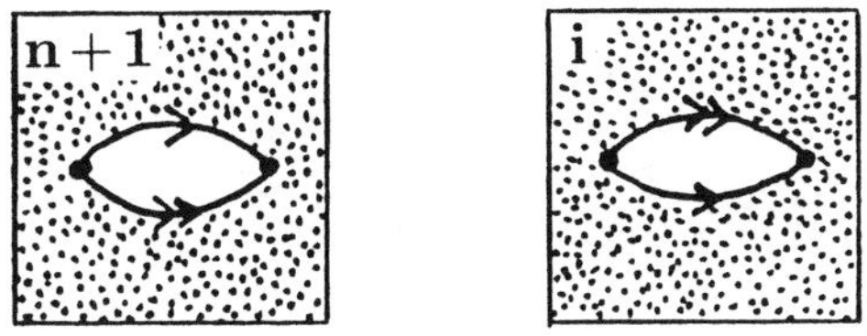

FIGURE 24.10. Joining the ith and $(n+1)$st sheets

D^2_{n+1}, and then reglue the boundaries thus created as shown in Fig 24.10.

From the topological point of view, this operation consists in cutting M^3 along a 2-disk and attaching a 3-disk along the boundary 2-sphere created by the cut. Obviously the manifold M^3_1 thus constructed is homeomorphic to M^3. For the manifold M^3_1, an $(n+1)$-fold branched covering $p_1 : M^3_1 \to S^3$ is naturally defined. In a neighborhood of any point of the boundary circles of the disks D^2_i and D^2_{n+1} the map p_1 is the 2-fold branched covering shown in Fig.24.11.

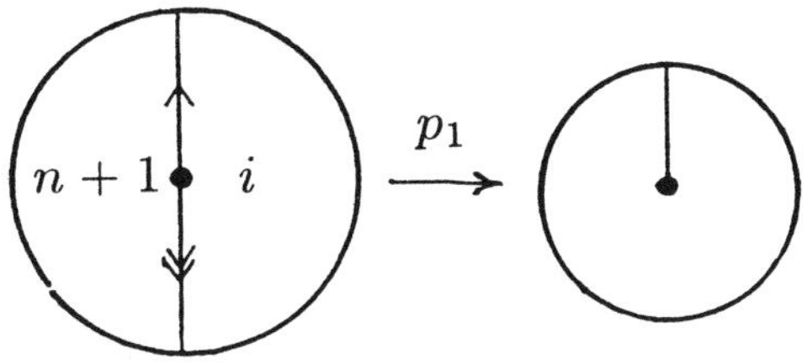

FIGURE 24.11. The map p_1 near a boundary circle

Therefore the corresponding transformation of the colored diagram of the branch set consists in adding to the branching link L an unknotted curve of color $(i, n+1)$ (that can be separated from the other components of L by a 2-sphere). Symbolically this is shown in Fig.24.12.

This completes our study of admissible transformations of branched coverings $p : M^3 \to S^3$.

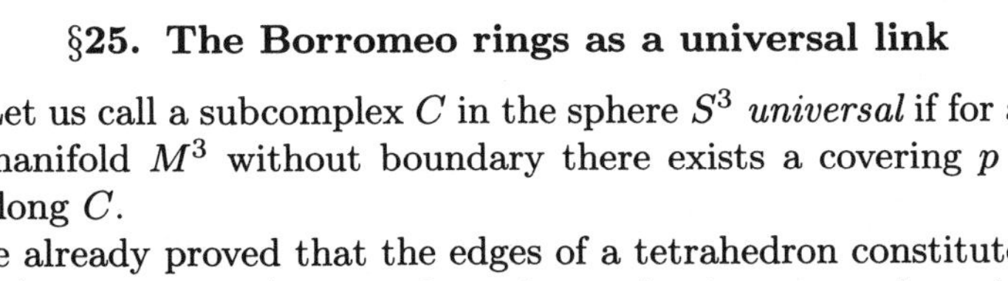

FIGURE 24.12. Addition of a trivial sheet

§25. The Borromeo rings as a universal link

25.1. Let us call a subcomplex C in the sphere S^3 *universal* if for any compact orientable manifold M^3 without boundary there exists a covering $p : M^3 \to S^3$ branching along C.

We have already proved that the edges of a tetrahedron constitute a universal subcomplex (Theorem 22.3). But this subcomplex is not a submanifold. In this section we shall give an example of a universal submanifold; namely, we shall prove the following statement.

25.2. THEOREM. *The Borromeo rings constitute a universal link, i.e., any compact orientable* 3*-manifold* M^3 *without boundary can be presented as the covering space of the* 3*-sphere with branch set the Borromeo rings.*

The proof of this theorem takes up the rest of this section. The general strategy is the following. Given the manifold M^3, we first apply the Hilden–Montesinos theorem (23.1, see also 24.2) to obtain a 3-fold covering $p : M^3 \to S^3$ branching along a (colored) link L. Then, in five steps, we carry out admissible transformations modifying L, first making it more and more complicated but more uniform (Steps 1–4), and then simplifying it to get the Borromeo rings (Step 5).

25.3. Step 1. *Braiding the link* L. By the Alexander Braiding Theorem 6.5, we can assume that the colored link $L \subset S^3$ (the branch set of the covering map $p : M^3 \to S^3$) is represented as the closure of some braid, which we can also assume colored. All the arcs of the braided link diagram (which we still denote by L) cannot be of the same color, because otherwise the covering would be 2-fold.

Moreover, we can assume that at all the crossings of the braided link L, there are arcs of different colors. Indeed, we can get rid of monochromatic crossings by using the isotopy pictured in Fig.25.1.

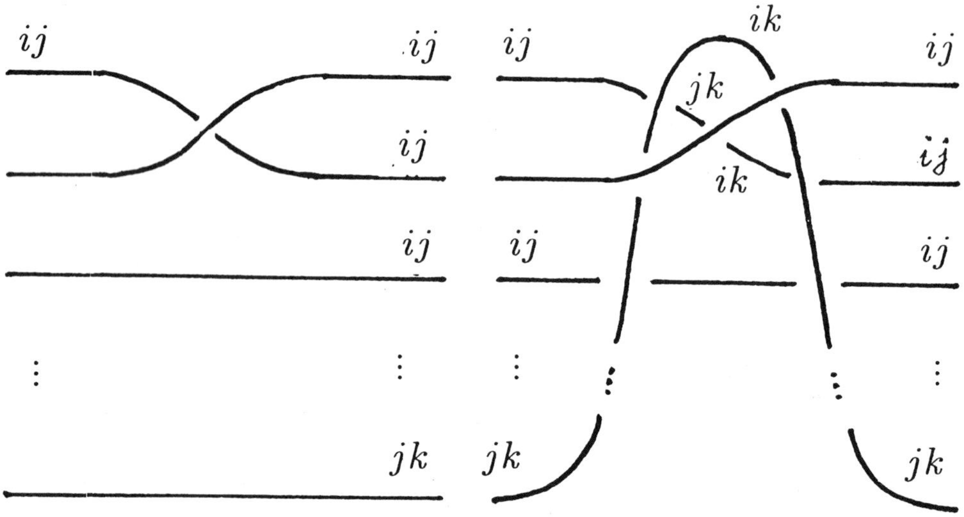

FIGURE 25.1. Isotopy destroying monochromatic crossings

Here we must avoid moving the strand of color (j,k) above other strands of that same color, but this can be ensured if we move the strand of color (j,k) nearest to the crossing point.

25.4. Step 2. *Making all the crossings positive.* We shall now need the admissible transformation shown in Fig.25.2 (see Figure 24.4 and Problem 24.3).

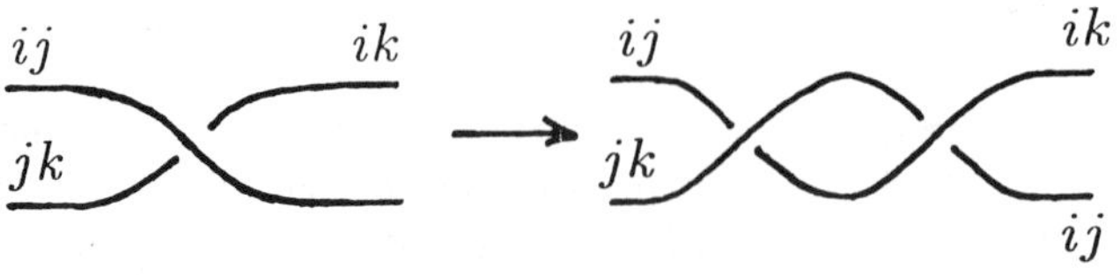

FIGURE 25.2. Admissible transformation

Using the transformation depicted in Fig.25.2, we can arrange for all the braid crossings to be positive, i.e., to be as pictured in Fig.25.3,a, rather than negative as in Fig.25.3,b. Indeed, an application of this transformation replaces one negative crossing by two positive ones. Moreover, it does not destroy the coloring properties of the crossings obtained at Step 1.

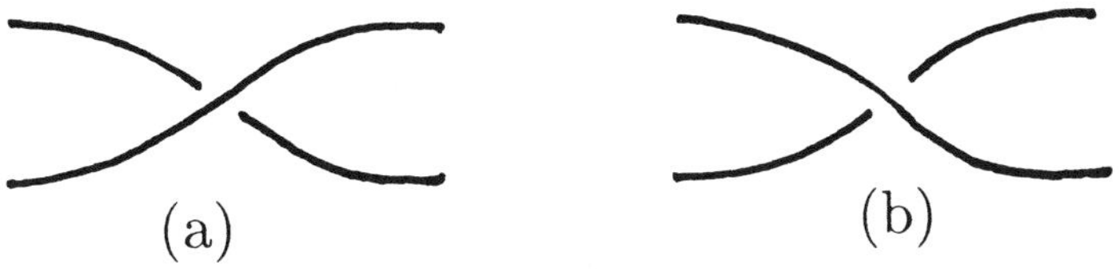

FIGURE 25.3. Positive and negative braid crossings

25.5. Step 3. *Trivializing the braid.* The transformations constituting this step are shown in Fig.25.4,a–f.

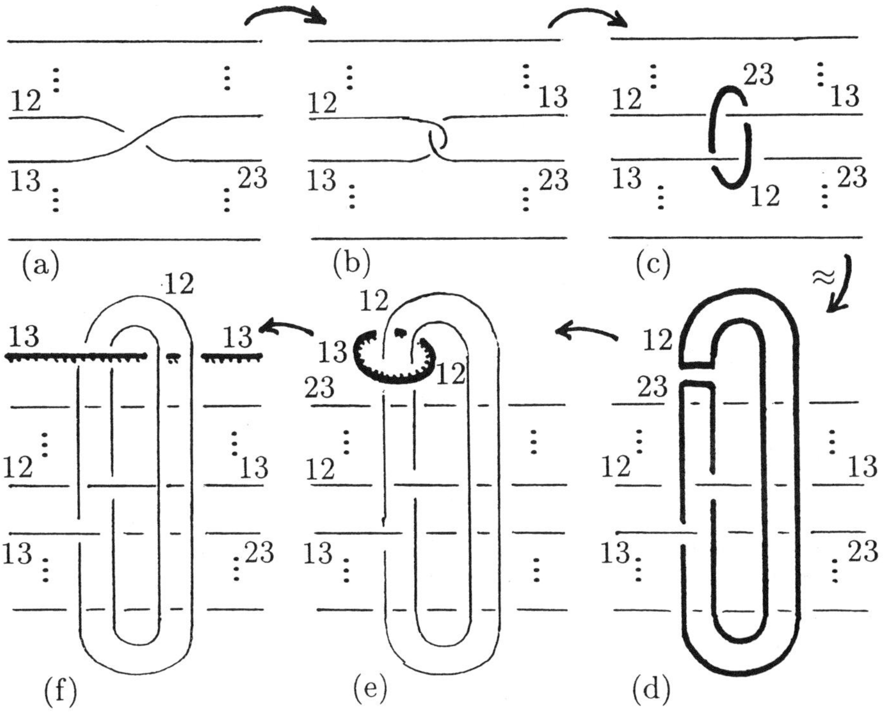

FIGURE 25.4. Trivializing the braid

To pass from (a) to (b) and (c), and also from (d) to (e), we apply the result of Problem 24.3. The passage from (c) to (d) is simply an isotopy of the thick circle. The thick circle in (e), which appears as the result of applying the last transformation from Fig.24.4 to (d), becomes an extra strand of the braid in (f).

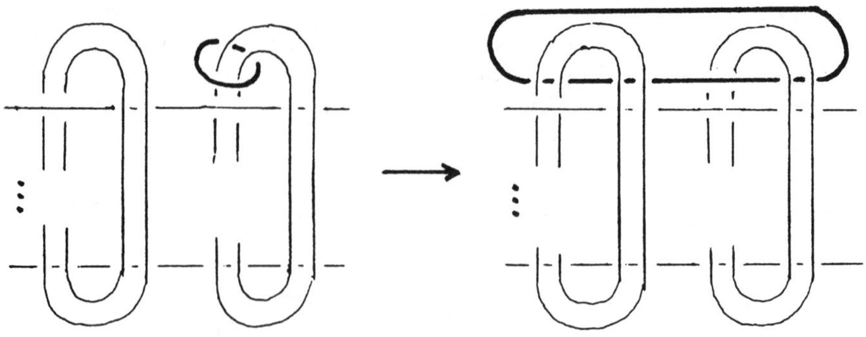

FIGURE 25.5. Stretching a circle into a strand

Note that in this last process we must move this thick circle *under* the big circles that appeared in the previous steps (Fig.25.5).

When all the q crossings of our braid are destroyed, we get a trivial braid with $2q$ big circles linking the parallel strands of the braid as shown in Fig.25.6. Note that the right half of each linking circle is situated *above* the "interior" strands of the braid.

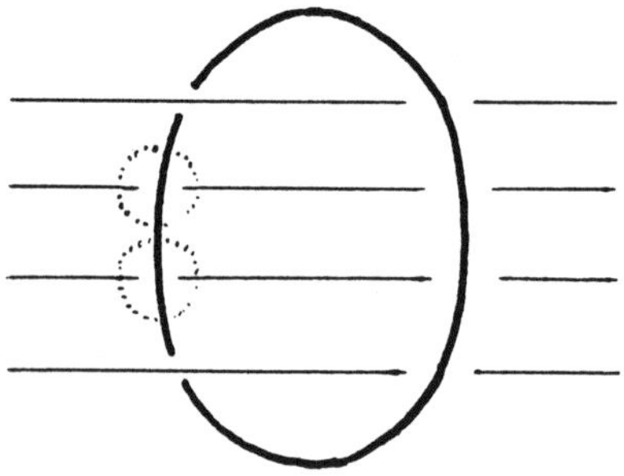

FIGURE 25.6. The result of Step 3

25.6. Step 4. *Uniformization of the link on the torus.* Now let us (admissibly) transform our link so that the left side of the big circle will pass *under* the interior strands of the braid. To do that, we must perform crossing changes at the points shown by the dotted circles in Fig.25.6. These crossing changes can be carried out by means of the operation of adding trivial sheets (see 24.7) and the operation shown in Fig.25.7 (where the integers i, j, k and l must be pairwise distinct). The crossing that must be changed can be either be monochromatic or be constituted by arcs of different colors. These two cases must be treated separately.

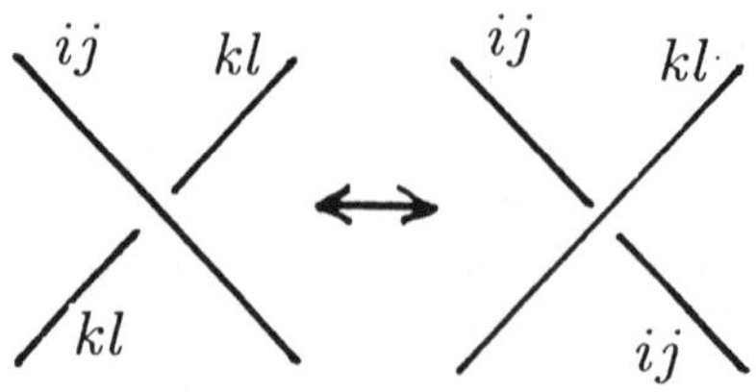

FIGURE 25.7. An admissible transformation

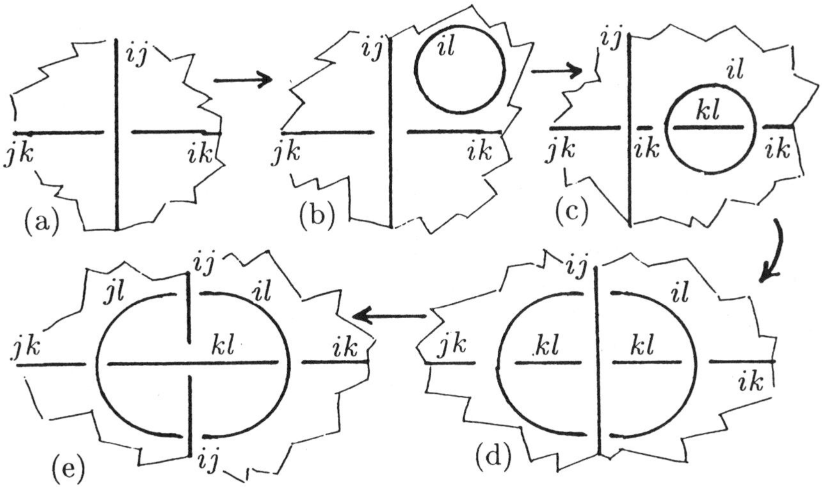

FIGURE 25.8. Multi-colored crossing change

In the case of a multi-colored crossing, the necessary transformations are shown in Fig.25.8. First we add a trivial sheet, i.e., we add a little circle of color (i, l), then perform Reidemeister moves, and finally carry out the transformation shown separately in Fig.25.7.

In the case of a monochromatic crossing, a similar addition of a trivial sheet produces a crossing painted (i, j), (i, l) and (j, l). The transformation from Fig.25.7 can be applied only to a crossing of type (i, j), (k, l), (k, l), where the numbers i, j, k, l are pairwise distinct, and therefore cannot be applied to the crossing under consideration. In order to get a crossing with the required coloring, we first add *two* trivial sheets, using the same little circle twice (Fig.25.9,b). Its color will be $(i, l)(j, l + 1)$.

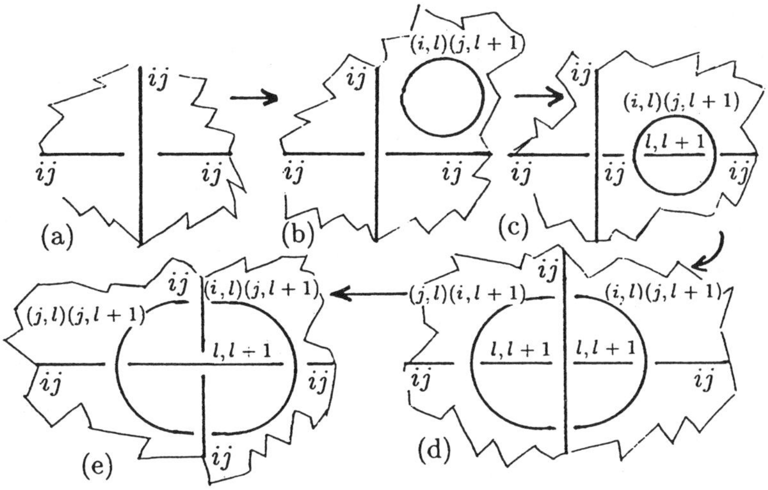

FIGURE 25.9. Monochromatic crossing change

Then we perform the sequence of transformations shown in Fig.25.9,c-e. The changes in the coloring of the diagram under the Reidemeister moves are determined by the following relations in the permutation group:

$$(i,l)(j,l+1)(i,j)(i,l)(j,l+1) = (l,l+1),$$
$$(i,j)(i,l)(j,l+1)(i,j) = (j,l)(i,l+1).$$

When we will have performed the needed crossing changes by means of the indicated transformations, the branching set will look globally as depicted in Fig.25.10. In this figure the parallels of the torus are the strands of the trivialized braid, the meridians are the big circles that appeared in Step 3, and the little circles are the ones that appeared just now when we performed the crossing changes. (Of course the number of circles of each type is not necessarily the same as in the picture!) Note that not all of the little circles are part of the branch set (recall the crossing points in Fig.25.6 that were not marked by dotted circles).

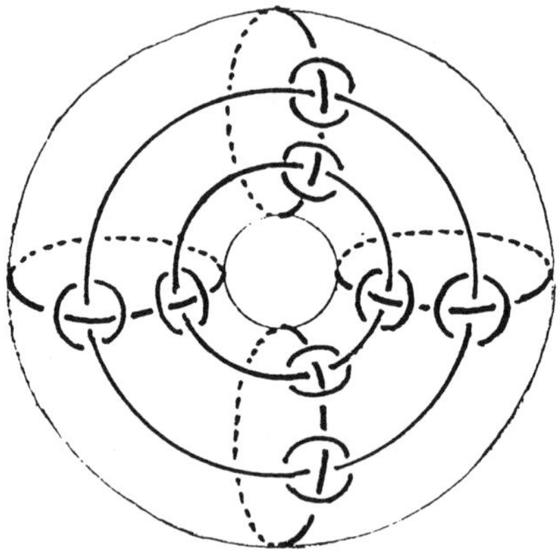

FIGURE 25.10. The result of Step 4

25.7. Step 5. *Final simplifications.* First let us recall the structure of an n-fold cyclic branched covering $p : S^3 \to S^3$. The sphere S^3 may be represented as $\mathbb{R}^3 \cup \infty$. Let us identify points obtained from each other by a rotation of $2\pi/n$ about the axis l. The result is an n-fold branched covering $p : S^3 \to S^3$ called cyclic. The branch set of the covering p is the circle $l \cup \infty$ of color $(12\ldots n)$.

Let p_1 and p_2 be the cyclic coverings branching along the central circle S_1 of the solid torus and along its complementary circle S_2, respectively (Fig.25.11); the branching indices are equal to the number of parallels and meridians (respectively) constituting the branch set obtained after Step 4 (Fig.25.10).

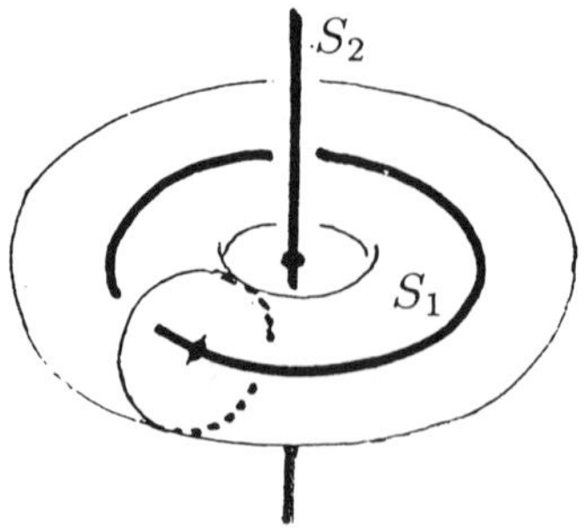

FIGURE 25.11. Two cyclic branched coverings

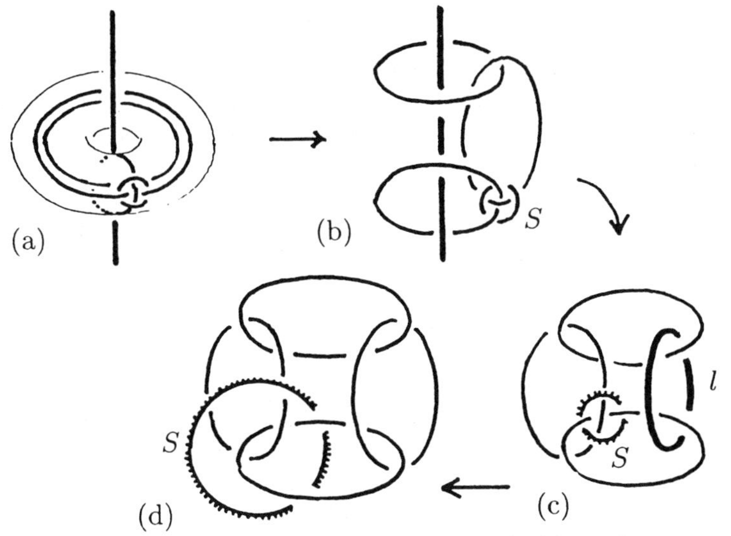

FIGURE 25.12. Isotopies of the simplified branch set

We can assume that these parallels and meridians are arranged in a symmetric way on the torus. Then the branch set of the composition of maps

$$M^3 \xrightarrow{p} S^3 \xrightarrow{p_1} S^3 \xrightarrow{p_2} S^3$$

will be as shown in Fig.25.12,a. This set may be transformed by a sequence of isotopies to appear as in Fig.25.12,d.

The component denoted S in this picture may be taken as a symmetry axis and then used to define a 2-fold cyclic branched covering $p_3 : S^3 \to S^3$. The resulting branch set of the composition of maps

$$M^3 \xrightarrow{p} S^3 \xrightarrow{p_1} S^3 \xrightarrow{p_2} S^3 \xrightarrow{p_3} S^3$$

is shown in Fig.25.13,a. It is easy to see that it is none other than the Borromeo rings (Fig.25.13,b). □

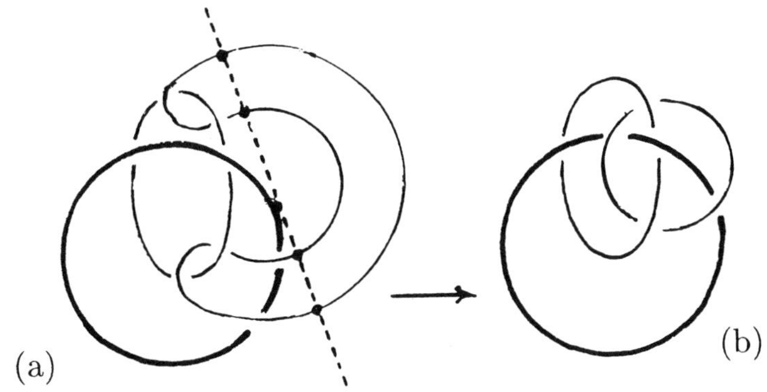

FIGURE 25.13. Two-fold covering producing the Borromeo rings

Comments

The concluding material of this chapter originated in the pioneering and technically rather difficult work carried out in the seventies by J. M. Montesinos and H. M. Hilden. In the eighties, together with their collaborators, these authors produced complete and sharpened versions of this work, on which much of this chapter is based. In particular, our exposition of the proof of the beautiful result asserting that the Borromeo rings B are a universal link (for producing any 3-manifold via an appropriate covering branching along B) follows along the lines of [**HLMW**] (1987).

Of course the construction of branched coverings of three-dimensional manifolds generalizes and uses branched coverings of manifolds in dimension two. This is a classical topic, originating in the 19th century in complex analysis, but very topological in nature. We were unable to resist the temptation of interspersing our exposition of this material with digressions in various directions, in the hope of sharing a little of the sheer beauty of the subject with the reader. We should perhaps mention here that the bridge between dimension two and dimension three, like so many innovations in three-dimensional topology, was the result of work by J. W. Alexander.

Our treatement of the three-dimensional part of the theory is of course far from complete. For example, we have not included the results asserting that the figure eight knot is universal ([**HLM**], 1983), but the trefoil is not ([**GH**], 1972), and, among the applications, the new proof of the fact that all compact orientable 3-manifolds are parallelizable ([**HMT**],1976).

The material of this chapter is not used further in this book. We have included it, however, not only for aesthetic reasons, but because we feel there are some nice open problems remaining in the theory, as well as potential applications lurking in the background. (Let us warn the ambitious reader that a proof of the Poincaré conjecture or a counterexample on the basis of this theory, if they exist at all, are not anywhere on the surface: many good mathematicians have tried, unsuccessfully.)

Let us conclude this commentary with two open questions.

- Is there some way of distinguishing knots and links that are universal (in the sense of producing all 3-manifolds via branched coverings) from those that are not?
- Is there a nice algorithm (computer implementable) for passing from surgery presentations to branched covering presentations of 3-manifolds and vice versa?

CHAPTER VIII

Skein Invariants of 3-Manifolds

There are many different invariants of 3-manifolds. Classical examples are the homology and homotopy groups, as well as various numerical characteristics derived from them (e.g. the Betti numbers or the Euler characteristic). Here we present the recently discovered complex-valued invariants based on surgery presentations of 3-manifolds and on skein algebras (which come from the bracket polynomial). The main idea of the construction is simple and quite natural: represent the given manifold M^3 by a framed link in the sphere S^3 and define a complex number $W(M^3)$ (related to values of a polynomial of the Jones-Kauffman type) that would be invariant with respect to both the Reidemeister moves and the Kirby moves. Unfortunately, the implementation of this program turns out to be rather complicated: the definition of $W(M^3)$ will only appear near the end of this chapter, in §29. Working towards this definition, we shall establish the necessary properties of the auxiliary objects that appear on the way.

The notation W for the invariant is in honor of E. Witten, although his original definition was based on totally different ideas coming from quantum field theories and presented "on the physical level of rigor". (For more historical background, see the Comments at the end of this chapter and §32).

§26. The Temperley–Lieb algebra and other skein algebras

26.1. The Temperley–Lieb algebra is a classical object in the theory of operator algebras, not far removed from the Hecke algebra. However, in this chapter it appears in a different guise, as one of the so-called skein algebras. The latter, roughly speaking, are algebras of linear combinations of certain little diagrams factorized by the skein relation coming from the bracket polynomial (§3). Their basic role in our exposition is to ensure invariance of various constructed objects with respect to the Reidemeister moves (the reader may recall that the bracket polynomial is Ω_2- and Ω_3-invariant, but not Ω_1-invariant).

26.2. The skein space S. We begin with a surgery presentation of our 3-manifold M^3 in the form of a link diagram without any framing indices (see the end of §19). Recall that such a presentation is obtained by making the twisted ribbons defining the surgery "horizontal" (replacing twists by kinks as in the belt trick) and then projecting on the horizontal plane. We obtain a link diagram (which encodes the framing information by kinks, just as the ribbon presentation encodes it by twists) presenting our manifold. The manifold M^3 does not change if we perform the Reidemeister moves Ω_2, Ω_3, and the double twist move Ω_1' (Fig.26.1).

In §3 we assigned to each diagram D of a link L its bracket polynomial, denoted $\langle D\rangle$, in the variables a and a^{-1}. This polynomial is Ω_2- and Ω_3-invariant and

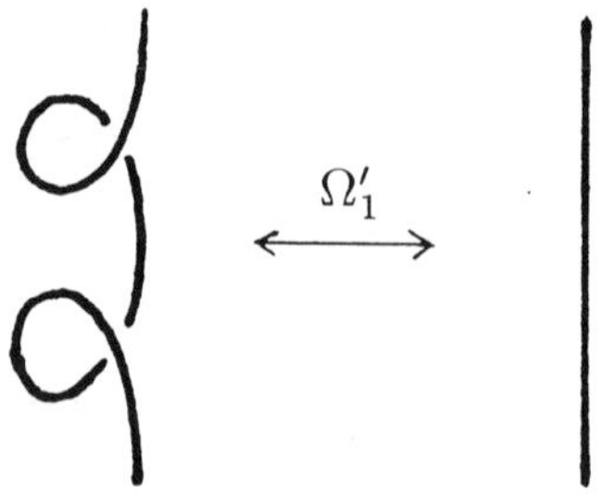

FIGURE 26.1. The double twist move Ω_1'

satisfies the following defining relations:

(1) 

(2) $$\langle D \cup \bigcirc \rangle = (-a^{-2} - a^2)\langle D \rangle;$$

(3) $$\langle \bigcirc \rangle = 1.$$

For a fixed complex number $a = a_0 \in \mathbb{C}$, the value of the polynomial $\langle D \rangle$ may be computed in the following way. Consider a set $\{D_i\}$ of pairwise nonisotopic link diagrams. Let V be the vector space over $\mathbb{C}$ whose elements are finite linear combinations of the diagrams D_i. In the space V consider the subspace V_0 generated by vectors of the form

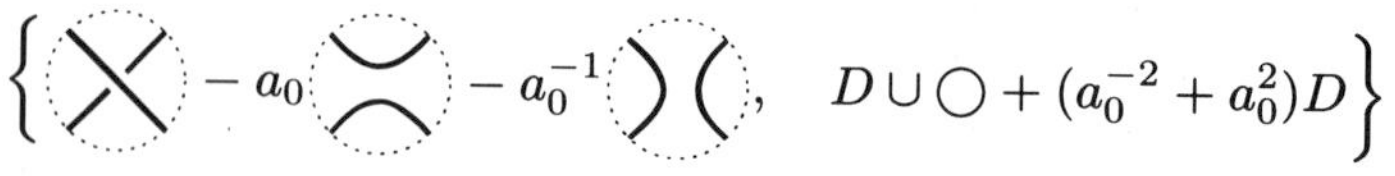

(the diagram D is understood as the vector $1 \cdot D \in V$). Put $S = V/V_0$; let us call S the *skein space*. Then under the quotient map $p : V \to V/V_0 = S$, the element D is taken to λe_1, where λ is the value of the polynomial $\langle D \rangle$ at the point a_0, while e_1 is the image in S of the link diagram consisting of one circle. Indeed, suppose that E_i is the link diagram consisting of i disjoint circles (in the plane). It follows from the proof of the existence and uniqueness theorem for the bracket polynomial $\langle D \rangle$ (Theorem 3.4) that

$$D = \sum \lambda_i E_i + w_1 \quad \text{and} \quad \sum \lambda_i E_i = \lambda E_1 + w_2,$$

where $w_1, w_2 \in V_0$ and λ is the value of the polynomial $\langle D \rangle$ at the point a_0.

Note that the image under the quotient map of any diagram generates the skein space S, and hence $\dim S = 1$, i.e., $S \cong \mathbb{C}$.

Now let e be any nonzero element of the space S. Then $p(D) = f(D)e$, where $f(D) \in \mathbb{C}$. In the role of e, it is convenient to take the element e_0 corresponding to the empty diagram. For such a choice we have the following multiplicativity relation:

(26.1) $$f(D_1 \cup D_2) = f(D_1)f(D_2).$$

Indeed, let $D_1 = \sum \lambda_i E_i + w_1$, $D_2 = \sum \mu_j E_j + w_2$, where $w_1, w_2 \in V_0$. For nonnegative integers m and n, we have the relation $mE_s \cup nE_t = E_{ms} \cup E_{nt} =$

E_{ms+nt}. Hence,

$$p(D_1 \cup D_2) = p\Big(\big(\sum_i \lambda_i E_i\big) \cup \big(\sum_j \lambda_j E_j\big)\Big) = p\big(\sum_{i,j} E_{i\lambda_i + j\mu_j}\big) = \big(\sum_{i,j} (c^i)^{\lambda_i} (c^j)^{\mu_j}\big) e_0,$$

where $c = (-a_0^2 - a_0^{-2})$. Therefore,

$$f(D_1 \cup D_2) = \sum_{i,j} (c^i)^{\lambda_i} \times (c^j)^{\mu_j} = \sum_i (c^i)^{\lambda_i} \times \sum_j (c^j)^{\mu_j} = f(D_1) f(D_2),$$

which proves relation (26.1).

26.3. The skein space S may be generalized as follows. Suppose F is a surface, i.e., an oriented 2-manifold with (possibly empty) boundary ∂F. Choose $2n$ points on the boundary (if $\partial F = \emptyset$, then $n = 0$). We shall consider *diagrams* on the surface F whose intersection with ∂F is exactly the set of $2n$ distinguished points (see Fig.26.2). Fix a number $a_0 \in \mathbb{C}$ and consider the vector space $V(F, 2n)$ over $\mathbb{C}$ constituted by finite linear combinations of (isotopy classes of) diagrams. Let $V_0(F, 2n)$ be the subspace of $V(F, 2n)$ generated by vectors of the form

$\{$  $- a_0$  $- a_0^{-1}$ , $D \cup \bigcirc + (a_0^{-2} + a_0^2) D \}$.

Put $S(F, 2n) = V(F, 2n)/V_0(F, 2n)$. For brevity the space $S(F, 0)$ will be denoted by $S(F)$.

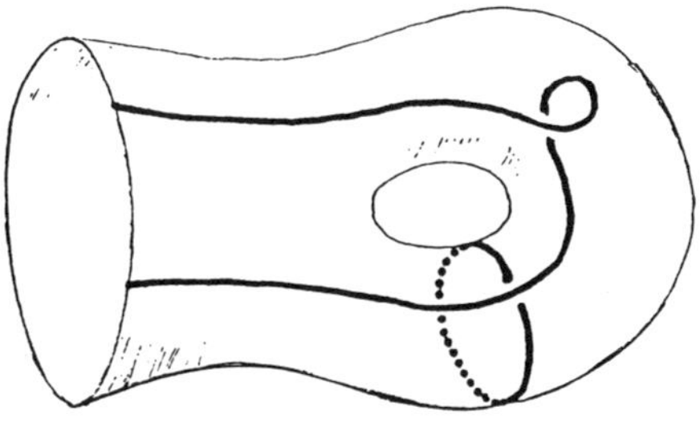

FIGURE 26.2. An element of $V(F, 2n)$

Remark. The assumption that the surface F be orientable is essential, because on a nonorientable surface it is impossible to distinguish between the crossing types  and , and so it is not clear where to put the coefficients a_0 and a_0^{-1}.

26.4. THEOREM. *The image $p(D)$ under the quotient map p of the diagram D in $S(F, 2n)$ is invariant with respect to the double twist move Ω_1' and to the second and third Reidemeister moves Ω_2 and Ω_3.*

Proof. If $w \in V_0$, then $p(w) = 0$. Therefore,

(26.2) $p($  $) = a_0 p($  $) + a_0^{-1} p($  $)$,

(26.3) $p(D \cup \bigcirc) = -(a_0^{-2} + a_0^2) p(D)$.

In §3 we proved (see 3.1) that (26.2) and (26.3) imply the relations

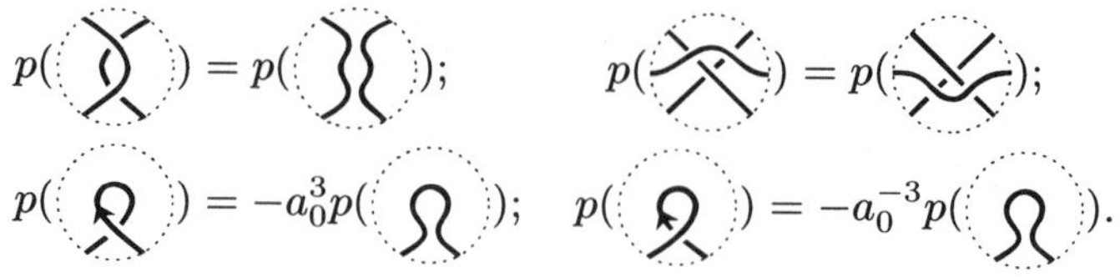

The first two relations express invariance with respect to the second and third Reidemeister moves, while the last two imply Ω_1'-invariance. Indeed,

because $(-a_0^{-3})(-a_0^3) = 1$. □

26.5. THEOREM. *A basis of the space $S(F, 2n)$ is constituted by the images under the quotient map of all the (isotopy classes of) diagrams D_i that do not contain any crossings or any contractible closed curves.*

Proof. By (26.2) we can represent any diagram in $S(F, 2n)$ as a linear combination of diagrams without crossings. By (26.3) we can represent any diagram without crossings in the form λe_i, where e_i is the image of D_i under the quotient map. Therefore the vectors e_i generate the space $S(F, 2n)$.

It remains to prove that the vectors e_i are linearly independent. To this end let us recall that (26.2) and (26.3) can be applied in any order, i.e., the representation of a diagram D from $S(F, 2n)$ as a linear combination of the vectors e_i does not depend on the order in which we destroy the crossings and the contractible closed curves (see the proof of Theorem 3.4). Thus to each diagram D we can assign a well-defined sum $\sum \lambda_i D_i$. This assignment induces a linear map $f : V \to S'$, where S' is the set of all finite linear combinations of the vectors D_i. Further, it is obvious that $f(V_0) = 0$. Hence the map f induces a linear map $\widetilde{f} : V/V_0 \to S'$, where $\widetilde{f}(e_i) = f(D_i) = D_i$. The elements $e_1, e_2, \dots$ are taken by the linear map $\widetilde{f}$ to basis elements of the space S'; therefore they must be linearly independent. □

Now let us consider some simple particular cases of spaces $S(F, 2n)$.

26.6. PROPOSITION. $S(\mathbb{R}^2) \cong \mathbb{C}$.

Proof. By definition, the space $S(\mathbb{R}^2)$ coincides with the skein space S, which we know is one-dimensional. □

26.7. PROPOSITION. $S(S^2) \cong S(\mathbb{R}^2) \cong \mathbb{C}$.

Proof. Represent the sphere S^2 as $\mathbb{R}^2 \cup \infty$. To each diagram on S^2 not passing through the point ∞ we can assign a diagram on the plane $\mathbb{R}^2$. In any isotopy class of diagrams on S^2 there exists a diagram not passing through the point ∞. Thus we obtain a correspondence between isotopy classes of diagrams on S^2 and isotopy classes of diagrams on $\mathbb{R}^2$. To any class of plane diagrams there corresponds exactly one spherical class, but to one class of diagrams on the sphere there may correspond several plane classes. However, we claim that when we pass to the quotient spaces $S(\mathbb{R}^2)$ and $S(S^2)$, the correspondence becomes one-to-one.

First we shall show that to isotopic diagrams in S^2 corresponds one and the same element of $S(\mathbb{R}^2)$. To do this, it suffices to show that if we "throw a strand around infinity" (recall the "sphere trick" from §2) the image (under the quotient map) of the diagram in $S(\mathbb{R}^2)$ does not change (Fig.26.3,a). On the plane $\mathbb{R}^2$, this transformation looks as in Fig.26.3,b. Using the second and third Reidemeister move, we can obtain the diagram shown in Fig.26.3,c. This diagram can be transformed into the initial one by the move Ω'_1. All these transformations do not change the corresponding element of $S(\mathbb{R}^2)$.

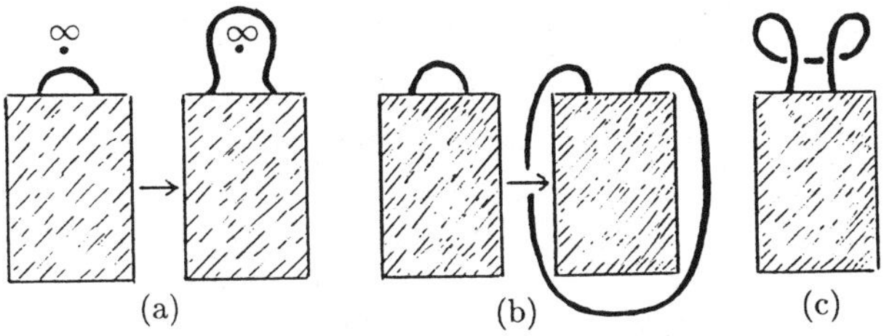

FIGURE 26.3. Going around infinity

Using the fact that isotopic diagrams on S^2 are assigned one and the same element of the space $S(\mathbb{R}^2)$, it is easy to prove that $S(S^2) \cong S(\mathbb{R}^2)$. Indeed, formulas (26.2) and (26.3) for diagrams on S^2 and diagrams on $\mathbb{R}^2$ differ only in that a closed curve C going around ∞ on the sphere (Fig.26.4,a) can be contracted to a point within the shaded region that it bounds, while this cannot be done in the plane. But the curve C can be replaced by an isotopic curve (by slipping it off ∞) that can be contracted in the plane in the corresponding region (Fig.26.4,b).

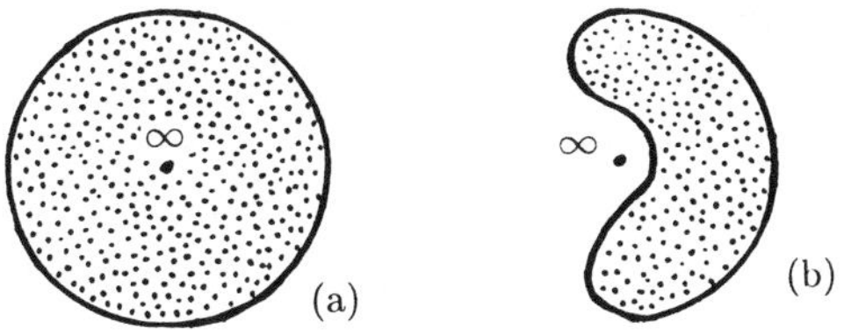

FIGURE 26.4. Contracting curves on the sphere and plane

26.8. PROPOSITION. *The space $S(I \times S^1)$ has the natural structure of an algebra over $\mathbb{C}$. This algebra is isomorphic to $\mathbb{C}[\alpha]$ (the algebra of polynomials over $\mathbb{C}$ in one variable α).*

Proof. According to Theorem 26.5, for the basis of $S(I \times S^1)$ we can take the family of diagrams each of which consists of n circles parallel to the base of the cylinder $I \times S^1$; here $n = 0, 1, 2, \ldots$; for $n = 2$, see Fig.26.5,a.

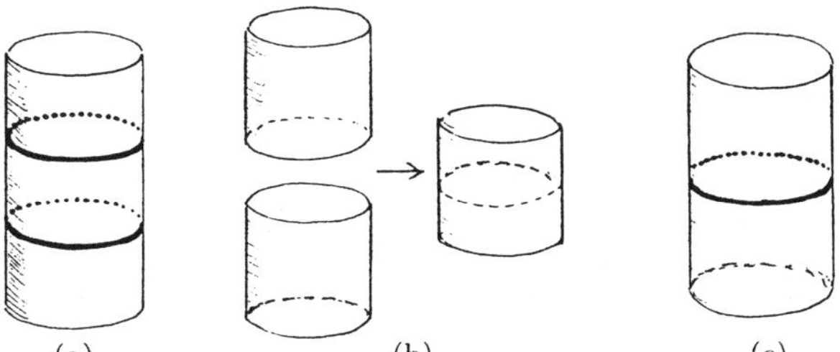

FIGURE 26.5. Generators and product operation in $S(I \times S^1)$

Define the product of two diagrams by attaching the top of one cylinder to the bottom of the other and then contracting the resulting cylinder to half its height (Fig.26.5,b). This operation endows the space $S(I \times S^1)$ with the structure of an algebra over $\mathbb{C}$.

Let α be the image (under the quotient map) in $S(I \times S^1)$ of the diagram shown in Fig.26.5,c. Then the basis of the space $S(I \times S^1)$ will be $\alpha^0, \alpha^1, \alpha^2, \ldots$, where α^0 is the image of the empty diagram. Therefore $S(I \times S^1) \cong \mathbb{C}[\alpha]$. □

26.9. The Temperley-Lieb algebra. We now consider the space $S(D^2, 2n)$, which will be quite important in the sequel. According to Theorem 26.5, the basis of such spaces consists of isotopy classes of diagrams not containing any crossings or contractible closed curves; but in the case $F = D^2$, all the closed curves can be successively destroyed, beginning with the innermost ones. So we are left with nonintersecting arcs with endpoints on the boundary of D^2.

Problem 26.1. Let $c_n = \dim(S(D^2, 2n))$. Prove that

$$\text{(a)}\quad c_{n+1} = \sum_{i=0}^{n} c_i c_{n-i}; \qquad \text{(b)}\quad c_n = \frac{1}{n+1}\binom{2n}{n}.$$

The space $S(D^2, 2n)$ can be supplied with the structure of an algebra over $\mathbb{C}$, but not in a canonical way. To do that we must represent our disk D^2 with $2n$ distinguished points as a square with n distinguished points on each of two opposite sides. (Such a representation is obviously not unique.) Then the product of two diagrams in the square with n distinguished points on opposite sides is defined in the natural way: as the juxtaposition of the two squares followed by a contraction of the resulting rectangle back to a square (Fig.26.6). This operation transforms the linear space $S(D^2, 2n)$ into an algebra, denoted TL_n and called the *Temperley-Lieb algebra*.

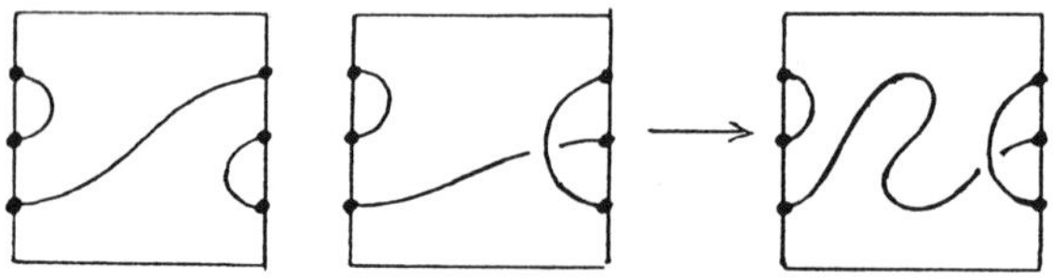

FIGURE 26.6. The product operation in TL_3

To simplify our diagrammatic notation, let us agree to denote k parallel arcs by one arc marked by the number k. Let e_i be the element of the algebra TL_n corresponding to the diagram shown in Fig.26.7,a. (For $n = 4$ all such elements are depicted in their entirety in Fig.26.7,b.) Further denote by 1_n the element of the algebra TL_n corresponding to the diagram with n parallel arcs.

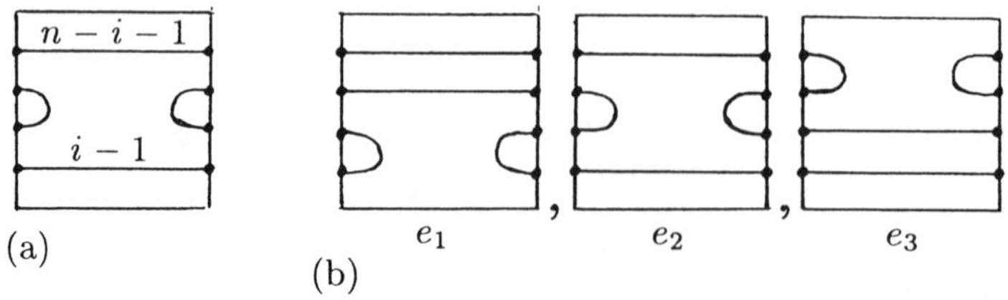

FIGURE 26.7. Basis elements of the Temperley–Lieb algebra

26.10. THEOREM. *The elements $1_n, e_1, \dots, e_{n-1}$ generate the Temperley–Lieb algebra TL_n.*

Proof. Consider an arbitrary diagram corresponding to a basis element of the space TL_n, say the element shown in Fig.26.8,a. Modify the diagram by an isotopy so that all its arcs consist of horizontal segments and semicircles with vertical diameters (Fig.26.8,b). Such a modification is possible because of the following property: any arc joining points of the same side of the square moves up (or down) by an odd number of steps, whereas any arc joining opposite sides of the square moves up (or down) by an even number of steps. To prove this property, the main idea is to successively destroy all arcs joining pairs of neighboring points on the same side of the square (this does not change the parity of the remaining points) until only parallel arcs remain. This idea can easily be transformed into a proof by induction.

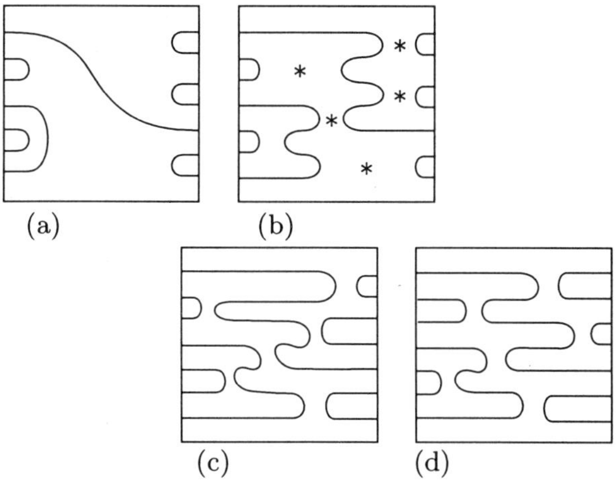

FIGURE 26.8. Expressing an element of TL_7 in the generators

To each little star in Fig.26.8,b there corresponds one of the generators e_i. Perform an isotopy (moving points only horizontally) so that the projections of the stars are uniformly distributed on the base of the square. This can be done in different ways (compare Fig.26.8,c and Fig.26.8,d). The diagram thus obtained can easily be expressed in the generators e_i. □

Remark. It is not difficult to find relations defining the Temperley–Lieb algebra as an abstract algebra over $\mathbb{C}$ (see [**Jon2**]), but we shall not do so because these relations are not needed in the sequel.

§27. The Jones–Wentzl idempotent

27.1. In order to construct the skein invariants of 3-manifolds, we shall define a special element $f^{(n)}$ of the Temperley–Lieb algebra TL_n, known as the *Jones-Wentzl idempotent.* Let A_n be the subalgebra of TL_n generated by the elements $e_1, \dots, e_{n-1}$ (the only missing generator is 1_n); the element f_n is defined by the two following relations:

(1) $f^{(n)} A_n = A_n f^{(n)} = 0;$

(2) $1_n - f^{(n)} \in A_n$.

Such an element actually exists only under a certain condition on the element $a_0 \in \mathbb{C}$ which appears in the definition of the algebra TL_n. Recall that this element is involved in the defining relations of the subspace $V_0 \subset V$ (see 26.2).

27.2. THEOREM. (a) *Suppose* $a_0^{4k} \neq 1$ *for* $k = 1, \dots, n-1$. *Then there exists a unique element* $f^{(n)} \in TL_n$ *satisfying conditions* (1) *and* (2).

(b) *The element* $f^{(n)}$ *is an idempotent, i.e.,* $f^{(n)} f^{(n)} = f^{(n)}$.

Proof. First let us note that the existence of $f^{(n)}$ implies both its uniqueness and its idempotence. This is because $1_n - f^{(n)}$ is the unit element in the algebra A_n. Indeed, according to condition (1), for any $x \in A_n$ we have $(1_n - f^{(n)})x = x$ and $x(1_n - f^{(n)}) = x$, while $1_n - f^{(n)} \in A_n$ by condition (2). Uniqueness now follows from the uniqueness of the unit in any algebra, and its idempotence from the relation $(1_n - f^{(n)})(1_n - f^{(n)}) = 1_n - f^{(n)}$.

Now let us construct the element $f^{(n)}$ by induction on n. The algebra TL_1 is generated by the element 1_1; hence $A_1 = 0$. Thus $f^{(1)} = 1_1 \in TL_1$.

An element of the algebra TL_n is a linear combination of the images (under the quotient map $V \to V/V_0$) of diagrams (from V). However, when we define certain operations on the elements of this algebra, it is convenient to think of these images as actual diagrams (rather than classes of diagrams). In such situations we shall suppose that all the diagrams appearing in the linear combination are subjected to the same operation. In accordance to this, let us define an element $\Delta_n \in S(\mathbb{R}^2)$ by means of the *closure* operation shown in Fig.27.1. Under this operation, for each diagram appearing in the decomposition of the element $f^{(n)}$, we take its closure, and then take the image in $S(\mathbb{R}^2)$ of the diagram thus obtained in $\mathbb{R}^2$. Then the linear combination of diagrams becomes a linear combination of elements of $S(\mathbb{R}^2)$. Recall that $S(\mathbb{R}^2) \cong \mathbb{C}$, so that Δ_n may be regarded as a complex number (for the basis element of the space $S(\mathbb{R}^2)$ we choose the image of the empty diagram).

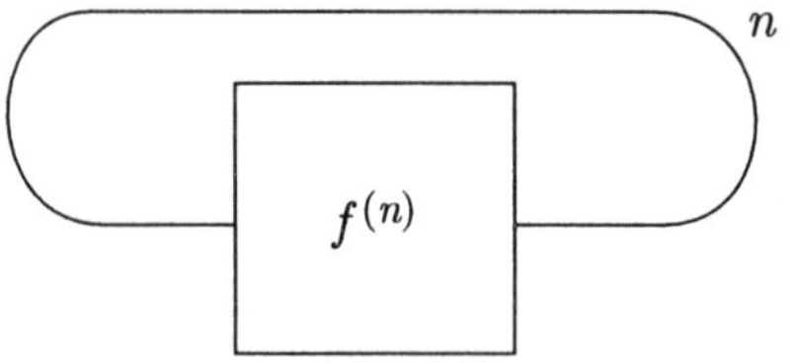

FIGURE 27.1. The closure of $f^{(n)}$

For example, the closure of the element $f^{(1)} = 1_1$ is simply the circle, so $\Delta_1 = -a_0^{-2} - a_0^2$.

The equality $f^{(n)} f^{(n)} = f^{(n)}$, in accordance with our agreement, may be pictured as in Fig.27.2. Let us prove that it has the generalization shown in Fig.27.3.

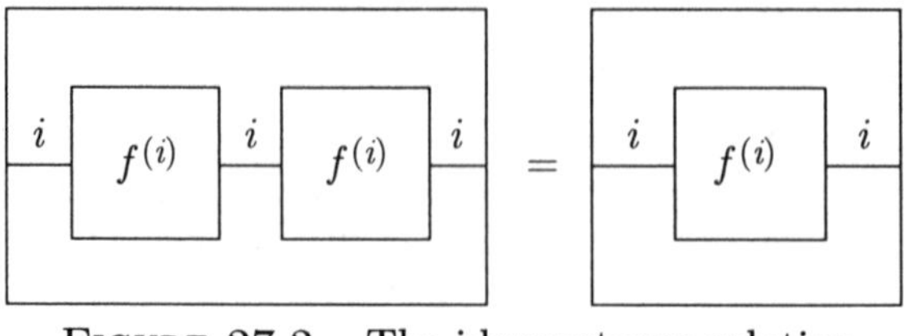

FIGURE 27.2. The idempotence relation

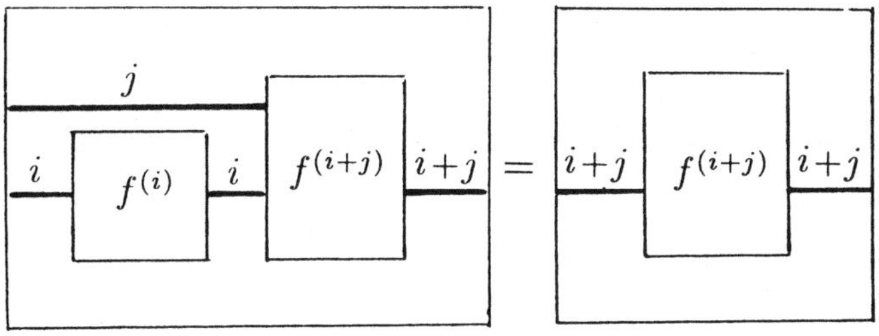

FIGURE 27.3. The generalized idempotence relation

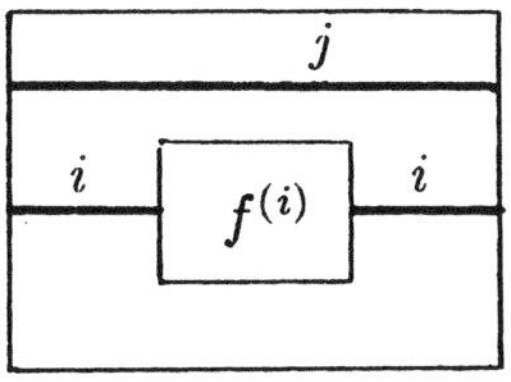

FIGURE 27.4. The addition of j parallel strands

Suppose that $f_j^{(i)}$ is the element of TL_{i+j} obtained from the element $f^{(i)} \in TL_i$ by means of the transformation pictured in Fig.27.4.

Under this transformation, the element $1_i \in TL_i$ is mapped to $1_{i+j} \in TL_{i+j}$, while the generator $e_p \in TL_i$ becomes the generator $e_p \in TL_{i+j}$. And since we have

$$f^{(i)} = 1_i + \sum \lambda_{i_1,\ldots,i_s} e_{i_1} \ldots e_{i_s},$$

it follows that

$$f_j^{(i)} = 1_{(i+j)} + \sum \lambda_{i_1,\ldots,i_s} e_{i_1} \ldots e_{i_s}.$$

Therefore,

$$f_j^{(i)} f^{(i+j)} = \left(1_{(i+j)} + \sum \lambda_{i_1,\ldots,i_s} e_{i_1} \ldots e_{i_s}\right) f^{(i+j)} = f^{(i+j)}.$$

Now let us put $i = n-1$ and $j = 1$. Then the relation in Fig.27.3 obviously implies the one in Fig.27.5. Let x be the element of the algebra TL_{n-1} that appears in the right-hand side of this relation. A simple isotopy shows that in the left-hand side we actually have $f^{(n-1)}x$. Hence $f^{(n-1)}x = x$. On the other hand, any element

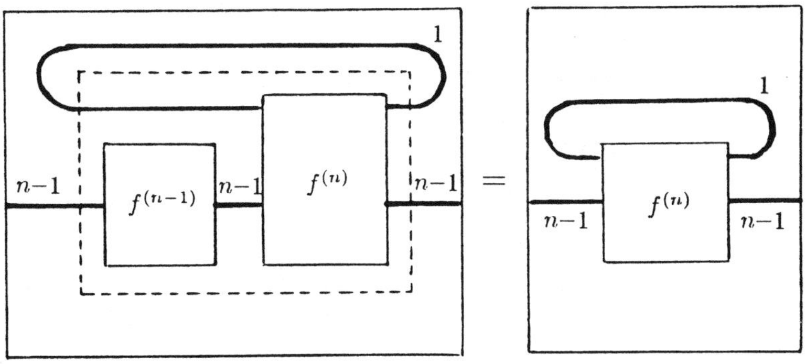

FIGURE 27.5. An auxiliary relation

$x \in TL_{n-1}$ may be presented in the form $x = \lambda 1_{n-1} + x'$, where $x' \in A_{n-1}$. Hence $f^{(n-1)}x = \lambda f^{(n-1)}$, and therefore $f^{(n-1)} = \lambda^{-1} f^{(n-1)} x = \lambda^{-1} x$.

In order to compute λ, take the closures of the diagrams x and $f^{(n-1)}$ (Fig.27.6), and then consider the images of the resulting diagrams in $S(\mathbb{R}^2)$.

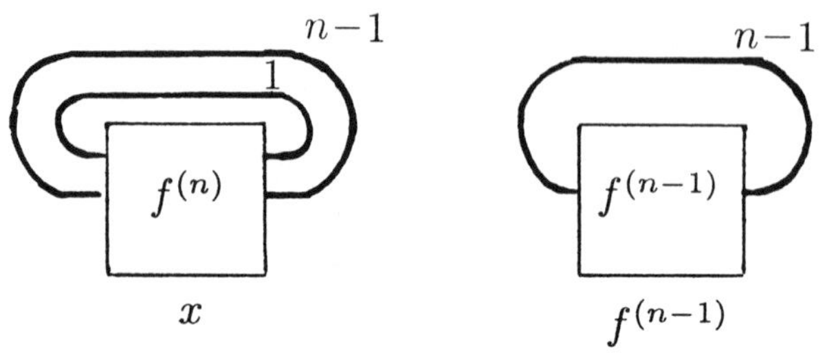

FIGURE 27.6. Closures of X and $f^{(n-1)}$

As the result x will be taken to Δ_n and $f^{(n-1)}$ to Δ_{n-1}. Hence $\Delta_n = \lambda \Delta_{n-1}$, and therefore

$$f^{(n-1)} = \frac{\Delta_{n-1}}{\Delta_n} f^{(n-1)} x;$$

this last relation is pictured in Fig.27.7.

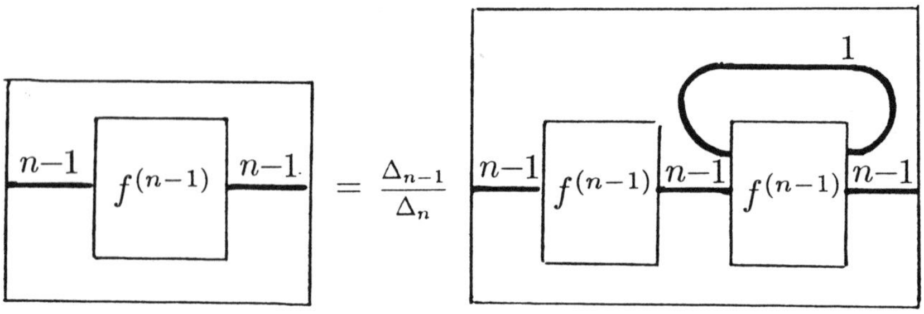

FIGURE 27.7. Another auxiliary relation

The fact that $\Delta_n \neq 0$ whenever $a_0^{4(n+1)} \neq 1$ will be the object of Lemma 27.3 below. In the meantime, resuming our inductive argument, we assume that the element $f^{(n)}$ has been constructed (the inductive assumption) and $\Delta_n \neq 0$. Then let us prove that (when $n \geqslant 1$) for $f^{(n+1)}$ we can take

$$f^{(n+1)} = f_1^{(n)} - \frac{\Delta_{n-1}}{\Delta_n} f_1^{(n)} e_n f_1^{(n)};$$

see Fig.27.8. To validate this formula when $n = 0$, we shall assume that $f^{(0)}$ is the empty diagram (then we have $\Delta_0 = 1$) and that $\Delta_{-1} = 0$.

For the element $f^{(n+1)}$ thus defined we must prove that $f^{(n+1)} e_i = e_i f^{(n+1)} = 0$ for $i = 1, \dots, n$ and $1_{n+1} - f^{(n+1)} \in A_{n+1}$. For $i = 1, \dots, n-1$, the relation $f^{(n+1)} e_i = e_i f^{(n+1)} = 0$ immediately follows from the $n-1$ relations $f^{(n)} e_i = e_i f^{(n)}$. It is also clear that $1_{n+1} - f_1^{(n)} \in A_{n+1}$ and $f_1^{(n)} e_n f_1^{(n)} \in A_{n+1}$; therefore

$$1_{n+1} - f^{(n+1)} = (1_{n+1} - f_1^{(n)}) + \frac{\Delta_{n-1}}{\Delta_n} f_1^{(n)} e_n f_1^{(n)} \in A_{n+1}.$$

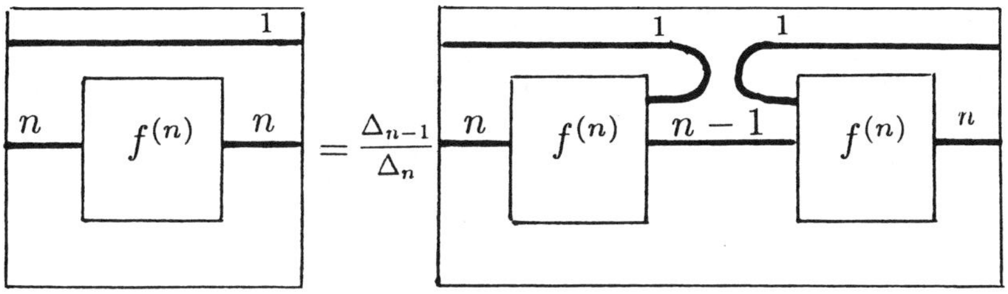

FIGURE 27.8. Another intermediate relation

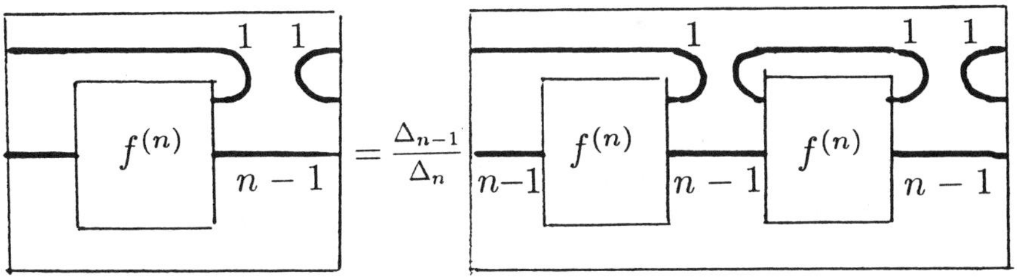

FIGURE 27.9. The intermediate relation $f^{(n+1)}e_n = 0$

Some more efforts are required to prove that $f^{(n+1)}e_n = e_n f^{(n+1)} = 0$. We shall only prove that $f^{(n+1)}e_n = 0$ (Fig.27.9), since the proof of the other equality is similar.

The relation pictured in Fig.27.7 implies the one in Fig.27.10.

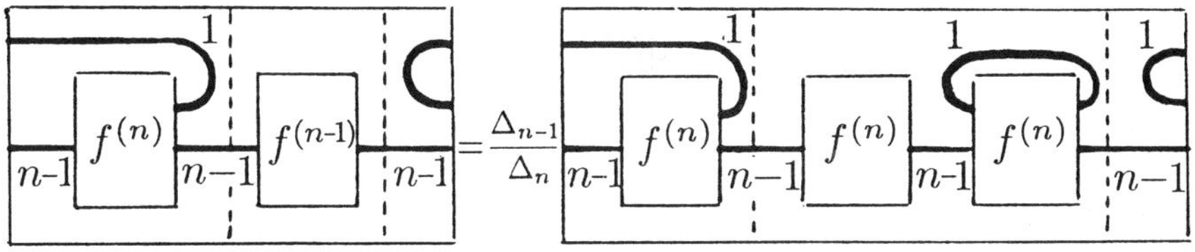

FIGURE 27.10. A consequence of the relation in Fig.27.7

An isotopy yields the relation pictured in Fig.27.11. Just as we proved the relation $f_j^{(i)} f^{(i+j)} = f^{(i+j)}$, we can show that $f^{(i+j)} f_j^{(i)} = f^{(i+j)}$. In Fig.27.11 we have enclosed the diagram $f^{(i+j)} f_j^{(i)}$, where $i = n-1$, $j = 1$; replacing it by $f^{(i+j)}$, we obtain the required relation.

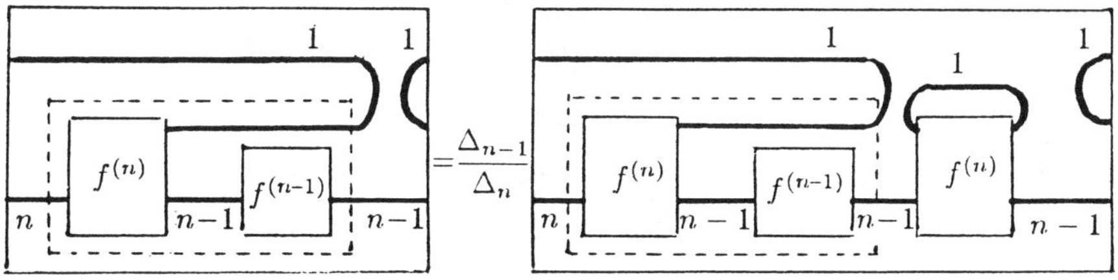

FIGURE 27.11. A relation isotopic to the previous one

To conclude the proof of Theorem 27.2, it remains to prove that $\Delta_1 \dots \Delta_n \neq 0$.

27.3. Lemma. *If $a_0^{4(n+1)} \neq 1$, then $\Delta_n \neq 0$.*

Proof. We shall prove a more precise statement, namely

$$\Delta_n = \frac{(-1)^n(a_0^{2(n+1)} - a_0^{-2(n+1)})}{a_0^2 - a_0^{-2}}. \tag{27.1}$$

From the definition of the element $f^{(n+1)} \in TL_{n+1}$ we get the relation, pictured in Fig.27.12, between elements of the algebra $S(I \times S^1)$. Suppose $S_n(\alpha)$ is the image in $S(I \times S^1)$ of the closure of the element $f^{(n)}$. Then Δ_n is the image of the element $S_n(\alpha)$ under the map $S(I \times S^1) \to S(\mathbb{R}^2)$ induced by the inclusion of $I \times S^1$ in $\mathbb{R}^2$. Under this map the element α is taken to the complex number $(-a_0^{-2} - a_0^2)$.

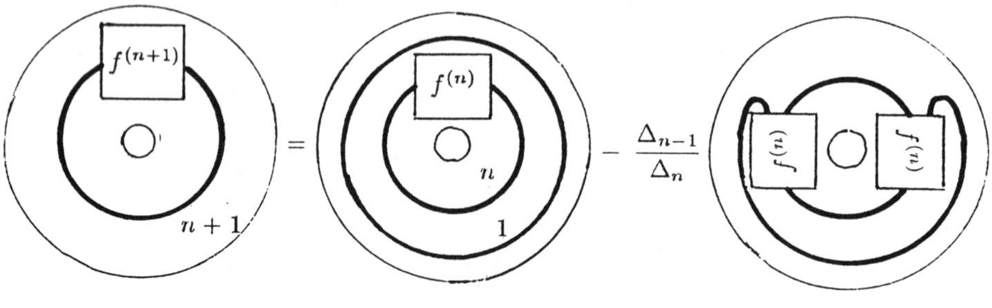

FIGURE 27.12. A relation in $I \times S^1$

Let us prove that the relation pictured in Fig.27.12 becomes

$$S_{n+1}(\alpha) = \alpha S_n(\alpha) - S_{n-1}(\alpha). \tag{27.2}$$

Only the term $S_{n-1}(\alpha)$ needs to be justified. The corresponding element of the space $S(I \times S^1)$ may be transformed as shown in Fig.27.13. The first of the equalities pictured there is based on the idempotence relation $f^{(n)}f^{(n)} = f^{(n)}$, while the second follows from the equalities in Figures 27.5 and 27.7.

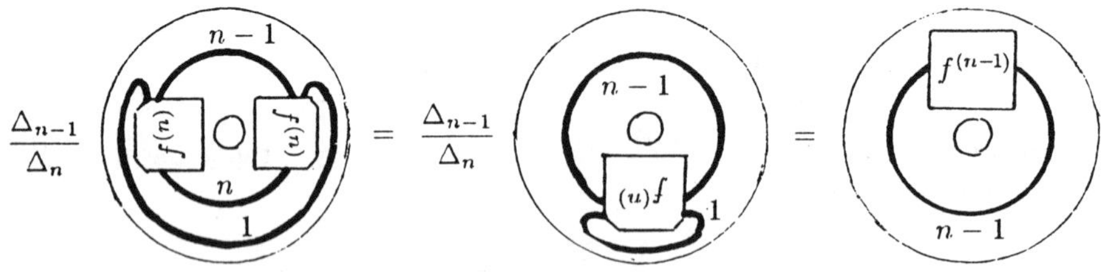

FIGURE 27.13. Transformations in $S(I \times S^1)$

Consider what happens to relation (27.2) under the map $S(I \times S^1) \to S(\mathbb{R}^2)$, i.e., substitute the number $(-a_0^{-2} - a_0^2)$ for α. The result will be

$$\Delta_{n+1} = (-a_0^{-2} - a_0^2)\Delta_n - \Delta_{n-1}. \tag{27.3}$$

It now remains to note that the numbers Δ_n defined by formula (27.1) satisfy both the reccurence relations (27.3) and the initial conditions $\Delta_0 = 1$ and $\Delta_1 = (-a_0^{-2} - a_0^2)$. □ □

§28. Invariance with respect to the second Kirby move

28.1. We now define, for any framed diagram with components $K_1, \dots, K_n$, a (poly)linear map

$$\langle \cdot, \dots, \cdot \rangle_D : S_1 \times \dots \times S_n \to S(\mathbb{R}^2),$$

where $S_i \cong S(I \times S^1)$. In order to do this, it suffices to specify all elements of the form $\langle \alpha_1^{k_1}, \dots, \alpha_n^{k_n} \rangle_D \in S(\mathbb{R}^2)$, where α_i is the generator of S_i corresponding to $\alpha \in S(I \times S^1)$.

Let us recall once more how the framed link diagrams that we are considering are obtained. The ribbon B_i defining the ith framed component of the diagram D is positioned horizontally above the plane and projected vertically down on it. For the diagram of the ith curve K_i take one of the two boundary components of the projection of B_i (of course the under-overcrossing information is encoded in the usual way).

In order to obtain a diagram corresponding to the element $\langle \alpha_1^{k_1}, \dots, \alpha_n^{k_n} \rangle_D \in S(\mathbb{R}^2)$, on each ribbon B_i we must draw k_i curves parallel to its boundary. For example, in the case of the framed diagram D shown in Fig.28.1,a, the diagram $\langle \alpha_1^3, \alpha_2^2, \alpha_3^0 \rangle_D$ is represented in Fig.28.1,b.

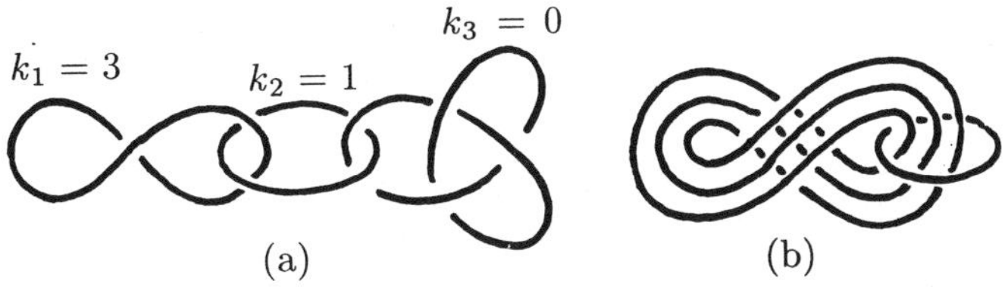

FIGURE 28.1. Constructing the element $\langle \alpha_1^3, \alpha_2^1, \alpha_3^0 \rangle_D$

28.2. Assume that the framed diagrams D and D' are equivalent, i.e., they can be obtained from each other by plane isotopies, the transformation Ω_1', and the second and third Reidemeister moves. Then the diagrams

$$\langle \alpha_1^{k_1}, \dots, \alpha_n^{k_n} \rangle_D \quad \text{and} \quad \langle \alpha_1^{k_1}, \dots, \alpha_n^{k_n} \rangle_{D'}$$

can also be obtained from each other by plane isotopies, the transformation Ω_1', and the second and third Reidemeister moves. Therefore, according to Theorem 26.4 the images of these two diagrams in $S(\mathbb{R}^2)$ coincide. This means that *the polylinear map* $\langle \cdot, \dots, \cdot \rangle_D$ *that we have constructed is an isotopy invariant of the framed link* D.

This makes our map an appropriate tool for constructing 3-manifold invariants (no more worries about isotopy invariance), but it doesn't solve all our problems yet: we need invariance with respect to the Kirby moves. The element Δ_n constructed in the previous section will help us in this respect.

To ensure invariance with respect to the second Kirby move (the goal of this section), we shall use the element

$$\omega = \sum_{n=0}^{r-2} \Delta_n S_n(\alpha) \in S(I \times S^1),$$

where $r \geqslant 3$ is an integer, while the polynomial $S_n(\alpha)$ and the complex number Δ_n were defined in 27.3. The element ω depends on r; this dependence does not appear explicitly in our notation, but is nevertheless implied.

28.3. THEOREM. *Suppose the complex number a_0 is chosen so that a_0^4 is a primitive rth root of unity, and the diagrams D and D' are obtained from each other by a second Kirby move. Then*

$$\langle \omega, \omega, \ldots, \omega \rangle_D = \langle \omega, \omega, \ldots, \omega \rangle_{D'}.$$

Proof. First we consider the simplest particular case of the second Kirby move; namely, we assume that to some ribbon we add a ribbon parallel to a circle S^1 with framing 0 (Fig.28.2).

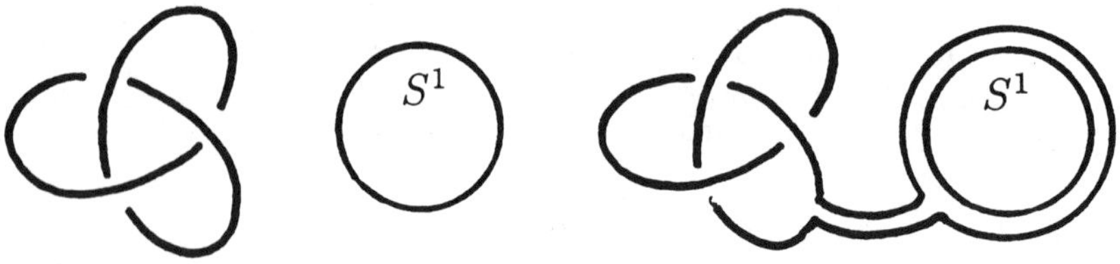

FIGURE 28.2. Simplest instance of the Kirby move

In a neighborhood of the circle S^1, this move will apear as shown in Fig.28.3,a. By an obvious isotopy the move can be made more symmetric (Fig.28.3,b). Let us call it an *arc switch.*

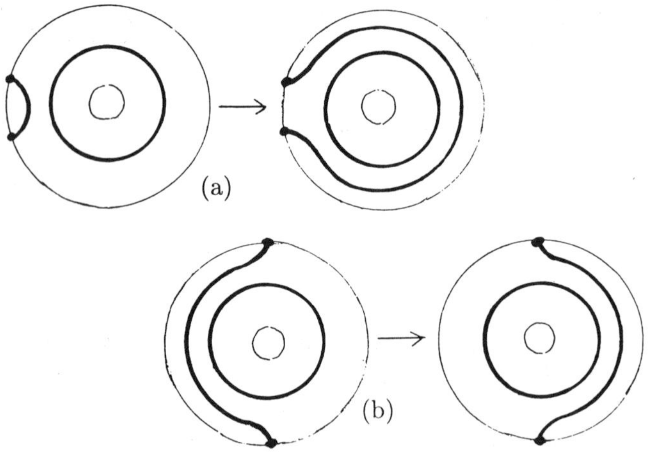

FIGURE 28.3. Two versions of the move from Fig.28.2

Therefore the passage from the element $\langle \omega, \ldots, \omega \rangle_D$ to the element $\langle \omega, \ldots, \omega \rangle_{D'}$ consists of several arc switches as shown in Fig.28.4. Note that the symbol ω next to the circle means that we have in mind the element

$$\omega = \sum_{n=0}^{r-2} \Delta_n S_n(\alpha) \in S(I \times S^1),$$

i.e., a linear combination of diagrams consisting of concentric circles. The arc switch is performed with respect to each of these concentric circles.

To prove Theorem 28.3, it now suffices to verify that the transformation of elements of the space $S(\mathbb{R}^2)$ shown in Fig.28.4 is in fact the identity. First we shall see what the corresponding transformation does on the level of elements of

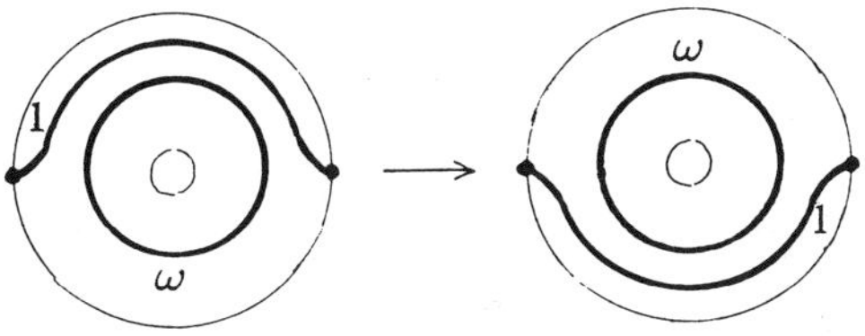

FIGURE 28.4. An arc switch with the element ω

$S(I \times S^1)$, and then check that after the map $S(I \times S^1) \to S(\mathbb{R}^2)$ we do indeed obtain the identical transformation.

28.4. LEMMA. *The difference between the elements of the space* $S(I \times S^1)$ *shown in* Fig.28.4 *may be represented as a linear combination of diagrams each of which contains (as a constituent) the diagram* $f^{(r-1)}$.

Proof. The recursive definition of the element $f^{(n+1)}$ implies the relation pictured in Fig.28.5.

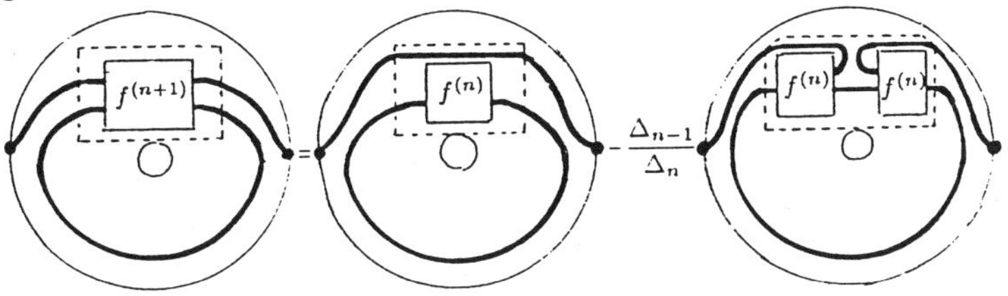

FIGURE 28.5. A consequence of the definition of $f^{(n+1)}$

The idempotence relation $f^{(n)}f^{(n)} = f^{(n)}$ (proved in Theorem 27.2) can be applied to the relation pictured in Fig.28.5. This allows us to deduce the relation shown in Fig.28.6.

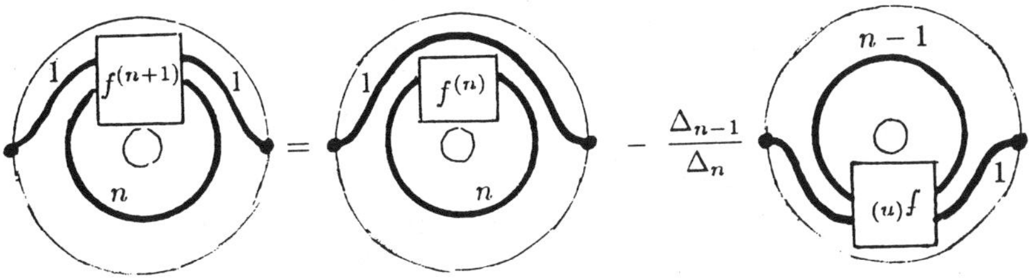

FIGURE 28.6. A modification of the relation in Fig.28.5

When $\Delta_n \neq 0$, this last relation is equivalent to the one shown in Fig.28.7.

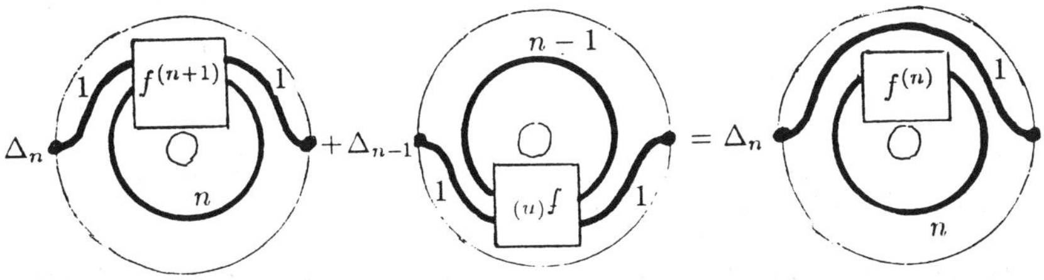

FIGURE 28.7. The previous relation after multiplication by Δ_n

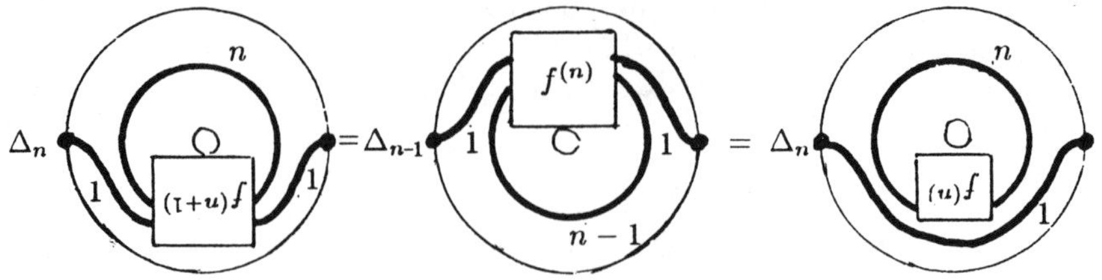

FIGURE 28.8. The previous relation after a 180° rotation

Rotating all the diagrams by 180°, we get the relation from Fig.28.8.

Consider the equalities expressed by Fig.28.7 for $n = 0, 1, \ldots, r-2$. The right-hand side of the sum of all these equalities will be the element shown in Fig.28.9,a. This is because the element ω is equal to the sum of elements shown in Fig.28.9,c for $n = 0, 1, \ldots, r-2$. For the same reason, a similar sum of right-hand sides of equalities expressed by Fig.28.8 will be the element shown in Fig.28.9,b.

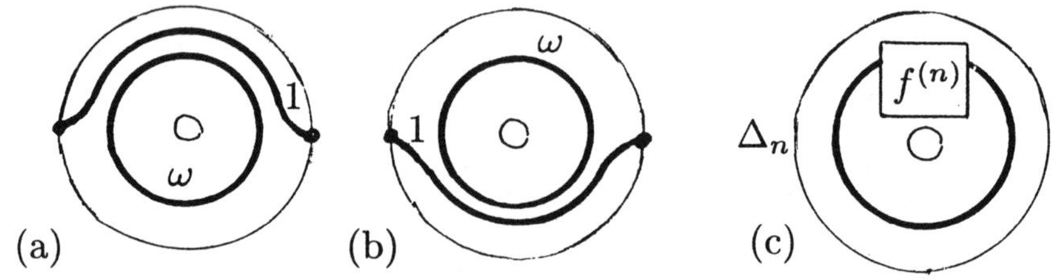

FIGURE 28.9. Three elements of the space $S(I \times S^1)$

The difference between the right-hand sides of the two summed relations is equal to the difference between their left-hand sides. It is easy to verify that the left-hand sides differ only by the elements shown in Fig.28.10 (in the proof of this statement one uses the fact that $\Delta_{-1} = 0$). Looking at these elements, we see that they both contain (as constituents) the element $f^{(r-1)}$, as claimed. □

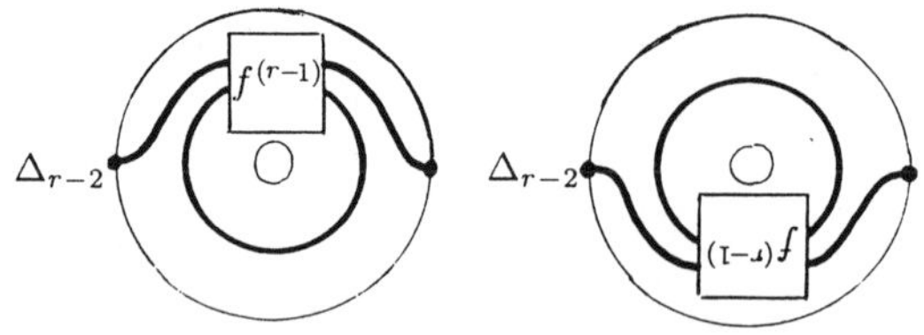

FIGURE 28.10. Difference between the left-hand sides

Returning to the proof of the theorem, recall that by assumption the number a_0^4 is a primitive rth root of unity. The fact that it is primitive garantees that $\Delta_n \neq 0$ for $n = 0, 1, \ldots, r-2$, while the relation $a_0^{4r} = 1$ implies that $\Delta_{r-1} = 0$. To complete the proof of the theorem (in the particular case being considered) it now remains to prove the following statement.

28.5. LEMMA. *Suppose that $\Delta_{r-1} = 0$. Then any element d of the space $S(\mathbb{R}^2)$ containing the diagram $f^{(r-1)}$ vanishes.*

Proof. According to Proposition 26.7, $S(\mathbb{R}^2)$ may be identified with $S(S^2)$. Let us picture the sphere as a cylinder. We can assume that the lower and upper base of the cylinder are free from any arcs belonging to the diagram representing the element d. Then d is the image in $S(\mathbb{R}^2)$ of the closure of the element $f^{(r-1)}x$ under the embedding of the lateral surface of the cylinder in S^2; here x is some element of the algebra TL_{r-1} (Fig.28.11).

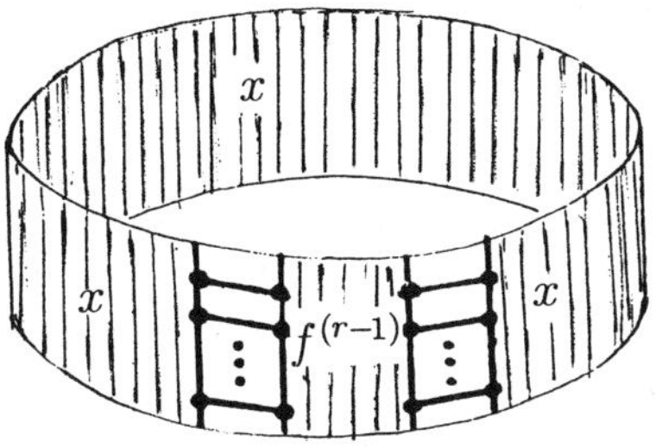

FIGURE 28.11. An element of the algebra TL_{r-1}

Since we have $x = \lambda 1_{r-1} + x'$, where $x' \in A_{r-1}$, it follows that $f^{(r-1)}x = \lambda f^{(r-1)}$. Now let us identify $S(S^2)$ and $S(\mathbb{R}^2)$. As the result we see that d is the image of the closure of the element $\lambda f^{(r-1)}$ under the embedding of $I \times S^1$ in $\mathbb{R}^2$. Therefore, $d = \lambda \Delta_{r-1} = 0$. □

28.6. The proof of Theorem 28.2 is complete in the particular case of the second Kirby move. Recall (see Fig.28.2) that the ribbon under consideration was the simplest possible (Fig.28.12,a). But the same arguments (with minor modifications) work with an arbitrary ribbon; for example, the ribbon may appear as in Fig.28.12,b. Indeed, all the relations between the diagrams drawn on the ribbon remain valid. For isotopies and operations destroying contractible circles this is obvious, while for the operations destroying crossings this follows from the fact that the orientation of any part of the (horizontal) ribbon agrees with the orientation of the plane where the diagrams live. □

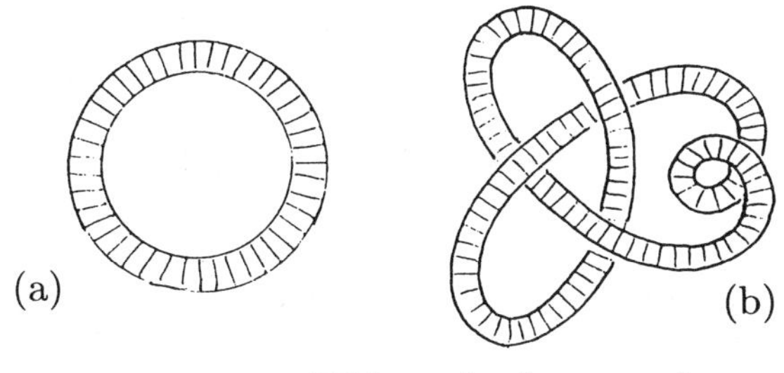

FIGURE 28.12. Ribbons in the general case

Our construction of the invariant of 3-manifolds based on the polylinear map defined in 28.1 would be complete if we could prove that this map is invariant with respect to the *first* Kirby move. Unfortunately, this is simply not true. In the next section, we shall have to introduce a "correction" to this map ensuring the required invariance.

§29. Invariance with respect to the first Kirby move

29.1. Linking number matrix. The expression $\langle \omega, \dots, \omega \rangle_D$ is invariant with respect to isotopies of the framed link D and with respect to the second Kirby move.

But we also need invariance with respect to the first Kirby move. To achieve that, we shall need the *linking number matrix* B of a framed link $L = \{K_1, \dots, K_n\}$, defined as follows. Outside the diagonal, the matrix elements are equal to the corresponding linking numbers, i.e., $b_{ij} = \mathrm{lk}(K_i, K_j)$, while the diagonal elements are equal to the framing indices of the corresponding components, i.e., $b_{ii} = \mathrm{lk}(K_i, K_i')$, where K_i' is the second boundary component of the ribbon that defines the framing of the component K_i.

The matrix B is symmetric, because $b_{ij} = \mathrm{lk}(K_i, K_j) = \mathrm{lk}(K_j, K_i)$. Hence all the eigenvalues of B are real. Let b_+ be the number of positive eigenvalues of B, and b_- the number of negative ones. To construct our invariant we don't really need to know the whole matrix B, but only the integers b_+ and b_-.

The matrix B depends on the choice of orientations for the components of L. However, the integers b_+ and b_- do not depend on this choice. For example, suppose we reverse the orientation of the component K_1. Then B will be replaced by the symmetric matrix B', where $b'_{11} = b_{11}$, $b_{1i} = -b_{1i}$ for $i \neq 1$, $b'_{ij} = b_{ij}$ for $i, j \neq 1$. It is easy to see that $B' = X^T B X$, where X is the diagonal matrix with diagonal $(-1, 1, \dots, 1)$. Therefore B and B' are matrices of the same quadratic form in different bases. It is a standard fact from linear algebra that the integers b_+ and b_- are well defined by the quadratic form and do not depend on the choice of basis.

Now we claim that *the integers b_+ and b_- are not changed by the second Kirby move.* To establish this claim, it suffices to prove the following statement.

29.2. THEOREM. *Under any Kirby move of the second kind, the linking number matrix B is replaced by the matrix $B' = X^T B X$, where X is a nondegenerate real matrix.*

Proof. It suffices to consider the case when to the ribbon defining K_1 we add a new part parallel to the ribbon K_2. Let us show that in this case

$$X = \begin{pmatrix} 1 & 0 & 0 & \dots & 0 \\ \pm 1 & 1 & 0 & \dots & 0 \\ 0 & 0 & 1 & \dots & 0 \\ \vdots & \vdots & \vdots & \ddots & \vdots \\ 0 & 0 & 0 & \dots & 1 \end{pmatrix},$$

and therefore

$$B' = \begin{pmatrix} b_{11} \pm 2b_{12} + b_{22} & b_{12} \pm b_{22} & \dots & b_{1n} \pm b_{2n} \\ b_{12} \pm b_{22} & b_{22} & \dots & b_{2n} \\ \vdots & \vdots & \ddots & \vdots \\ b_{1n} \pm b_{2n} & b_{2n} & \dots & b_{nn} \end{pmatrix}.$$

Recall that the second Kirby move consists in moving the ribbon K_1 up to the ribbon K_2 (Fig.29.1) and then replacing the small hatched part of K_1 by a long ribbon parallel to K_2 (part of which is shown by dotted lines in Fig.29.1). Actually there are two essentially different ways to move the ribbon K_1 up to K_2. By essentially different here we mean that in one case the orientation of the added ribbon is the same as that of K_2 (Fig.29.1,a), and in the other case is opposite (Fig.29.1,b).

The ribbon K_1 is moved up to K_2 by an isotopy. This does not change the linking number, which is only modified by the addition of the new part of the ribbon

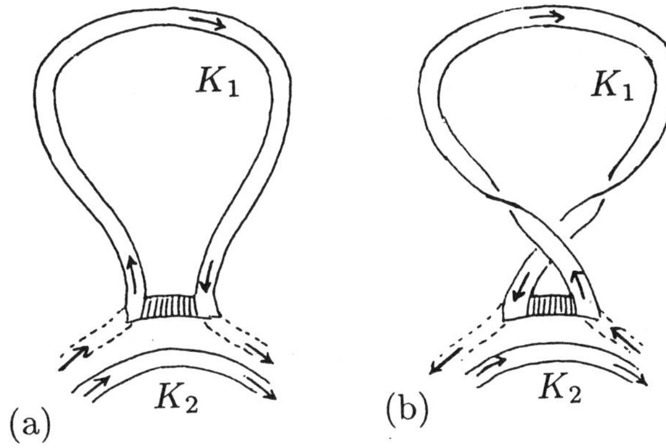

FIGURE 29.1. Two ways to do the second Kirby move

parallel to K_2. Therefore, up to a set that does not affect the linking numbers, we can assume that K_1 has been replaced by $K_1 \cup (\pm K_2)$. When $i \neq 1$, we get

$$b'_{1i} = \mathrm{lk}(K_1 \cup (\pm K_2), K_i) = \mathrm{lk}(K_1, K_i) \pm \mathrm{lk}(K_2, K_i) = b_{1i} \pm b_{2i}.$$

In order to compute the element b'_{11}, we must consider the curves K'_1 and K'_2: the second components of the boundaries of the ribbons defining the framings of K_1 and K_2. It is now obvious that

$$b'_{11} = \mathrm{lk}(K_1 \cup (\pm K_2), K'_1 \cup (\pm K'_2)) = b_{11} \pm 2b_{12} + b_{22}. \quad \square$$

29.3. The invariant. Consider the three standard framed diagrams U_+, U and U_- shown in Fig.29.2. They are simply circles with framing indices equal to 1, 0 and -1, respectively. In the case when $\langle\omega\rangle_{U_+}\langle\omega\rangle_{U_-} \neq 0$, for each framed link diagram D we can consider the complex number

$$I(D) = \langle\omega, \dots, \omega\rangle_D \langle\omega\rangle_{U_+}^{-b_+} \langle\omega\rangle_{U_-}^{-b_-}. \tag{29.1}$$

As we shall see, this number is the invariant that we require.

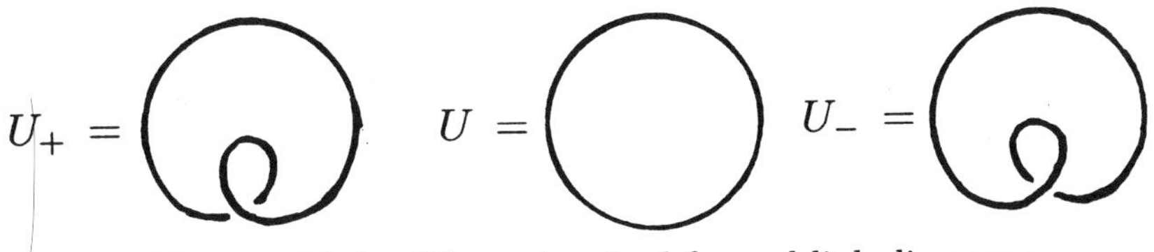

FIGURE 29.2. Three standard framed link diagrams

29.4. PROPOSITION. *The complex number $I(D)$ defined by* (29.1) *is a topological invariant of the* 3*-manifold presented by the framed link diagram D provided that $\langle\omega\rangle_{U_+}\langle\omega\rangle_{U_-} \neq 0$.*

Remark. Actually the last condition of the theorem can always be satisfied by an appropriate choice of the parameter a_0, as we shall show.

Proof of Proposition 29.4. Under isotopies and Kirby moves of the second kind none of the three factors in the expression for $I(D)$ changes. The factors do change under the first Kirby move, but as we shall see, these changes cancel each other out. Recall that the first Kirby move (see 19.1) consists in adding or deleting the diagram $U_\pm$.

Since we have chosen the empty diagram for the basis element of the space $S(\mathbb{R}^2)$, we have $p(D_1 \cup D_2) = p(D_1)p(D_2)$, where $p(D_i)$ is the image of D_i in $S(\mathbb{R}^2)$. This relation implies

$$\langle x_1, \dots, x_{n+m}\rangle_{D_1 \cup D_2} = \langle x_1, \dots, x_n\rangle_{D_1} \langle x_{n+1}, \dots, x_{n+m}\rangle_{D_2}.$$

In particular,

$$\langle \omega, \dots, \omega, \omega\rangle_{D \cup U_\pm} = \langle \omega, \dots, \omega\rangle_D \langle \omega\rangle_{U_\pm}$$

It is also clear that the addition of the diagram $U_\pm$ results in the increase by 1 of the integer $b_\pm$. Therefore the changes in the factors $\langle \omega, \dots, \omega\rangle_D$ and $\langle \omega\rangle_{U_\pm}^{-b_\pm}$ compensate each other. □

29.5. In order to ensure that $\langle \omega\rangle_{U_+}\langle \omega\rangle_{U_-} \neq 0$, we shall have to impose a supplementary conditions on a_0. Namely, we shall assume in the sequel that the parameter a_0 satisfies one of the following conditions:

(1) a_0 is a primitive root of degree $4r$ of unity;

(2) a_0 is a primitive root of degree $2r$ of unity and r is odd.

In both cases a_0^4 will be a primitive root of 1 of degree r. Indeed, in case (1) we have $a_0 = \exp(2\pi i k/4r)$ and $(k, 4r) = 1$; therefore $a_0^4 = \exp(2\pi i k/r)$ and $(k, r) = 1$. In case (2), $a_0 = \exp(2\pi i k/2r)$ and $(k, 2r) = 1$; therefore $a_0^4 = \exp(2\pi i(2k)/r)$ and $(2k, r) = 1$ since r is odd.

Note that conditions (1) and (2) do impose essential supplementary conditions on the parameter a_0. For example, the number $a_0 = \exp(2\pi i/3)$ is such that a_0^4 is a primitive root of unity of degree $r = 3$; but this number satisfies neither condition (1) nor condition (2).

29.6. PROPOSITION. *If the number a_0 satisfies either condition* (1) *or condition* (2), *then* $\langle \omega\rangle_{U_+}\langle \omega\rangle_{U_-} \neq 0$

Proof. Consider the framed diagram $D = U_+ \cup U_-$ (Fig.29.3,a). By means of the second Kirby move the diagram D may be transformed into the diagram D' (Fig.29.3,b). Using plane isotopies and the move Ω_1', we can represent D' in the form shown in Fig.29.3,c.

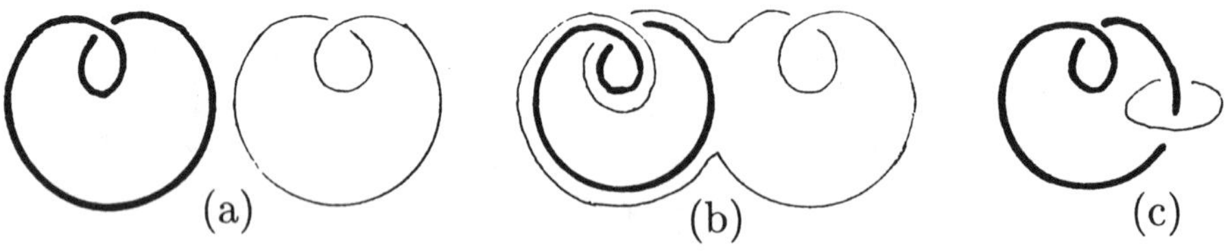

FIGURE 29.3. Transformations of the diagram D

According to Proposition 29.4,

$$\langle \omega, \omega\rangle_{D'} = \langle \omega, \omega\rangle_D = \langle \omega\rangle_{U_+}\langle \omega\rangle_{U_-}.$$

Therefore we must prove that the expression $\langle \omega, \omega\rangle_{D'}$ is nonzero. To compute this expression, recall that

$$\omega = \sum_{n=0}^{r-2} \Delta_n S_n(\alpha).$$

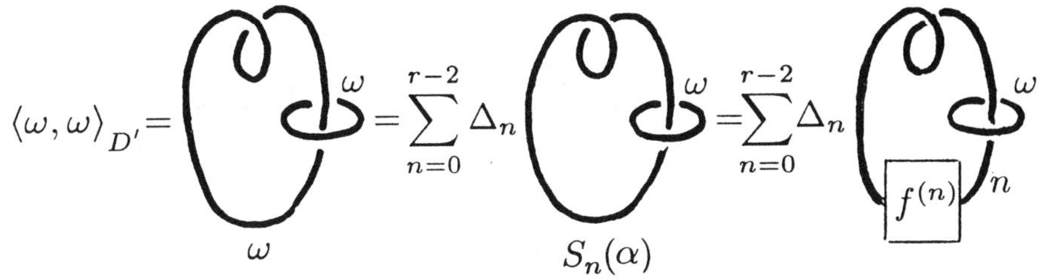

FIGURE 29.4. Diagrammatic sum for $\langle\omega,\omega\rangle_{D'}$

This definition and that of the polynomial $S_n(\alpha)$ (see 27.3) immediately imply the series of equalities pictured in Fig.29.4.

In order to compute the sum, we shall need the diagrammatic relation shown in Fig.29.5.

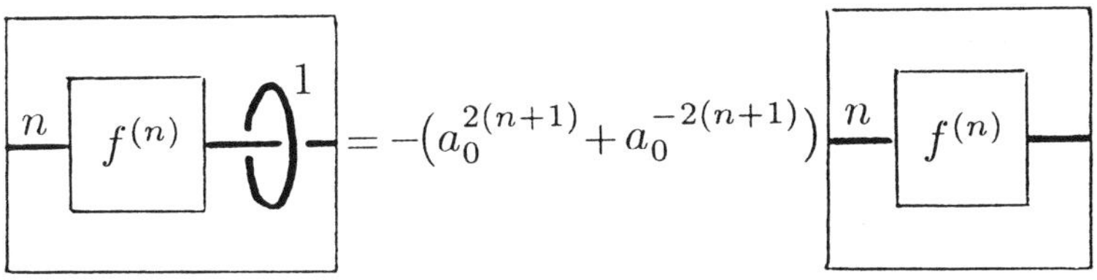

FIGURE 29.5. An auxiliary relation

Problem 29.1. Prove the relation contained in Fig.29.5.

Note that the diagram D_n shown in Fig.29.6 appears in each summand of the sum in Fig.29.4 that we are computing.

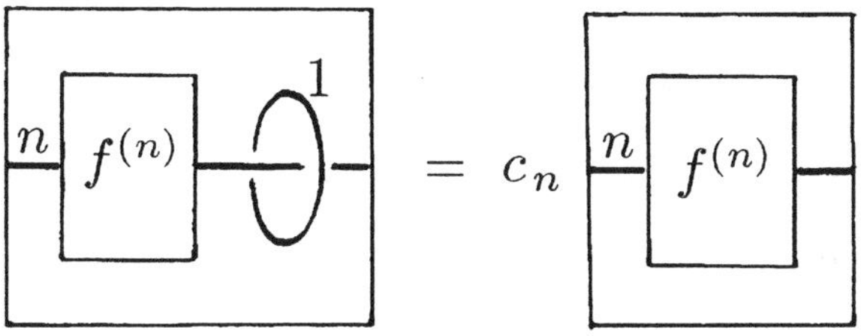

FIGURE 29.6. A recurring expression

29.7. LEMMA. *Suppose the complex number a_0 satisfies either condition (1) or condition (2). Then the following situations may occur:*

(i) *when $1 \leqslant n \leqslant r-3$, any diagram containing D_n is equal to zero;*

(ii) *when $n = r-2$, in the case of condition (1), any diagram containing D_n is still zero, while in the case of condition (2), it may be replaced by $\langle\omega\rangle_U f^{(n)}$.*

(Here by "diagrams" we actually mean their images in $S(\mathbb{R}^2)$.)

Proof. According to Theorem 28.3, the second Kirby move does not change the image in the space $S(\mathbb{R}^2)$ of any diagram all of whose components are marked by the symbol ω. But actually we proved a stronger statement there: only the component which is encircled (i.e., the one for which a parallel ribbon is added to another component) need be marked by the symbol ω; the other components may

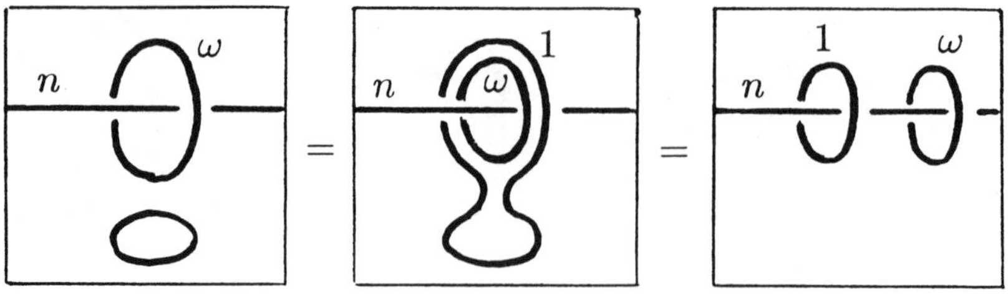

FIGURE 29.7. Another auxiliary diagrammatic relation.

have arbitrary markings. Using this consideration, we obtain the relations shown in Fig.29.7.

Suppose that D_n is the diagram depicted in Fig.29.6. For the image in $S(\mathbb{R}^2)$ of any diagram containing D_n, we get the relations shown in Fig.29.8. The first of these relations involves the operations of adding and deleting a circle, the second follows from Fig.29.7, while the third is a consequence of the relation depicted in Fig.29.5.

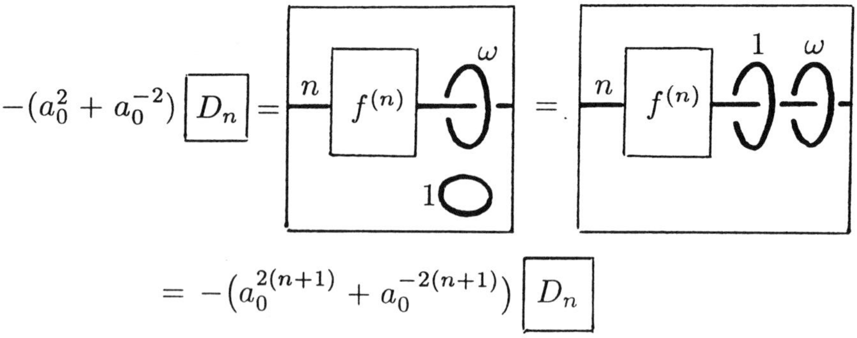

FIGURE 29.8. Relations involving D_n

Therefore, any diagram containing D_n can be nonzero only if

$$a_0^2 + a_0^{-2} = a_0^{2(n+1)} + a_0^{-2(n+1)}, \quad \text{i.e.,} \quad a_0^2(a_0^{2n} - 1)(a_0^{-2(n+2)} - 1) = 0.$$

This last relation holds only if $a_0^{2n} = 1$ or $a_0^{2(n+2)} = 1$. In the case when we have condition (1) for $1 \leqslant n \leqslant r-2$ neither of these equalities can hold. In the case when we have condition (2) for $1 \leqslant n \leqslant r-3$, it still follows that neither equality can hold, but if $n = r-2$, we get $a_0^{2(n+2)} = 1$. Let us consider this last situation. Since $a_0^{2r} = 1$, we get $a_0^{2(n+1)} = a_0^{2(r-1)} = a_0^{-2}$. Therefore the relation shown in Fig.29.5 acquires the form represented in Fig.29.9.

But the multiplication of a diagram by $-(a_0^2 + a_0^{-2})$ is equivalent to adding a circle. This gives us the relation shown in Fig.29.10.

In other words, one of the circles may be unlinked from the strands. But then this can be done for several concentric circles, so that we obtain the relation shown in Fig.29.11. Therefore D_{r-2} may be replaced by the diagram $\langle\omega\rangle_U f^{(r-2)}$. □

29.8. Lemma 29.7 will allow us to complete the computation of the sum appearing in Fig.29.4. In case (1) only one nonzero summand remains, the one corresponding to $n = 0$. In case (2) there are two, corresponding to $n = 0$ and $n = r-2$.

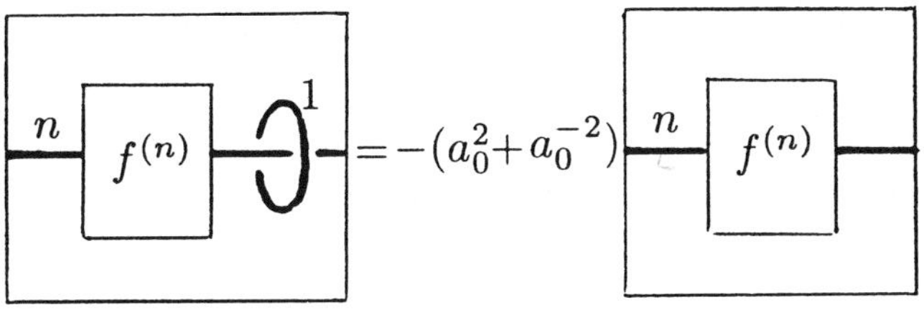

FIGURE 29.9. Another form of Fig.29.5

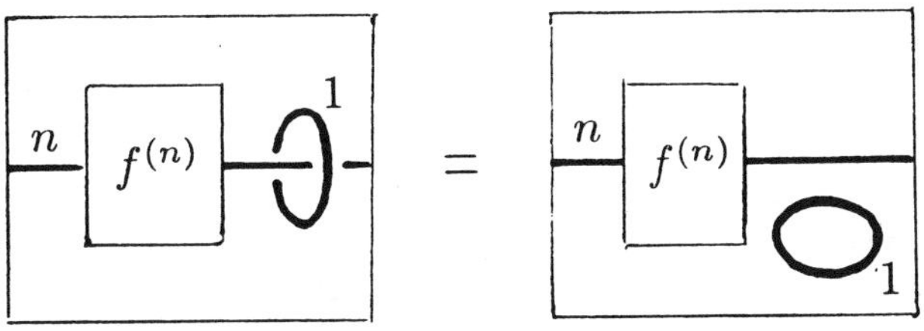

FIGURE 29.10. Unlinking one circle

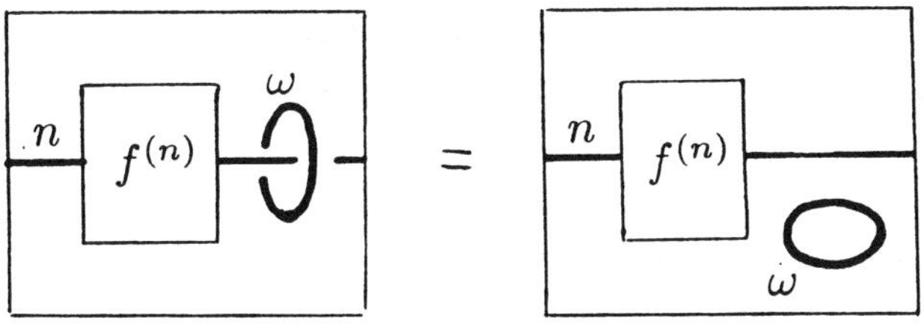

FIGURE 29.11. Unlinking several circles

The summand corresponding to $n = 0$ is the diagram U marked by the symbol ω (recall that $\Delta_0 = 1$). The image of this diagram in $S(\mathbb{R}^2)$ is $\langle\omega\rangle_U$.

It is easy to check that if $a_0^{2r} = 1$ and $a_0^2 \neq 1$, then

$$\Delta_{r-2} = -\frac{a_0^{2(r-1)} - a^{-2(r-1)}}{a_0^2 - a_0^{-2}} = 1.$$

Hence in case (2) the summand corresponding to $n = r-2$ can be represented by the diagram shown in Fig.29.12,a. According to Lemma 29.7, this diagram is equivalent to the one in Fig.29.12,b. Hence the summand under consideration is equal to the product of $\langle\omega\rangle_U$ by the image in $S(\mathbb{R}^2)$ of the diagram shown in Fig.29.12,c.

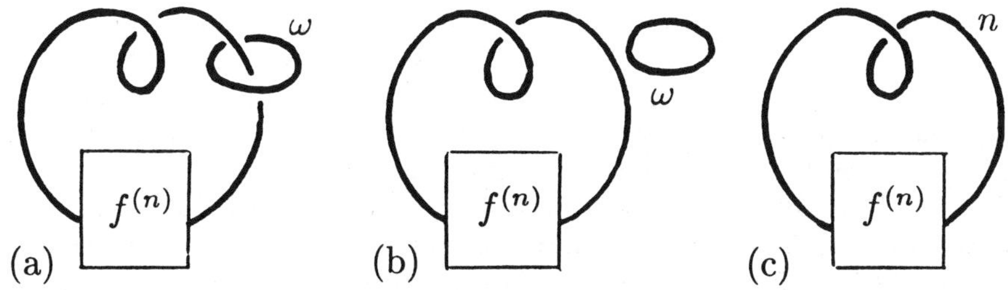

FIGURE 29.12. Modifications of one of the remaining summands

To complete the proof of Proposition 29.6, it remains to solve two fairly easy problems.

Problem 29.2. Prove that if a_0 is a primitive root of unity of degree $4r$, $r \geqslant 3$, then

$$\langle \omega \rangle_U = -\frac{2r}{(a_0^2 - a_0^{-2})^2} \neq 0.$$

Problem 29.3. Prove that in case (2), the image in $S(\mathbb{R}^2)$ of the diagram shown in Fig.29.12,c is equal to 1.

This completes our proof of Proposition 29.6. □

Comparing Propositions 29.4 and 29.6, we obtain the main result of this chapter.

29.9. THEOREM. *If either of the conditions* (1) *or* (2) *from* 29.5 *on the parameters* $a_0 \in \mathbb{C}$ *and* $r \in \mathbb{N}$ *hold, then the formula* (*already appearing in* 29.3 *above*)

$$W(M^3) = I(D) = \langle \omega, \dots, \omega \rangle_D \langle \omega \rangle_{U_+}^{-b_+} \langle \omega \rangle_{U_-}^{-b_-} \tag{29.1}$$

in which D *is any framed link diagram presentation of* M^3, *defines a topological invariant for any compact oriented 3-manifold without boundary* M^3.

29.10. Examples. Now that we have established the topological invariance of the number

$$I(D) = \langle \omega, \dots, \omega \rangle_D \langle \omega \rangle_{U_+}^{-b_+} \langle \omega \rangle_{U_-}^{-b_-}$$

characterizing the manifold M^3 presented by the framed link diagram D, we can compute its values for concrete 3-manifolds.

The sphere S^3 is presented by the empty diagram; therefore $I(S^3) = 1$.

The manifold $S^1 \times S^2$ is presented by the diagram U. For this diagram the linking number matrix has the form $B = (b_{11})$, where $b_{11} = 0$. Therefore $b_+ = b_- = 0$ and so $I(S^1 \times S^2) = \langle \omega \rangle_U$.

Comments

The first generalization of the Jones polynomial from links in the 3-sphere to links in an arbitrary 3-manifold appeared in the remarkable paper of E. Witten [**Wit2**]. A brief description of Witten's approach appears in subsection 32.4, below. However, mathematicians were not satisfied by the level of rigor in Witten's paper, pointing out (with reason) that the integral appearing in Witten's main definition is mathematically undefined. (For the sake of fairness, let us note that physicists have been working with some success with "integrals" of this type for nearly two decades, in particular by using the so-called Feynman path integrals.)

A mathematically correct definition of this invariant was obtained in the work of the Saint Petersburg school (N. Reshetikhin, V. Turaev, O. Viro). Essentially, our exposition in this chapter is based on their quantum group approach, yielding the so-called $SU_q(2)$ 3-manifold invariants; however, following [**Lic4**], we present a simplified version that avoids quantum groups altogether.

It should be noted that one of the main ingredients of the definition is the skein algebra, which comes, as its name indicates, from the skein relation (in our case, from Kauffman's version of this relation). Some mathematical physicists, and some mathematicians, feel that this relation, as well as, say, Kauffman's bracket polynomial, have physical significance and must be included in the foundations of topological quantum field theories (see subsection 32.5 below, or, for a more substancial account, [**Ati2**] or [**BHMV**]).

We should perhaps mention here that the skein relation is also quite meaningful in biology: there is in fact a specific enzyme that "cuts" long molecules (e.g. DNA) and "glues" them back together, thereby effecting crossing changes. In this connection, see [**Sum**].

CHAPTER IX

Invariants of Links in 3-Manifolds

This concluding chapter consists of three sections, which are related only by the fact that in each we are concerned with invariants of links (and hence of knots) in 3-manifolds.

The first one, §30, is rather elementary, and can be read directly after Chapter II. In that section, we give an exposition of the results of Yu. Drobotukhina, who constructed Jones-type polynomial invariants for links in $\mathbb{R}P^3$ by an appropriate modification of the Kauffman approach (compare 3.5 and Theorem 3.6).

The second section of this chapter, §31, is the culmination of this book, and makes use (directly or indirectly) of most of the preceding material. Its goal is to present a rigorous definition of the invariants of links (and knots) in arbitrary compact oriented 3-manifolds. These invariants, which generalize the Jones polynomial $V_L(q)$ (more precisely, its numerical values for fixed values of q), were discovered by E. Witten in the late eighties. However, the ideas behind Witten's original definition, which was presented "on the physical level of rigor", have not been given a treatment that would satisfy mathematicians. For this reason, our exposition is based on a totally different approach, due to several mathematicians (see the Comments to the previous chapter and §32 for references). The brevity of §31 is due to the fact that most of the hard work (related to the use of the Kirby calculus) was carried out in the previous chapter, when we discussed skein invariants of 3-manifolds (which should now be regarded as "3-manifolds containing the empty link").

The third section of this chapter, §32, is an overview of the present state of affairs in the relationships between knot theory and physics. This is presently a very fashionable topic, with hundreds and hundreds of publications in the last few years, including a dozen or so monographs (e.g. [**Kau4**], [**Tur3**], [**Ati1**], [**Kas**], [**BM**], [**Fu**], [**CP**], [**Lus**]). We hope that our brief exposition will help the reader understand the sources of this (often mysterious) interaction of seemingly unrelated topics; and, should interest in this subject matter arise, §32 can perhaps serve as a motivating guide in the vast literature.

§30. Polynomial invariants of links in $\mathbb{R}P^3$

30.1. In §3 we described a construction defining an invariant of oriented links L in the sphere S^3, namely the Kauffman polynomial $X(L)$. This invariant can be carried over in a natural way to oriented links in $\mathbb{R}P^3$. We describe this natural construction following [**Dro**].

30.2. The manifold $\mathbb{R}P^3$ may be defined as the sphere S^3 with diametrically opposed points identified. The sphere S^3 consists of two hemispheres, each of which is homeomorphic to the disk D^3. Therefore, $\mathbb{R}P^3$ may be obtained from D^3

by identifying diametrically opposed points of the boundary $S^2 = \partial D^3$. Hence any link in $\mathbb{R}P^3$ is defined by a union of smooth closed curves and arcs lying in the disk D^3 provided that the set of all the endpoints of the arcs lies on the boundary S^2 and splits into pairs of diametrically opposed points. Thus a link in $\mathbb{R}P^3$ may be presented by a diagram in the plane disk D^2 in which the points where the curves reach the boundary circle $S^1 = \partial D^2$ are centrally symmetric. An example of such a diagram is pictured in Fig.30.1.

FIGURE 30.1. Diagram of a link in $\mathbb{R}P^3$

Problem 30.1. Let the diagram of a link in $\mathbb{R}P^3$ be given. Describe its preimage in the sphere S^3 under the 2-fold covering $p : S^3 \to \mathbb{R}P^3$.

30.3. Reidemeister moves. For links in $\mathbb{R}P^3$, just as for links in S^3, we shall only consider diagrams in general position. In particular, this means that the diagram is not allowed to be tangent to the boundary of D^2, nor can an overpass occur at points of this boundary. Recall that in the case of the sphere S^3, forbidden projections (see Fig.1.2) gave rise to the Reidemeister moves Ω_2 and Ω_3. In a similar way, the forbidden projections mentioned above (tangency or overpass at points of S^2) give rise to two new moves, also called *Reidemeister moves* and denoted by Ω_4 and Ω_5; these moves are pictured in Fig.30.2.

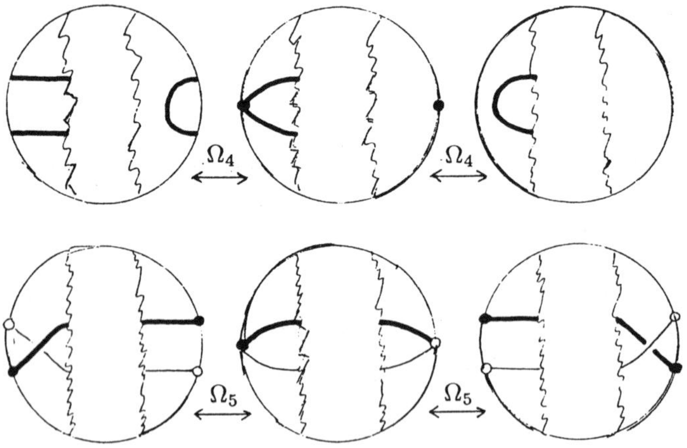

FIGURE 30.2. The additional Reidemeister moves Ω_4 and Ω_5

30.4. THEOREM. *Two link diagrams correspond to isotopic links in $\mathbb{R}P^3$ if and only if they can be obtained from each other by isotopies of the disk D^2 fixed on the boundary, by the classical Reidemeister moves Ω_1, Ω_2, Ω_3, and by the two new Reidemeister moves Ω_4 and Ω_5.*

The proof of this theorem is a general position argument similar to the one proving the classical Reidemeister theorem in Chapter I, and is omitted.

30.5. Definition of the bracket polynomial. For an oriented link in $\mathbb{R}P^3$, we shall define the polynomial invariant just like the Kauffman polynomial $X(L)$ for oriented links in S^3 (see §3); some additional arguments will be needed only to show its invariance with respect to the new moves Ω_4 and Ω_5.

First, given an unoriented link (in particular, a knot) diagram L in $\mathbb{R}P^3$, we define a polynomial $\langle L \rangle$ possessing the following properties:

(1) 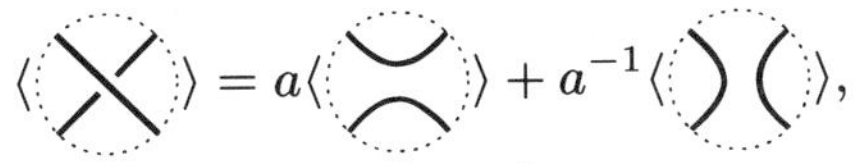

(2) $$\langle L \cup \bigcirc \rangle = (-a^{-2} - a^2)\langle L \rangle,$$

(3) $$\langle \bigcirc \rangle = 1.$$

The proof of the existence and uniqueness of the bracket polynomial $\langle L \rangle$ is exactly the same as that of Theorem 3.4; moreover, no modifications are needed to prove the invariance of the bracket polynomial with respect to the moves Ω_2 and Ω_3.

To prove the invariance of the polynomial $\langle L \rangle$ with respect to the move Ω_4, it is convenient to use formula (3.3) (the explicit definition of $\langle L \rangle$ given in Chapter II). Indeed, for any state s of the diagram L, the numbers $\alpha(s)$, $\beta(s)$, and $\gamma(s)$ are not changed by the transformation Ω_4 (for links in $\mathbb{R}P^3$ the integer $\gamma(s)$ is defined as the number of components of the link L, i.e., the number of closed curves of L in $\mathbb{R}P^2$.

The proof of the invariance of the bracket polynomial with respect to the move Ω_5 follows from the sequence of equalities shown in Fig.30.3.

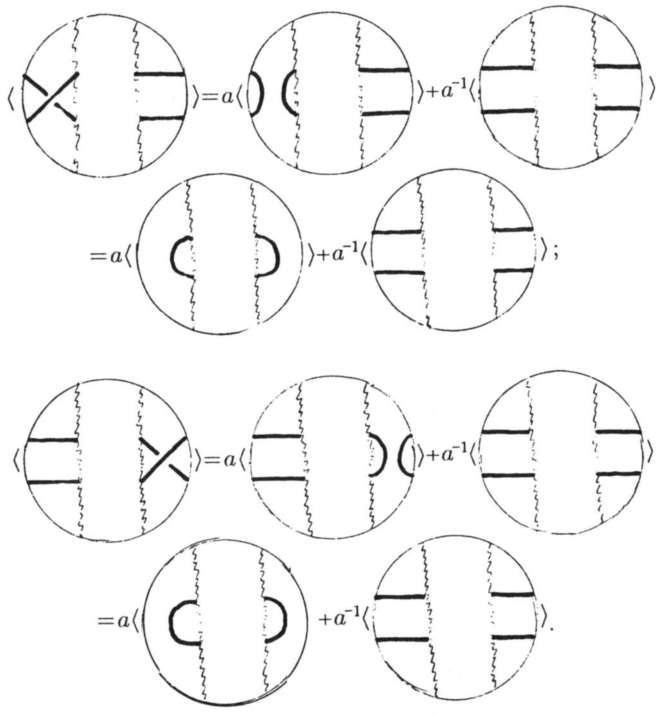

FIGURE 30.3. Invariance of $\langle L \rangle$ with respect to Ω_5

With respect to the move Ω_1 the bracket polynomial is not an invariant: it is multiplied by the factor $(-a)^{\pm 3}$, just like the bracket polynomial for links in S^3.

30.6. Definition of the polynomial $X(L)$. We continue exactly as in §3. We define the *writhe number* $w(L)$ of an *oriented* link L in $\mathbb{R}P^3$ in the same way as for oriented links in S^3. Then we define the polynomial invariant $X(L)$ by setting

$$X(L) = (-a)^{-3w(L)} \langle |L| \rangle, \tag{30.1}$$

where the diagram $|L|$ is obtained from the diagram L by erasing the arrows (forgetting the orientation).

30.7. THEOREM. *The polynomial $X(L)$ is an isotopy invariant for oriented links L in projective space $\mathbb{R}P^3$.*

The proof follows immediately from Theorem 30.4, the definition (30.1), and the invariance with respect to the moves $\Omega_2 - \Omega_5$ established above (compare with the proof of Theorem 3.6). □

Problem 30.2. Prove that for any link in $\mathbb{R}P^3$ only even powers of the variable a appear in the polynomial $X(L)$.

Problem 30.3. Prove that for the knot L shown in Fig.30.1, the variable a appears in the polynomial $X(L)$ not only in powers divisible by four (cf. Theorem 3.9).

§31. Invariants of framed links in three-manifolds

The construction used in Chapter VIII to define numerical invariants of compact oriented 3-manifolds without boundary will be used here to define invariants of more complicated topological objects, namely framed links embedded in such 3-manifolds. Geometrically, this means that we are defining invariants of ribbons in 3-manifolds (since framed links can be represented by ribbons, as explained in 16.3).

31.1. Consider a 3-manifold M^3 (oriented, compact, and without boundary) containing a framed link L. The manifold M^3 can be obtained from the sphere S^3 by surgery along a framed link $L_M \subset S^3$ (§16). This means that there exists a homeomorphism

$$f : S^3 - L_M \to M^3 - \widetilde{L}_M,$$

where $\widetilde{L}_M$ is a certain link in M^3. By general position considerations, the link $\widetilde{L}_M$ may be chosen so that it does not intersect the given link $L \subset M^3$ (although of course various components of L and $\widetilde{L}_M$ may link with each other). Suppose $L_L = f^{-1}(L)$ is the preimage of the link $L \subset M^3$ in the sphere S^3. Note that if L has a framing, then so does L_L. Thus a framed link L in the manifold M^3 can be represented by a pair of nonintersecting framed links (L_L, L_M) in the sphere S^3 such that L_L has the same number of components as L.

31.2. Now let us find under what conditions two pairs (L_L, L_M) and (L'_L, L'_M) determine the same framed link L in M^3. Surgery along the framed links L_M and L'_M yields the same manifold M^3; hence, by Theorem 19.3, the link L'_M can be obtained from L_M by means of the first and second Kirby moves and isotopies.

Note that in the process of these isotopies the link L_L never intersects L_M. This is because the isotopy taking L_M to L'_M induces the homeomorphism

$$f : S^3 - L_M \to M^3 - \widetilde{L}_M.$$

It is also obvious (by compactness and general position considerations) that we may assume the Kirby moves to be the identity in some neighborhood of the link L_L.

After performing the appropriate Kirby moves, we can assume that $L'_M = L_M$. In this process, the link $L_L \cup L_M$ will be transformed by the Kirby moves and isotopies, while the link L_L only undergoes isotopies. Thus it only remains to find the relationship between L_L and L'_L in the case when $L_M = L'_M$.

In this case, after surgery on the sphere S^3 along the link L_M, we obtain the manifold M^3 in which the images of the links L_L and L'_L are isotopic. In the process of this isotopy in M^3, the images of the links L_L and L'_L are allowed to intersect the image of L_M. Hence in the sphere S^3, the corresponding transformation does not reduce to an isotopy of the link $L_L \cup L_M$. More precisely, if in the manifold M^3 a component K_L of the image of the link L_L intersects a component K_M of the image of L_M during the isotopy, a second Kirby moves occurs in the sphere S^3; namely, a ribbon parallel to K_M is added to the ribbon K_L.

Let us note two obvious but important circumstances:

1. the number of components of the links L_L and L'_L is the same;
2. under the second Kirby moves, ribbons parallel to components of the link L_L are never added.

31.3. To the framed link $L_L \cup L_M$ we can assign the corresponding framed diagram $D_L \cup D_M$ (see 19.6). For the link L_M, consider the matrix B of the pairwise linking numbers of its components (see 29.1). Let b_+, and b_- the number of positive and negative eigenvalues of the matrix B, and n the number of components of L_L. As in §28, let us use the angle brackets $\langle \cdot \rangle_D$, where D is a framed link diagram, to denote the polylinear map from the product of the appropriate number of copies of the skein algebra $S(S^1 \times I)$ to the skein algebra $S(\mathbb{R}^2)$. The algebra $S(\mathbb{R}^2)$ is isomorphic to the complex numbers $\mathbb{C}$. Recall that the definition the algebras $S(\mathbb{R}^2)$ and $S(S^1 \times I)$ involves the choice of a complex number a. Thus the values of the angle brackets also depend on how we choose the complex number $a \in \mathbb{C}$. Further, let us choose fixed elements

$$p_1(\alpha), \dots, p_n(\alpha) \in S(I \times S^1) \cong \mathbb{C}[\alpha] \tag{31.1}$$

(see 26.8). Finally, recall that ω denotes the specific element of $S(S^1 \times I)$ defined in 28.2, while $U_\pm$ are the concrete link diagrams shown in Fig.29.2.

31.4. THEOREM. *Let r be an integer satisfying $r \geqslant 3$ and let the complex number $a = a_0$ be a primitive root of unity of degree $4r$, or let r be odd and a_0 a primitive root of unity of degree $2r$. Then the complex number*

$$W(M^3, L) = \langle p_1(\alpha), \dots, p_n(\alpha), \omega, \dots, \omega \rangle_{D_L \cup D_M} \langle \omega \rangle_{U_+}^{-b_+} \langle \omega \rangle_{U_-}^{-b_-} \tag{31.2}$$

is an isotopy invariant of the framed link L in the manifold M^3.

Proof. First note that the integer n is well defined: it is equal to the number of components of the given link L. The proof of the invariance of the number (31.2) with respect to the admissible transformations of $L_L \cup L_M$ mainly follows the proofs of Theorems 29.4 and 29.6. The only additional argument is the following. Under the second Kirby moves, we never add ribbons parallel to components of the link L_L. This means that we can mark them not only by the element ω, but by arbitrary elements $p_1(\alpha), \dots, p_n(\alpha) \in S(I \times S^1)$. □

31.5. Remark. Formula (31.2) defines not one invariant, but an infinite series of invariants parametrized by the following data:

1. the complex number a_0, which must satisfy one of the two conditions stated at the beginning of Theorem 31.4;
2. the n-tuple of elements (31.1) of the skein algebra $S(S^1 \times I)$.

At present little is known about the relative strength of these invariants and on how well they distinguish knots (or links) in a fixed 3-manifold. Their relationship to numerical values of the invariant $X(L)$ from the previous section (in the case $M^3 = \mathbb{R}P^3$) has not been studied either.

§32. Knots and physics

In this section, written more in the genre of an essay than a survey, we try to explain the relationship of the subject matter of this book to physics. There will be no precise formulations and proofs here, our aim being to display the underlying reasons for the truly fascinating and mysterious connections between various topological invariants and certain physical models, rather than to summarize the main achievements in this booming field. Before we begin our exposition, we would like to stress two points:

• We feel it is too early yet to assess the importance or even the relevance to reality of the models that have appeared here, e.g. in quantum field theory. The history of string theory (which, in our opinion, has produced some beautiful mathematics, but so far seems to have little to do with physical reality) shows that the expression "applications to physics" in reference to present day fashionable work (even to that of such outstanding researchers as Atiyah and Witten) should not be taken too literally.

• On the other hand, the absence of specific practical applications to experimental physics does not make the surprising interconnections that have appeared here (even if they are "only" between various branches of "pure" mathematics) less interesting.

32.1. The Yang-Baxter equation and topology. Undoubtedly, the starting point of the relationship between physics and the recent work on knot invariants is the observation that the Yang-Baxter equation (a fundamental law both in statistical physics and in quantum physics) also appears in three-dimensional topology under different guises. Namely, in a certain formal sense, the Yang-Baxter equation "coincides" with the braid relation (see §5) and with the third Reidemeister move (see §1). Before we discuss this coincidence, let us explain (from the mathematical viewpoint) what the Yang-Baxter equation is.

Suppose V is a finite-dimensional vector space with basis $e_1, \dots, e_n$ over the field F (actually one may consider the more general setting of a free module over a commutative ring). Let $R : V \otimes V \to V \otimes V$ be an automorphism. Consider the endomorphism

$$R_i = 1 \otimes 1 \otimes 1 \dots 1 \otimes R \otimes 1 \otimes \cdots \otimes 1 : V^{\otimes n} \to V^{\otimes n},$$

where $V^{\otimes n}$ stands for the n-fold tensor product (over F) and in the expression for R_i the automorphism R operates on the $(i, i+1)$th component. Then R is called a *Yang-Baxter operator* (or an *R-matrix* if the basis is fixed) if it satisfies the following relations:

$$\begin{aligned} R_i R_{i+1} R_i &= R_{i+1} R_i R_{i+1}, \quad i = 1, \dots, n-1 \\ R_i R_j &= R_j R_i, \quad |i - j| \geqslant 2. \end{aligned} \tag{32.1}$$

The first of these relations is known as the *Yang-Baxter equation*, while the second is the *far commutativity relation* for the endomorphisms $R_1, \dots, R_{n-1}$.

The physical meaning of Yang-Baxter operators will be briefly described in the following subsection, while here we shall discuss their relationship with three-dimensional topology. The reader will of course have noticed the formal coincidence between the relations (32.1) and the braid group relations (see (5.1) and (5.2)): changing the letter R to the letter b and vice versa, we pass from one to the other. The formal similarity between the Yang-Baxter equation and the third Reidemeister move Ω_3 (see Fig.1.11), which is more graphical than algebraic, is displayed in Fig.32.1.

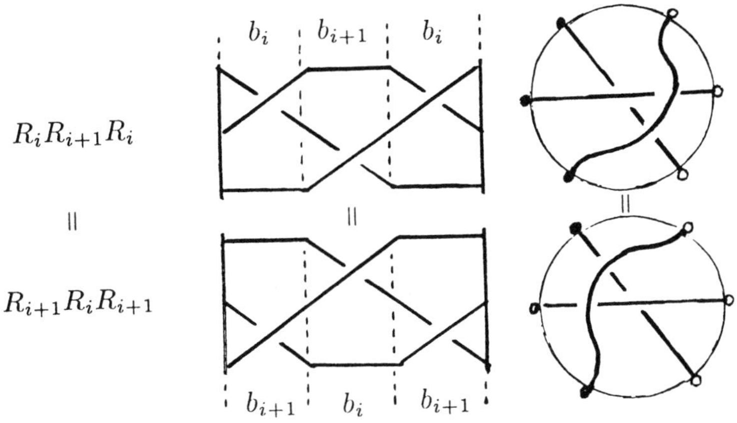

FIGURE 32.1. Yang-Baxter equation, braid relation, and Ω_3

It is natural to expect that the formal coincidences noted above lead to meaningful interactions between the theories from which they come. This is indeed the case, as we shall explain below. To conclude this subsection, let us note a significant connection between Yang-Baxter operators and knot invariants, which is purely mathematical in nature.

It can be shown (see [**Dri1-2**]) that any irreducible representation of a simple Lie algebra can be used to effectively produce an R-matrix. In its turn, any R-matrix $R : V^{\otimes 2} \to V^{\otimes 2}$ can be used to produce a link invariant in the following simple way. First we use R to define a representation of the braid group $\rho : B_n \to V^{\otimes n}$ by the natural formula $\rho(b_i) = R_i$, where the b_i are the standard braid generators. (This assignment is indeed a representation, precisely because the braid relations coincide with the equations (32.1) for Yang-Baxter operators.) Then, given a link L, we find a braid $b \in B_n$ whose closure is L (it exists by Alexander's theorem, see §6), and set $T(L) = \operatorname{trace}(\rho(b))$. It is easy to prove (using Markov's theorem, see §6) that the number $T(L)$ is well defined and is indeed an isotopy invariant of links. In particular, if the simple Lie algebra we start with is

$\mathfrak{sl}(2, \mathbb{R})$, then the Jones invariants (numerical values of the Jones polynomial) can be obtained in this way.

Of course the constructions described in the previous paragraph are purely mathematical, but it should be noted that the basic ingredients (the Yang-Baxter equation and its solutions obtained by Lie algebra representations) are due to physicists.

32.2. Two physical interpretations of the Yang-Baxter equation. The Yang-Baxter equation comes from two totally disconnected physical sources. One is statistical physics (more precisely, completely solvable two-dimensional state lattice models) and quantum mechanics (the simultaneous interaction of three particles in the one-dimensional quantum model).

For a detailed exposition of the theory of lattice models (of which the Ising model is the most popular instance) the reader is referred to the classical monograph [**Bax**]. An example of a lattice model (of sorts) was described above when we proved the existence of the Kauffman bracket (see §3): the shadow of any link is a planar graph whose vertices represent *particles* of a two-dimensional medium (the *model*) and whose edges represent *interactions* between the particles, the choice of an A or B angle at a vertex is the assignment of a *spin* to the particle that the vertex represents, a choice of spins at all vertices is the *state* of the lattice model, and the Kauffman bracket represents the state model's *partition function* (which is the main tool used by physicists to compute the *total energy* of the model, its *phase transfers*, etc.).

Of course this model does not describe any known physical reality, but only resembles (not too closely) certain models actually studied by physicists. To make this resemblance more visual, in Fig.32.2 we show how a square lattice in the plane (the two-dimensional state model) is related to links. It should be noted that the idea of joining far-away points (e.g. the points A and A' in Fig.32.2,b), which seems physically unnatural, was put forward by physicists (in order to avoid unpleasant "boundary effects") long before connections with knot theory were established. Thus a square lattice becomes a special kind of knot (or link) to which topological invariant theory may be applied, possibly in a physically meaningful way.

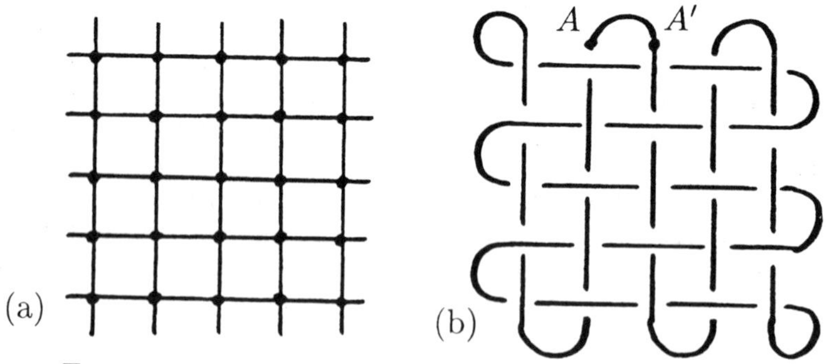

FIGURE 32.2. Square lattice and the corresponding link

Looking at things in the opposite direction (from physics to knot theory), we must note that one of the classical two-dimensional state models from statistical physics can in fact be used (as we used the Kauffman model in §3) to define the

Jones polynomial. This is the *Potts model*, which describes "two-dimensional ice"[1] (in this model phase transfer is simply the freezing of water or the melting of ice). For details, the reader is refered to [**Jon3**].

The other physical interpretation of the Yang-Baxter equation has to do with three quantum particles interacting on a line (one-dimensional space or two-dimensional space-time). We shall describe this situation very briefly. Now we must consider the *quantum Yang-Baxter equation*, which has the form

$$R_{12}R_{13}R_{23} = R_{23}R_{13}R_{12}. \tag{32.2}$$

Here the matrices R_{ij} are obtained from a fixed automorphism matrix $R : V^{\otimes n^2} \to V^{\otimes n^2}$ (e.g. $R_{12} = R \otimes I_n$, where I_n is the unit $n \times n$ matrix). A solution of this equation is called a *quantum R-matrix*. The matrix R_{ij} describes the interaction of the ith and jth particles, and the entire equation describes what happens when the first (counting from the left) particle changes places with the third one, the second remaining between them. This can happen in two ways:

1. first $\leftrightarrow$ second, then first $\leftrightarrow$ third, and then second $\leftrightarrow$ third, or
2. second $\leftrightarrow$ third, then first $\leftrightarrow$ third, and then first $\leftrightarrow$ second.

The equation says that the result of these two processes is the same (actually, according to the Heisenberg indeterminacy principle, we cannot ever learn which of the two processes actually occured).

Comparing equation (32.2) with Fig.32.1, we see that this equation is just another way to write the braid relation (for $n = 3$): the first strand passes over the second (R_{12}), then the first passes over the third (R_{13}), etc. Just as in the case of the classical Yang-Baxter equation (32.1), this is not only a formal similarity: the quantum Yang-Baxter equation (32.2) can be effectively used to produce knot and link invariants (see, for example, [**Haz**]).

At this point it is logical to pass to the fundamental notion of *quantum group*, but we precede that with some basic notions of classical and quantum mechanics that the reader may not be familiar with.

32.3. Invariants as observables: the classical and quantum cases. The basic mathematical model of classical mechanics is a symplectic $2n$-dimensional manifold M (the *phase space*). The points of M are the *states* of the system (in the simplest case a state is the position of a moving point together with its velocity $s = (x(t), \dot{x}(t))$, and so M is the total space $T(N) = M$ of the tangent bundle of an n-dimensional manifold N, the *configuration space*). The symplectic structure induces a multiplication (the *Poisson bracket*) in the space of functions on M, and the *observables* are functions on M that form a commutative algebra with respect to this bracket. The evolution of the system in time depends on the choice of the Lagrangian L (a functional, depending on the states, that must be minimized) and obeys the *Euler-Lagrange equations*:

$$\frac{\partial L}{\partial x_i(t)} - \frac{d}{dt}\frac{\partial L}{\partial \dot{x}_i(t)} = 0.$$

An equivalent (dual) construction is obtained by considering *momenta* $p_i = m\dot{x}_i$, i.e., covectors (elements of the cotangent space $T^*(N)$) instead of (velocity) vectors,

[1] Of course real ice is solid, i.e., three-dimensional, but only the two-dimensional model has been successfully studied in statistical physics and, besides, link shadows are also planar.

and replacing the Euler-Lagrange equations by *Hamilton's equations*:

$$\dot{x}_i = \frac{\partial H}{\partial p_i}, \quad \dot{p}_i = -\frac{\partial H}{\partial x_i}.$$

In quantum mechanics the *states* no longer form a finite-dimensional manifold, but constitute a Hilbert space $\mathcal{H}$. Physically, a state is no longer a pair (point, velocity) but a density function describing the probability of a particle having a certain location and a certain velocity. The *observables* are now operators from $\mathcal{H}$ to $\mathcal{H}$ forming a noncommutative algebra with respect to the usual commutator of operators $[P, Q] = P \circ Q - Q \circ P$. The choice of $\mathcal{H}$ and of the algebra of observables depends on the specifics of the physical system under consideration.

In the simplest situation we have $\mathcal{H} = L_2(\mathbb{R}^3)$, the algebra of observables satisfies certain simple anticommutativity relations, and the evolution in time of a state φ must satisfy the *Schrödinger equation*:

$$i\frac{d\varphi}{dt} = \hat{H}\varphi,$$

where $\hat{H}$ is an operator related to the energy function H (the *Hamiltonian*, a concrete function depending on the specifics of the physical problem considered).

This approach (and its generalizations), dealing with observables and based on the choice of a Hamiltonian, may be called the *Hamiltonian approach*, in contrast with the dual *Lagrangian approach* which deals with states and is based on the minimization of a a certain functional, the *Lagrangian*. Quantum groups are a refinement of the algebra of observables in quantum theory, and thus follow the ideology of the Hamiltonian approach. (We shall come back to the Lagrangian approach after briefly discussing quantum groups).

32.4. Quantum groups and other bialgebras. Quantum groups appeared in the work of L. Faddeev and his school as the result of quantizing the inverse scattering method (see [**FRT**]), which led to the appearance of the quantum Yang-Baxter equation. A crucial contribution was the work of M. Jimbo (see [**Jim1-2**]. The algebraic theory of quantum groups was developed by Vladimir Drinfeld, who also introduced the notion of *quasitriangular* quantum group, which stresses the connection between quantum theory and the theory of knot invariants via the Yang-Baxter equation. We shall not give the formal definition of the notion of quantum group, nor seriously develop the corresponding theory, referring the reader to any of several monographs of this topic that have recently appeared (e.g. [**Kau4**], [**Tur3**], [**Ati1**], [**Kas**], [**BM**], [**Fu**], [**CP**], [**Lus**], [**Mont**]). We only give a rough sketch, concentrating on the relationships with knot invariants.

Quantum groups are not groups, but algebras, in fact *Hopf algebras*, i.e., algebras A (with unit) supplied with an additional structure: a comultiplication (a map $A \to A \otimes A$), a *counit*, and an antihomomorphism $S : A \to A$ (the *antipode*), which satisfy certain natural commutativity relations. Roughly speaking, to construct a quantum group, one starts with a Lie group G (whose elements are the *states* of the model) and a Poisson bracket on the algebra A of complex-valued functions on G (the *observables*). The multiplication in G defines, by duality, a comultiplication in A. For the quantum group to be quasitriangular, its elements must satisfy the *triangle relation*, which implies the quantum Yang-Baxter equation. In this case concrete representations of the group G not only provide us with meaningful information on states and observables, but lead to new link invariants along the lines of the general construction explained in 32.1 above.

It should also be noted that the Vassiliev invariants, although they do not come from any quantum groups, do have a natural bialgebra structure (the comultiplication coming from connected summing of knots). There are also many interesting bialgebras related to Vassiliev invariants, in particular those appearing in connection with the Kontsevich integral and with Tutte polynomials (see [**CDL**], [**BN**]).

Thus the algebraic formalism of quantum groups, motivated by physics, turns out to be useful for three-dimensional topology. But, conversely, does three-dimensional topology help physics? Before we try to answer this question, we briefly discuss an approach to physical models, mainly due to E. Witten, and in a certain sense dual to Drinfeld's Hamiltonian approach.

32.5. Witten's Lagrangian approach. To really understand Witten's work, one needs to know a lot of physics and mathematics (string theory, Riemann surfaces and geometric quantization, nonabelian gauge theories and affine connections in principal fiber bundles, instantons and magnetic monopoles as solutions of Yang-Mills equations, quantum chromodynamics and the Atiyah-Singer index theorem, conformal field theory, etc.). Here we merely sketch Witten's original "physical" definition of invariants of links in 3-manifolds, a mathematical version of which was described in the previous section (§31).

Consider an oriented link $L = \bigcup_i K_i$ in an orientable 3-manifold M (the manifold is $(2+1)$-dimensional *space-time*, its points are the *states*, and the components K_i of the link are called *Wilson lines*). Choosing a *connection* A (a certain 1-form on M; see, for example, [**KN,Ch.III,§3**]), Witten considers the following *Chern-Simons Lagrangian*:

$$L_{CS}(A) = \frac{k}{4\pi} \int_M \operatorname{Tr}\big(A \wedge dA + \frac{2}{3} A \wedge A \wedge A\big),$$

where k is a fixed positive integer and Tr is a trace (given by the Killing form). It turns out that $L_{CS}(A)$ is well defined (does not depend on the choice of the metric); the motivation behind the choice of the 3-form $A \wedge dA + \frac{2}{3} A \wedge A \wedge A$, is that it is the only gauge-invariant 3-form on M.

Now Witten defines his invariant $W_L(M,k)$ by setting

$$W_L(M,k) = \int_A \left(\exp(ikL_{CS}(A)) \prod_i \operatorname{Tr}\exp \int_{K_i} A\right) DA. \tag{32.3}$$

In the case of the sphere $M = S^3$, we have the following remarkable relationship with the Jones polynomial $V_L(q)$:

$$W_L(M,k) = V_L(e^{2\pi i/k}).$$

Unfortunately, from the mathematician's viewpoint, the definition (32.3) does not make sense, because there is no measure on the set of all connections A on M, so the outer integral "DA" is not mathematically defined. However, this "integral" can be computed (by assymptotic methods or by using *Feynman path integrals*), so it is good enough for the physicists.

We shall not try to explain the physical meaning of definition (32.3). Let us note, however, that Witten's integral is an averaging "over all connections", so that $W_L(M,k)$ is a kind of partition function (standardly denoted by $Z_L(M,k)$). The reader should note a certain similarity with the discrete partition function used to define the Kauffman bracket in §3, although of course this similarity is far from obvious.

32.6. Topological quantum field theories. In the evolution of the formalisms underlying field theories, one of the noteworthy features is the progressive generalization of the group of admissible transformations. Striving for the appropriate generality, transformations evolved from those preserving the metric or the pseudometric (the Lorentz group in special relativity) to symplectomorphisms and contactomorphisms in classical mechanics, to various Poisson maps (that preserve various versions, including quantum mechanical ones, of the Poisson bracket) or maps preserving superstructure, to maps preserving affine connections (this was a crucial step), and, more recently, to various types of gauge transformations. Topological quantum field theories (TQFT's) go even further, allowing arbitrary diffeomorphisms. Therefore the invariants in those theories (which are supposed to have an intrinsic physical meaning) must be diffeomorphism invariants. Since topology (at least differential topology) is precisely about such invariants, it is not surprising that the physicists started looking for things like knot invariants, or various kinds of invariant traces, discrete and continuous partition functions, and the like. And were apparently satisfied with whatever it was they found.

The main ideas underlying TFQT's come from the work of Witten (see [**Wit1**], but the formal mathematical treatment first appeared in M. Atiyah's papers [**Ati2-3**], and a more complete description (but by no means easy reading!) can be found in his book [**Ati1**]. According to this approach, a TQFT in dimension d is a functor Z which to each $(d+1)$-dimensional manifold Y with boundary $\Sigma = \partial Y$ (the "space") assigns a pair $Z(Y) \in Z(\Sigma)$, where $Z(Y)$ is a vector (the "partition function") and $Z(\Sigma)$ is a finite-dimensional complex vector space (the "Hilbert space obtained by quantization"). The axioms, which we do not list here (see [**Ati1**], p.12), involve an involution in the dual space and a multiplication (generated by disjoint union), and imply, in particular, that for a closed $(d+1)$-manifold Y the vector $Z(Y)$ is one-dimensional, i.e., is simply a complex number. This is in fact the situation considered in §29, where $d = 2$ and $Z(Y)$ is the Jones–Witten invariant for 3-manifolds ("containing" the empty link).

For the more general situation in which a (nonempty) link is given in a manifold without boundary (as in §31 above), there is also an appropriate axiomatization, the "relative" version of Atiyah's functor. This functor is now defined on triples (Y, Σ, L), where Y and Σ are as before, while L (in the case $d = 2$) is a link in Y whenever Σ is empty, and is a system of closed curves and arcs in Y with endpoints in $\Sigma = \partial Y$ satifying certain transversality conditions when $\Sigma \neq \varnothing$. These axioms are satisfied by the Jones-Witten invariant (see §31) for $\Sigma = \varnothing$.

Thus we see how the invariants studied in this book are viewed by many physicists not only as helpful, but as fundamental notions in contemporary mathematical physics.

The subject of TQFT's is presently in a state of very intense development, and it is hardly possible to give an up-to-date bibliography at this stage. We would only like to mention two papers: one by P. Vogel and his collaborators [**BHMV**], and another by L. Crane and D. Yetter [**CY**]: the first, because the specific functor considered is based on the Kauffman bracket polynomial (with which we began our study of knot invariants back in §3), and the second, partly because it involves a simplicial construction very similar to the one with which we began our study of 3-manifolds (compare our Fig.8.3 with Figure 1 in [**CY**]), but mainly because the authors succeed in constructing a meaningful theory in the case $d = 3$, i.e., for the four-dimensional space-time that we are supposed to live in.

Appendix

This Appendix contains some basic material usually included in introductory topology courses and freely used in this text. We suggest that the reader refer to the Appendix in the process of reading the main text whenever he or she comes across a notion that may be insufficiently familiar. The exposition in the Appendix is very succinct (no examples, no proofs); it does not pretend to play the role of a textbook. Let us note, however, that the material contained here is presented in detail in the three following introductory textbooks, to which we refer the reader.

1. Czes Kosniowski, *A first course in algebraic topology*, Cambridge University Press, 1980.

(This text contains practically all the topological prerequisites for our book.)

2. J. R. Munkres, *Elementary differential topology*, Ann. of Math. Studies, no. 54, Princeton University Press, 1966.

(This classical book contains, in particular, a proof of the fact that smooth manifolds can be triangulated.)

3. V. Prasolov, *Intuitive topology*, Amer. Math. Soc., 1995.

(This is an informal introduction to the theory of knots and surfaces.)

§1. Topological spaces

A *topological space* is a set X with a distinguished family τ of subsets possessing the following properties:

1) the empty set and the whole set X belong to τ;
2) the intersection of a finite number of elements of τ belongs to τ;
3) the union of any subfamily of elements of τ belongs to τ.

The family τ is said to be the *topology* on X. Any set belonging to τ is called *open*. A *neighborhood* of a point $x \in X$ is any open set containing x. Any set whose complement is open is called *closed*. The minimal closed set (with respect to inclusion) containing a given set $A \subset X$ is called the *closure* of A and is denoted by $\bar{A}$ (or sometimes by $[A]$). The maximal open set contained in a given set $A \subset X$ is called the *interior* of A and is denoted by $\operatorname{Int} A$.

Any subset $Y \subset X$ becomes a topological space in the *induced topology* if its open sets are defined as the intersections with Y of all the open sets of X.

The map of one topological space into another is called *continuous* if the preimage of any open set is open. A map $f : X \to Y$ is said to be a *homeomorphism* if it is bijective and both f and f^{-1} are continuous; the spaces X and Y are then called *homeomorphic* or *topologically equivalent.*

The notion of topological space includes a vast and varied class of objects. But in our book we shall only be concerned with sufficiently nice topological spaces, namely, manifolds. Their definition will be given below.

A topological space X is called *compact* if any *open covering* of X (i.e., any collection of open sets of X whose union is X) has a finite subcovering.

A topological space is said to be a *Hausdorff space* if any two distinct points of the space have nonintersecting neighborhoods.

A topological space is called *connected* if it cannot be presented as the union of two nonintersecting nonempty sets each of which is simultaneously open and closed. When a topological space is presented as the union of nonintersecting nonempty open and closed subsets each of which is connected, these subsets are said to be the *connected components* of the given space.

An *equivalence relation* on the set X is a subset $R \subset X \times X$ possessing the following properties

1) $(x, x) \in R$ for all $x \in X$;
2) if $(x, y) \in R$, then $(y, x) \in R$;
3) if $(x, y) \in R$ and $(y, z) \in R$, then $(x, z) \in R$.

When $R \subset X \times X$ is an equivalence relation, two elements x and y for which $(x, y) \in R$ are called *equivalent*; equivalence of elements is usually denoted by $x \sim y$.

For a given equivalence relation R on X, the set X splits into pairwise nonintersecting sets of equivalent elements; these sets are called *equivalence classes.* The set $X/_\sim$ whose elements are the equivalence classes is said to be the *quotient set* of X by the relation $\sim$. By assigning to each element $x \in X$ the equivalence class that contains it, we obtain a map $p : X \to X/_\sim$, called the *projection.* If X is a topological space, then the quotient set $X/_\sim$ becomes a topological space if we declare open all of its subsets whose inverse image under p is an open set in the space X; this topological space is said to be the *quotient space* of X by the equivalence relation $\sim$.

An often used method for defining an equivalence relation on a set X is derived from the action of a group on this set. We say that the group G *acts* on the set X if to each pair $(g, x) \in G \times X$ an element $gx \in X$ is assigned so that we have $g(hx) = (gh)x$ and $1x = x$ for all $g, h \in G$ and $x \in X$, where 1 is the unit element in G. The *orbit* of the element x is the set of all elements gx, $g \in G$. The action of a group G on the set X defines the following equivalence relation on X: two elements are called *equivalent* iff they belong to the same orbit. In the case when X is a topological space, the corresponding quotient space is called the *quotient space* of X with respect to the action of G or the *orbit space* of the action of G on X.

Suppose X and Y are topological spaces without common elements, A is a subset of X, and $f : X \to Y$ is a continuous map. In the set $X \cup Y$, let us introduce the relation $a \sim f(a)$. The resulting quotient space $(X \cup Y)/\sim$ is denoted by $X \cup_f Y$; the procedure of constructing this space is called *gluing* or *attaching* Y to X along the map f.

If $X \times Y$ is the Cartesian product of the topological spaces X and Y (regarded as sets), then $X \times Y$ becomes a topological space (called the *product* of the spaces X and Y) if we declare open all the products of open sets in X and in Y and all possible unions of these products.

Two maps f_0, $f_1 : X \to Y$ are called *homotopic* if there exists a continuous map $F : X \times [0,1] \to Y$ such that $F(x,0) = f_0(x)$ and $F(x,1) = f_1(x)$ for all $x \in X$ (in other words, if the maps f_0 and f_1 can be joined by a family of continuous maps $f_t : X \to Y$, $t \in [0,1]$, continuously depending on t). This family of maps is called a *homotopy* joining f_0 to f_1.

Two topological spaces are called *homotopy equivalent* if there exist continuous maps $f : X \to Y$ and $g : Y \to X$ such that $g \circ f$ and $f \circ g$ are homotopic to the identical maps of X and Y, respectively.

An *isotopy* of the subspaces A and B of the space X is a homotopy $f_t : A \to X$ such that f_0 is the identical inclusion $A \hookrightarrow X$, $f_1(A) = B$, and f_t is a homeomorphism of A onto $f_t(A)$ for all $t \in [0,1]$.

§2. Manifolds, simplicial spaces, and cell spaces

If (X, τ) is a topological space, a *base* of the space X is a subfamily $\tau' \subset \tau$ such that any element of τ can be represented as the union of elements of τ'. In other words, τ' is a family of open sets such that any open set of X can be represented as the union of sets from this family. In the case when at least one base of X is countable, we say that X is a space with *countable base.*

To define the topology τ, it suffices to indicate a base of the space. For example, in the space $\mathbb{R}^n = \{(x_1, \dots, x_n \,|\, x_i \in \mathbb{R}\}$, the standard topology is given by the base

$$U_{a,\varepsilon} = \{x \in \mathbb{R}^n \big| \, |x - a| < \varepsilon\},$$

where $a \in \mathbb{R}^n$ and $\varepsilon > 0$. We can additionally require that all the coordinates of the point a, as well as the number ε, be rational; in this case we obtain a countable base.

To the set $\mathbb{R}^n$ let us add the element ∞ and introduce in $\mathbb{R}^n \cup \{\infty\}$ the topology whose base is the base of $\mathbb{R}^n$ to which we have added the family of sets

$$U_{\infty,R} = \{x \in \mathbb{R}^n \big| |x| > R\} \cup \{\infty\}.$$

The topological space thus obtained is called the *one-point compactification* of $\mathbb{R}^n$; it can be shown that this space is homeomorphic to the n-dimensional sphere

$$S^n = \{x \in \mathbb{R}^{n+1} \big| |x| = 1\}.$$

Manifolds. A Hausdorff space M^n with countable base is said to be an *n-dimensional topological manifold* if any point $x \in M^n$ has a neighborhood homeomorphic to $\mathbb{R}^n$ or to $\mathbb{R}^n_+$, where

$$\mathbb{R}^n_+ = \{(x_1, \dots, x_n) \,|\, x_i \in \mathbb{R}, x_1 \geqslant 0\}.$$

The set of all points $x \in M^n$ that have no neighborhoods homeomorphic to $\mathbb{R}^n$ is called the *boundary* of the manifold M^n and is denoted by ∂M^n. When $\partial M^n = \varnothing$, we say that M^n is a *manifold without boundary.* It is easy to verify that if the boundary of a manifold M^n is nonempty, then it is an $(n-1)$-dimensional manifold.

An *atlas* on a topological manifold M^n is a covering of M^n by open sets U_α homeomorphic to $\mathbb{R}^n$. A set U_α together with a homeomorphism $\varphi_\alpha : U_\alpha \to \mathbb{R}^n$ is called a *chart.* A topological manifold is said to be a *smooth manifold of class* C^r if M^n possesses an atlas such that all the compositions $\varphi_\beta \circ \varphi_\alpha^{-1}$ are homeomorphic maps of class C^r on the subsets of $\mathbb{R}^n$ where they are defined, namely on $\varphi_\alpha(U_\alpha \cap U_\beta) \subset \mathbb{R}^n$. Here r denotes a natural number or ∞.

Suppose M^m and N^n are smooth manifolds of class C^r, while $\{(U_\alpha, \varphi_\alpha)\}$ and $\{(V_\beta, \psi_\beta)\}$ are their atlases (of class C^r). A map $f : M^m \to N^n$ is called *smooth of class C^r* if all the compositions $\psi_\beta \circ f \circ \varphi_\alpha^{-1}$ are differentiable maps of class C^r.

Suppose $f : M^m \to N^n$ is a smooth map (of class C^r) of smooth manifolds of class C^r, and $m \leqslant n$. The map f is called an *immersion* if the rank of f at each point of M^m (more precisely, the rank of the map $\psi_\beta \circ f \circ \varphi_\alpha^{-1}$) is equal to m. An immersion $f : M^m \to N^n$ which maps M^m onto $f(M^m) \subset N^n$ homeomorphically is called a *smooth embedding*.

Simplicial spaces and cell spaces. To each vector v in the vector space $\mathbb{R}^n$ we shall assign a point in the affine space $\mathbb{R}^n$, namely the endpoint (also denoted by v) of the vector v with initial point at the origin. We shall say that the points $v_0, v_1, \dots, v_m$ are *linearly independent* if they do not lie in any $(m-1)$-dimensional plane, i.e., if the vectors $v_1 - v_0, \dots, v_m - v_0$ are linearly independent.

If $v_0, v_1, \dots, v_m$ are linearly independent points in $\mathbb{R}^n$, then the set of endpoints of the vectors $x = \sum t_i v_i$, where $t_i \geqslant 0$ and $\sum t_i = 1$ for all i, is called an *m-dimensional simplex*. The endpoint of the vector $\left(\sum v_i\right)/(m+1)$ is said to be the *barycenter* of the simplex $[v_0, \dots, v_m]$. A *face* of the simplex $[v_0, \dots, v_m]$ is a simplex of the form $[v_{i_0}, \dots, v_{i_k}]$, where $\{v_{i_0}, \dots, v_{i_k}\}$ is any subset of $\{v_0, \dots, v_m\}$.

A *finite simplicial complex* K in $\mathbb{R}^n$ is a finite set of simplices in $\mathbb{R}^n$ possessing the following properties:

1. any face of a simplex from K belongs to K;
2. the intersection of any two simplices is a face of each (we agree that the empty set is a simplex of dimension -1 and is a face of all simplices).

The union of all the simplices of the simplicial complex K is said to be a (*simplicial*) *polyhedron* and is denoted by $|K|$.

A *cell* in $\mathbb{R}^n$ is any bounded subset of $\mathbb{R}^n$ determined by a system of equalities and inequalities of the form $\sum a_i x_i \geqslant b$. The *dimension* of a cell is the minimal dimension of an affine plane containing the cell. The *face* of a cell is its subset obtained by replacing one or more inequalities $\sum a_i x_i \geqslant b$ determining the cell by the corresponding equalities $\sum a_i x_i = b$.

A *finite cell complex* K in $\mathbb{R}^n$ is a finite number of cells in $\mathbb{R}^n$ possessing the following properties:

1. any face of a cell from K belongs to K;
2. the intersection of any two cells is a face of each (we agree that the empty set is a cell of dimension -1 and is a face of all cells).

The union of all the cells of the cell complex K is said to be a (*cellular*) *polyhedron* and is denoted by $|K|$.

Suppose K is a simplicial complex and M is a smooth manifold of class C^r. A map $f : |K| \to M$ is called *differentiable* or *smooth* of class C^r if the restriction of f to any simplex Δ of K is a smooth map of class C^r. The map f is called *nondegenerate* if the rank of its restriction to any simplex Δ of K is equal to the dimension of this simplex.

A nondegenerate C^r-smooth map $f : K \to M$ which is a diffeomorphism of the simplicial polyhedron $|K|$ onto the manifold M is called a *C^r-triangulation* of the manifold M. It can be proved that any C^r-smooth manifold without boundary possesses a C^r-smooth triangulation, and that any C^r-smooth triangulation of the boundary of a C^r-smooth manifold can be extended to the entire manifold (see [**Mun**]).

Any nondegenerate C^r-smooth map $f : K \to M$ which is a homeomorphism of the simplicial polyhedron $|K|$ onto its image in the manifold M is called an *embedding*. The image of $|K|$ in M is then called an *embedded subcomplex.*

Orientation. An atlas $\{(U_\alpha, \varphi_\alpha)\}$ of a smooth manifold is called *orienting* if the Jacobians of all the maps $\varphi_\beta \circ \varphi_\alpha^{-1}$, where $U_\beta \cap U_\alpha \neq \varnothing$, are positive. A manifold possessing an orienting atlas is called *orientable.*

If an orienting atlas of the manifold M is chosen, we say that an *orientation* is given on M. For two oriented atlases $\{(U_\alpha, \varphi_\alpha)\}$ and $\{(V_\beta, \psi_\beta)\}$ the Jacobians of all the maps $\varphi_\beta \circ \psi_\alpha^{-1}$ are either all positive or all negative. In the first case we say that these atlases *determine the same orientation*; in the second, that they *determine opposite orientations.*

For a manifold M with a fixed triangulation $f : K \to M$, there is another (equivalent) definition of orientability. We shall say that a simplex is *oriented* if its vertices are ordered (numbered); here we assume that two numberings give the same orientation if they can be obtained from each other by an even permutation, while two numberings that differ by an odd permutation determine opposite orientations. For the faces of dimension $n-1$ of a simplex of dimension n, the *induced orientation* is defined by leaving the vertices of the face obtained by omitting the first vertex of the n-simplex in the same order. But for the first vertex we may take any vertex, so this definition implies that the sign of the orientation of the face $[x_0, \dots, \hat{x}_i, \dots, x_n]$ (this notation means that the ith vertex is omitted) differs from the sign of the induced orientation by $(-1)^{i-1}$.

We shall say that an *orientation* is given on a triangulated n-dimensional manifold if all its n-dimensional simplices are oriented so that for any $(n-1)$-dimensional face the induced orientations coming from the two n-dimensional simplices containing this face are opposite. A triangulated n-dimensional manifold is called *orientable* if an orientation can be chosen on it.

Surfaces. By a *surface* M we mean a compact two-dimensional manifold (possibly with boundary). If the boundary ∂M is empty, we say that the surface M is *closed.* The *connected sum* of two surfaces M and N is the surface $M\#N$ obtained by removing small open disks D_1 and D_2 from M and N and attaching $M - D_1$ to $N - D_2$ along a homeomorphism $h : \partial D_1 \to \partial D_2$. When N is the torus $S^1 \times S^1$, the subset $N - D_2 \subset M\#N$ is called a *handle.*

CLASSIFICATION THEOREM FOR SURFACES. (a) *Closed orientable surfaces M are classified by their number of handles $g \in \{0, 1, \dots\}$, i.e.,*

$$M \in \{S^2, T^2 = S^1 \times S^1, T^2\#T^2, \dots, \mathop{\#}_{i=1}^{g} T_i^2, \dots\}.$$

(b) *Orientable surfaces M with boundary are classified by their number of handles $g \in \{0, 1, \dots\}$ and their number of holes (i.e., boundary components) $h \in \{1, 2, \dots\}$.*

(c) *Nonorientable closed surfaces N are connected sums of $g \in \{1, 2, \dots\}$ copies of the projective plane $\mathbb{R}P^2$, i.e.,*

$$N \in \{\mathbb{R}P^2, \mathbb{R}P^2\#\mathbb{R}P^2, \dots, \mathop{\#}_{i=1}^{g} \mathbb{R}P_i^2, \dots\}.$$

(d) *Nonorientable surfaces M with boundary are classified by the number $g \in \{1, 2, \dots\}$ of connected summands $\mathbb{R}P^2$ and the number of holes (i.e., boundary components) $h \in \{1, 2, \dots\}$.*

Tangent vectors. To any point x of a smooth manifold M^n without boundary one can assign the linear space $T_x M^n$ of all tangent vectors to M^n at the point x. A tangent vector can be understood as the velocity vector of a mobile point on the manifold. More precisely, let $\gamma : \mathbb{R} \to M^n$ be a smooth map such that $\gamma(0) = x$, i.e., let γ be a curve passing through x. To each smooth function $f : M^n \to \mathbb{R}$ we assign the number

$$D_\gamma(f) = \frac{d}{dt} f(\gamma(t))\bigg|_{t=0}.$$

We shall say that two curves γ_1 and γ_2 are equivalent if $D_{\gamma_1} = D_{\gamma_2}$. A class ξ of equivalent curves is called a *tangent vector* to the manifold M at the point x. The functional D_ξ corresponding to the tangent vector ξ is called the *derivative in the direction of the vector* ξ. The tangent vector corresponding to the functional $D_\xi + D_\eta$ is called the *sum* of the vectors ξ and η.

§3. Riemannian metric on a manifold

Let M be a C^∞-smooth manifold. Suppose that each tangent space $T_x(M)$ is supplied with a scalar product (v, w) smoothly depending on x, i.e., the assignment $v \mapsto (v, v)$ defines a C^∞-smooth map of the manifold TM of all tangent vectors to M into $\mathbb{R}$. In this situation we say that a *Riemanian metric* is given on M.

Suppose M is a manifold supplied with a Riemannian metric and $\gamma : [a, b] \to M$ is a smooth curve on M. The *length* of the curve γ is defined as the number

$$l = \int_a^b |\gamma'(t)|\, dt,$$

where $|\gamma'(t)|$ is the length (defined via the given scalar product) of the velocity vector of the curve γ at the point $\gamma(t)$. The *distance* between the points $x, y \in M$ is the greatest lower bound of the lengths of all smooth curves joining x and y.

Now an ε-neighborhood of a knot K in the manifold M^3 (i.e., of the image of a circle smoothly embedded in M^3) can be defined as the set of all points whose distance from K is less than ε.

§4. Coverings and the fundamental group

Suppose X is a topological space with a distinguished point $x_0 \in X$ (called the *base point*). A *loop* in such a space is a continuous map $f : I = [0, 1] \to X$ such that $f(0) = f(1) = x_0$. The *fundamental group* $\pi_1(X, x_0)$ is defined as the set of homotopy classes of loops (here it is assumed that homotopies are in the class of loops with base point x_0, i.e., $f_t(0) = f_t(1) = x_0$ for all $t \in [0, 1]$). The *product* of two loops f and g is the loop $h = fg$ defined by the formula

$$h(t) = \begin{cases} f(2t), & \text{for } 0 \leqslant t \leqslant 1/2, \\ g(2t - 1), & \text{for } 1/2 \leqslant t \leqslant 1. \end{cases}$$

This definition supplies the set $\pi_1(X, x_0)$ with a group structure. The unit element in this group is the constant loop $f : I \to x_0 \in X$, while the inverse element to the class of loops containing the loop $f(t)$ is the class of the loop $g(t) = f(1 - t)$.

If the space X is *arcwise connected* (i.e., any two points $x_1, x_2 \in X$ can be joined by a path $\alpha : I \to X$, $\alpha(0) = x_1$, $\alpha(1) = x_2$), then for any two points x_1 and x_2 the groups $\pi_1(X, x_1)$ and $\pi_1(X, x_2)$ are isomorphic, an isomorphism being defined by the formula $\gamma \mapsto \alpha\gamma\alpha^{-1}$, where α is any path joining the points x_1 and x_2. Therefore for arcwise connected spaces X the base point is usually omitted in the notation for the fundamental group $\pi_1(X)$.

If $\phi : X \to Y$ is a continuous map of arcwise connected spaces taking base point to base point, then it induces a homomorphism of the corresponding fundamental groups, $\phi_* : \pi_1(X) \to \pi_1(Y)$, specified by assigning to (the homotopy class of) the loop $f : [0,1] \to X$ (the homotopy class of) the loop $\phi f : [0,1] \to Y$.

It is often convenient to present the fundamental group by means of generators and relations. To give the formal definition of a group presented by generators and relations, we shall need the notion of *free group F_n in n generators* $a_1, \dots, a_n$; the group F_n consists of *words* $w = a_{i_1}^{\varepsilon_1} \dots a_{i_k}^{\varepsilon_k}$, where $\varepsilon = \pm 1$ and the word w does not contain occurences of the letters $a_i a_i^{-1}$ or $a_i^{-1} a_i$ ($i \in \{1, \dots, n\}$). The *product* of the words w_1 and w_2 is the word $w_1 w_2$ obtained by writing the word w_2 after w_1 and deleting all occurences of letters $a_i a_i^{-1}$ or $a_i^{-1} a_i$. The empty word is the unit element in the group F_n.

Suppose $w_1, \dots, w_k$ are words in the letters $a_1^{\pm 1}, \dots, a_n^{\pm 1}$. The *group with generators $a_1, \dots, a_n$ and relations $w_1, \dots, w_k$* is defined as the quotient group $G = F_n/H$, where F_n is the free group in the generators $a_1, \dots, a_n$ and H is the minimal normal subgroup of F_n containing the elements $w_1, \dots, w_k$. The group G defined in this way is usually denoted by

$$\langle a_1, \dots, a_n | \, w_1, \dots, w_k \rangle.$$

A continuous map $p : T \to X$ is said to be a *covering* if each point $x \in X$ possesses a neighborhood U whose inverse image $p^{-1}(U)$ is supplied with a homeomorphism h_U onto $U \times \Gamma$, where Γ is a discrete set and we have the commutativity relation $\pi \circ h_U = p$, the map π being the projection of the product $U \times \Gamma$ on the first factor U.

Suppose that T and X are arcwise connected. Then for any path $\gamma : I \to X$ and any point $\widetilde{x} \in p^{-1}(\gamma(0)) \subset T$ there exists a unique path $\widetilde{\gamma}$ originating at $\widetilde{x}$ and covering γ, i.e., $p(\widetilde{\gamma}(s)) = \gamma(s)$ for all $s \in I$. This statement is called the *path lifting property*. Further, if two paths γ_1 and γ_2 with common origin $x_0 = \gamma_1(0) = \gamma_2(0)$ are homotopic by a homotopy that is constant on the extremities of these two paths, two covering paths with common origin have the same terminal point, i.e., $\widetilde{\gamma}_1(1) = \widetilde{\gamma}_2(1)$.

When a finite group G acts without fixed points on a manifold M^n, it can be proved that the natural projection $p : M^n \to M^n/G$ on the orbit space is a covering, and the orbit space M^n/G is a manifold.

Solutions

Chapter I

1.1. See Fig.S1.1.

FIGURE S1.1

1.2–1.4. You can try transforming the knots (links) drawn in pencil on a piece of paper (or in chalk on a blackboard) with the help of an eraser, but we suggest using a piece of string or a shoelace (several for links) placed on your table to model the knot (link) diagrams and simply move it around to obtain the desired configuration.

1.5. A rotation of the oriented trefoil about the axis l (see Fig.S1.2) by $180°$ results in a change of orientation.

FIGURE S1.2

2.1. Six pairs of successive loops (of the 16 possible ones) cancel, namely those shown in Fig.S2.1. The first four (a–d) cancel via the Whitney trick, while the other two (e, f) cancel by the Reidemeister move Ω_2.

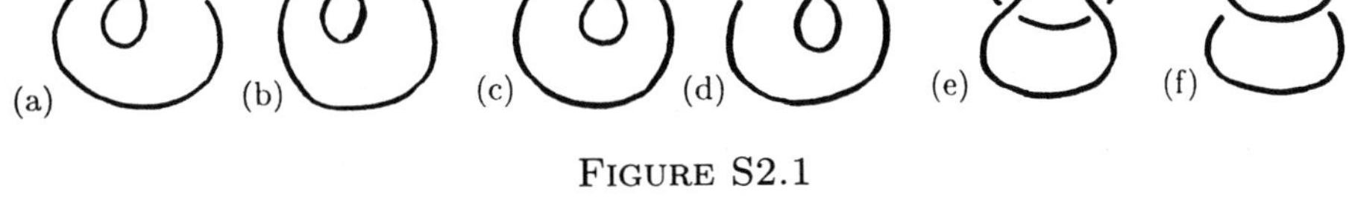

FIGURE S2.1

2.2. On the sphere, eight pairs cancel; they are the six that cancel on the plane, plus the two shown in Fig.S2.2. These two cancel by the sphere trick followed by the Reidemeister move Ω_2.

FIGURE S2.2

2.3. Note first of all that the boundary of the twisted strip is the trefoil knot. After the first cut, one gets a two-sided strip bounded by two trefoils (which are linked together). After the next cut, one obtains two two-sided trefoil-like strips linked together. Cutting these two, one gets four linked two-sided trefoil strips. At the nth step, there will be 2^{n-1} two-sided linked trefoil strips.

Remark. At this stage, we have no machinery to prove negative results, for example, to prove that some pairs of loops don't cancel (Problems 2.1 and 2.2) or that two ribbons don't unlink. This requires invariants. In Problems 2.1 and 2.2 some readers may succeed in inventing the appropriate invariants. Concerning linking (Problem 2.3), see the definition of linking number in §15.

Chapter II

3.1. (a) $a^8 - a^4 + 1 - a^{-4} + a^{-8}$.
(b) $-a^{11} + 2a^7 - a^3 + 2a^{-1} - a^{-5} + a^{-9}$.

3.2. The answers should of course be the same; in particular, for the knot (a),

$$q^{-2} - q^{-1} + 1 - q + q^2,$$

and, for the knot (c),

$$(-q^4 + q^3 + q)(-q^{-4} + q^{-3} + q^{-1}) = -q^3 + q^2 - q + 3 - q^{-1} + q^{-2} - q^{-3}.$$

3.3. The union of the sets of crossing points of the diagrams L_1 and L_2 coincides with the set of crossing points of the diagram $L_1\#L_2$ as well as with that of the diagram $L_1 \sqcup L_2$. Therefore to each pair of states s_1 and s_2 of the diagrams L_1 and L_2 we can assign the state $s_\#$ of $L_1\#L_2$ and the state $s_\sqcup$ of $L_1 \sqcup L_2$. For brevity, set $\alpha_1 = \alpha(s_1)$, $\alpha_\# = \alpha(s_\#)$, etc. It is clear that $\alpha_\# = \alpha_\sqcup = \alpha_1 + \alpha_2$ and $\beta_\# = \beta_\sqcup = \beta_1 + \beta_2$. Further, $\gamma_\# = \gamma_1 + \gamma_2 - 1$ and $\gamma_\sqcup = \gamma_1 + \gamma_2$. Therefore,

$$\langle L_1\#L_2\rangle = \sum_{s_\#} a^{\alpha_\#} b^{\beta_\#} c^{\gamma_\# - 1} = \sum_{s_1,s_2} a^{\alpha_1+\alpha_2} b^{\beta_1+\beta_2} c^{\gamma_1+\gamma_2-2}$$
$$= \Big(\sum_{s_1} a^{\alpha_1} b^{\beta_1} c^{\gamma_1 - 1}\Big)\Big(\sum_{s_2} a^{\alpha_2} b^{\beta_2} c^{\gamma_2 - 1}\Big) = \langle L_1\rangle\langle L_2\rangle.$$

Similarly,

$$\langle L_1 \sqcup L_2\rangle = c\sum_{s_1,s_2} a^{\alpha_1+\alpha_2} b^{\beta_1+\beta_2} c^{\gamma_1+\gamma_2-2} = c\langle L_1\rangle\langle L_2\rangle.$$

It is also clear that the writhe number w satisfies the relation

$$w(L_1\#L_2) = w(L_1 \sqcup L_2) = w(L_1) + w(L_2),$$

and therefore

$$X(L_1\#L_2) = X(L_1)X(L_2) \quad \text{and} \quad X(L_1 \sqcup L_2) = cX(L_1)X(L_2).$$

3.4. The links shown in Fig.3.16 can be represented in the form $L_1\#L_2$ and $L_2\#L_1$; therefore their Jones polynomials are equal to $V(L_1)V(L_2) = V(L_2)V(L_1)$.

3.5. Let us prove a more general statement, namely, if the part L_1 of a link or knot L is reflected in the line l (Fig.S3.1), then the link L' thus obtained has the same Jones polynomial as the given link L. Here we are assuming that the diagram L_1 is contained in a rectangle whose sides intersect L in four points positioned exactly as shown in Fig.S3.1 (the reader should have no trouble in finding this rectangle in Fig.3.17).

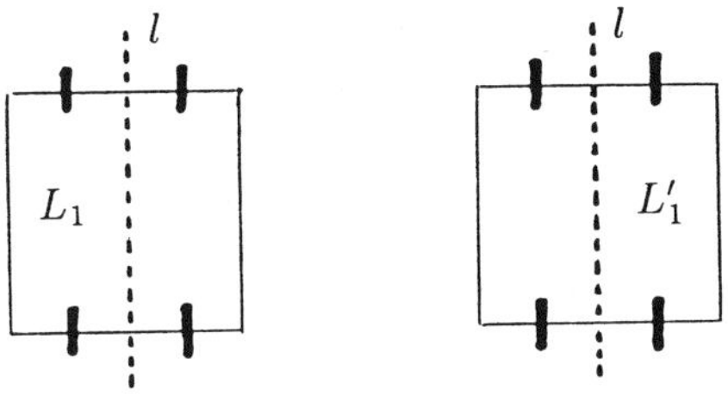

FIGURE S3.1

Under a symmetry with respect to a line, the numbers $\varepsilon_i = \pm 1$ whose sum is the winding number w do not change, therefore it suffices to prove that $\langle L \rangle = \langle L' \rangle$. To each crossing point of the diagram L corresponds a crossing point of L'; hence to every state of L we can assign the same state of L'. Then by eliminating the crossing points of L_1 and L'_1, we obtain diagrams symmetric with respect to the line l (Fig.S3.2).

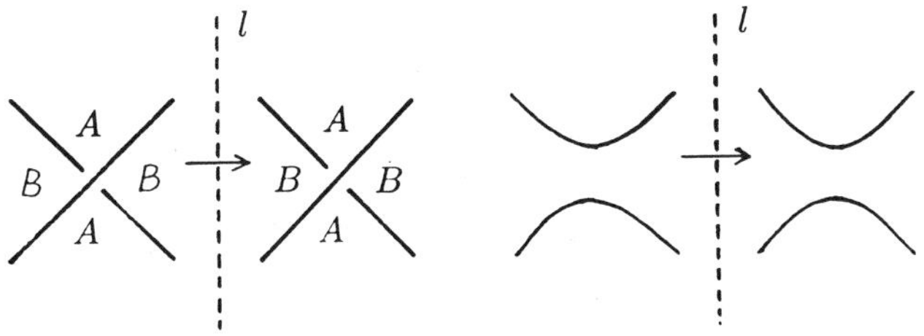

FIGURE S3.2

It remains to notice that under the symmetry with respect to l of the part of the link contained in the rectangle the number of components of the link does not change (Fig.S3.3)

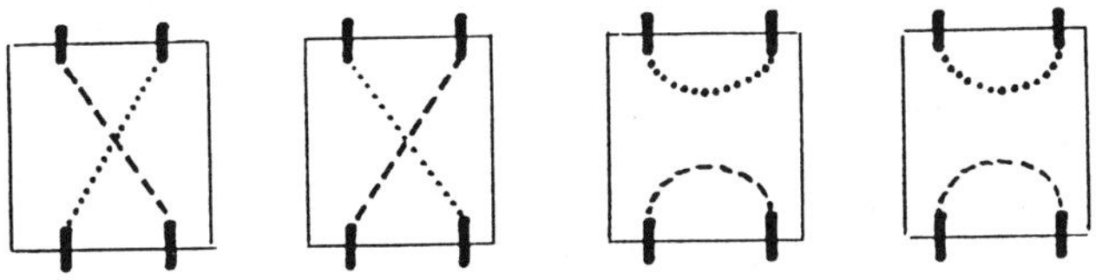

FIGURE S3.3

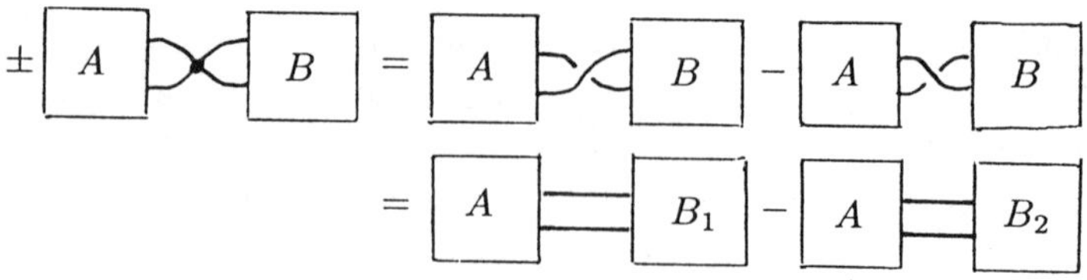

FIGURE S4.1

4.1. Relation (4.1) implies the equalities shown in Fig.S4.1.

The diagrams B_1 and B_2 are obtained from B by rotations by $\pm 180°$ in space, and so they coincide.

4.2. (a) As pictured in Fig.4.4 of the main text, all the chord diagrams of order 3 are clearly mirror symmetric with respect to the vertical diameter.

(b) See Fig.S4.2.

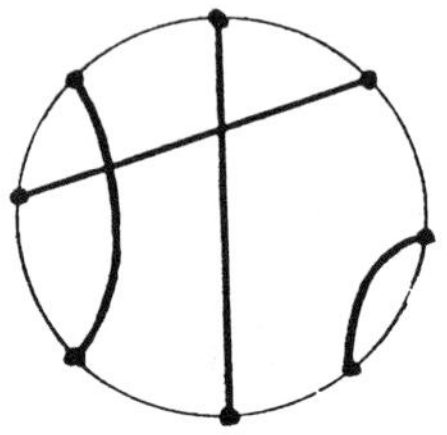

FIGURE S4.2

4.3. See Fig.S4.3.

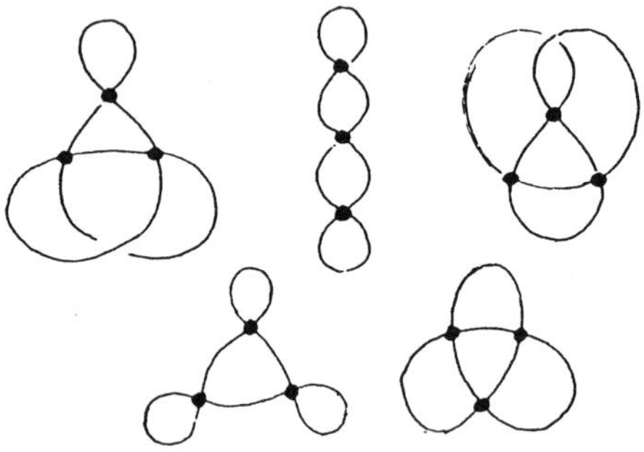

FIGURE S4.3

4.4. (a) To prove the required relation, we must write out the five four-term relations whose first terms are shown in Fig.S4.4; in this figure, the constant chords are shown by dotted lines.

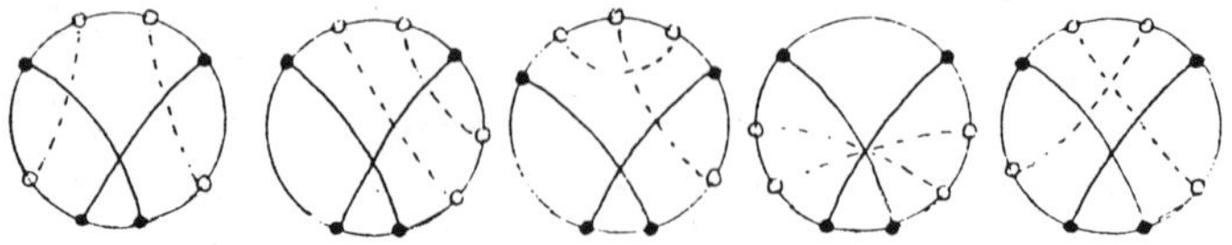

FIGURE S4.4

(b) It follows from part (a) that all the nonzero diagrams of order 4 can be expressed in terms of the three indicated diagrams. Reviewing all possible applications of the four-term relation shows that in the formulation of part (a) of the problem all the nontrivial relations between nonzero diagrams have been written out. Therefore $\dim \Delta_4 = 3$.

Chapter III

5.1. (a) Suppose G is the group with generators a, b and the relation $a^3 = b^2$. It is easy to verify that the assignments $a \mapsto xy$, $b \mapsto xyx$ and $x \mapsto a^{-1}b$, $y \mapsto ab^{-1}a$ can be extended to homomorphisms $G \to B_3$ and $B_3 \to G$ that are inverse ro each other. Indeed, the relation $a^3 = b^2$ becomes $xyxyxy = xyxxyx$, i.e., $yxy = xyx$, while the relation $xyx = yxy$ becomes $a^{-1}bab^{-1}aa^{-1}b = ab^{-1}aa^{-1}bab^{-1}a$, i.e., $a^3 = b^2$.

(b) The relations $x(xy)^3 = (xy)^3x$ and $y(xy)^3 = (xy)^3y$ are equivalent to the relation $(xyx)(yxy) = (yxy)(xyx)$.

5.2. Let us identify the plane $\mathbb{R}^2$ with the set of complex numbers $\mathbb{C}$. To the n-tuple of points $z_1, \dots, z_n \in \mathbb{C}$ we can assign the point $(z_1, \dots, z_n) \in \mathbb{C}^n$. The points $z_1, \dots, z_n$ are pairwise distinct iff the coordinates of the point $(z_1, \dots, z_n)$ are all different. In order to pass to unordered tuples of points, we simply identify all points of $\mathbb{C}^n$ obtained from each other by permutations of the coordinates.

5.3. Use the fact that $P(n, \mathbb{C})$ is homeomorphic to $C(n, \mathbb{R}^2)$.

5.4. (a) See the proof of Theorem 8.8.

(b) This immediately follows from item (a) and Problem 5.3.

6.1. (a) This is obvious, because the closures of the braids are isotopic to the unknot or to the Hopf link.

(b) The first two closures are the left and right trefoils, which are not isotopic (e.g. because they have different Jones polynomials). The closure of the braid $b_1b_2b_1^{-1}$ is the trivial two-component link, whereas one of the components of the other closure is a trefoil.

6.2. (a) The trefoil.

(b) The Borromeo rings.

(c) The figure eight knot.

6.3. *Hint.* Such a surface (which is known as a *Seifert surface*) may be obtained by joining disks bounded by the Siefert circles by twisted rectangles constructed at each crossing point, two parallel sides of the rectangles being attached to the two disks, while the two other sides are the two branches of the knot at the crossing points. For a more detailed description, see [**BZ**], pp. 17–18, or [**Pra**], pp. 17–22.

6.4. If the hints in the formulation of the problem are not sufficient for finding a proof that the Vogel algorithm terminates, see the original article [**Vog**].

Chapter IV

8.1. First choose an orientation on the common boundary N^2 of the manifolds M_1^3 and M_2^3 in the manifold M^3. To do that, let us specify a 2-frame e_1, e_2 at each point of N^2. To each one of these frames add a vector e_3 directed inside M_1^3. The orientation defined by these 3-frames can be extended from the boundary to the interior of the manifolds M_1^3 and M_2^3 .

8.2. Perform an inversion of the space $\mathbb{R}^3$ with respect to a point located outside the deleted sphere with handles.

8.3. $S^2 \times S^1$.

8.4. *Hint.* See the description of the Heegaard splitting of $L(p, q)$ and use the fact that $\mathbb{R}P^3 = L(2, 1)$.

9.1. See Fig.S9.1.

FIGURE S9.1

11.1. $\pi_1(L(p, q)) \cong \mathbb{Z}_p$.

Chapter V

12.1. Suppose that the handlebody M_1^3 has a center of symmetry, and let $s : \partial M_1^3 \to \partial M_1^3$ be the symmetry with respect to this point, restricted to the boundary. Clearly, s is an orientation-reversing homeomorphism of ∂M_1^3. Consider the homeomorphisms $h = f$ and $h' = s \circ f$. If we attach the boundary of one copy of the handlebody M_1^3 to the boundary of another copy along either of these homeomorphisms, we obtain the manifold M^3. One of the two homeomorphisms h and h' preserves the orientation of ∂M_1^3, while the other one reverses it.

12.2. The curves β and β' perform a different number of rotations in the meridional direction.

12.3. Let $n_{i,t}$ be an isotopy joining the homeomorphism n_i to the identity, i.e., $n_{i,0} = n_i$ and $n_{i,1} = \mathrm{id}_N$. Then the isotopy $n_{1,t} \circ h_1 \circ \dots n_{k,t} \circ h_k$ joins the homeomorphism $n_1 \circ h_1 \circ \dots n_k \circ h_k$ to the homeomorphism $h_1 \circ \dots \circ h_k$.

12.4. (a) The link L_1 contains one knotted component, while all the components of L_2 are unknotted.
(b) Use Lemma 12.5.

12.5. (a) We may assume that the solid torus T is embedded in $\mathbb{R}^3$. Then in $\mathbb{R}^3$ the knot K_1 is trivial, while K_2 is the trefoil. Therefore it suffices to prove that any isotopy $f_t : T \to T$ can be extended to an isotopy of $\mathbb{R}^3$. Consider an ε-neighborhood U of T in $\mathbb{R}^3$. The closure of the set $U - T$ is homeomorphic to $\partial T \times [0, 1]$. We can assume that the points of ∂T are identified with the points of $\partial T \times \{1\}$. Let g_t be the restriction of f_t to ∂T. Consider the homeomorphism $h_t : \partial T \times [0, 1] \to \partial T \times [0, 1]$ given by the formula $h_t(x, s) = (g_{st}(x), s)$. Since we have

$$h_t(x, 0) = (g_0(x), 0) = (x, 0) \quad \text{and} \quad h_t(x, 1) = (g_t(x), 1),$$

it follows that the homeomorphism h_t can be extended to a homeomorphism of $\mathbb{R}^3$ that coincides with f_t on T and is the identity outside of U. Thus we will have continued the isotopy $f_t : T \to T$ to the whole space $\mathbb{R}^3$.

(b) Cut the solid torus T along a meridional disk and give it a twist by 2π (Fig.S12.1). Pasting the solid torus back, we obtain a homeomorphism $f : T \to T$ taking K_1 to K_2. Then the restriction of f to $T - K_1$ is a homeomorphism taking $T - K_1$ to $T - K_2$.

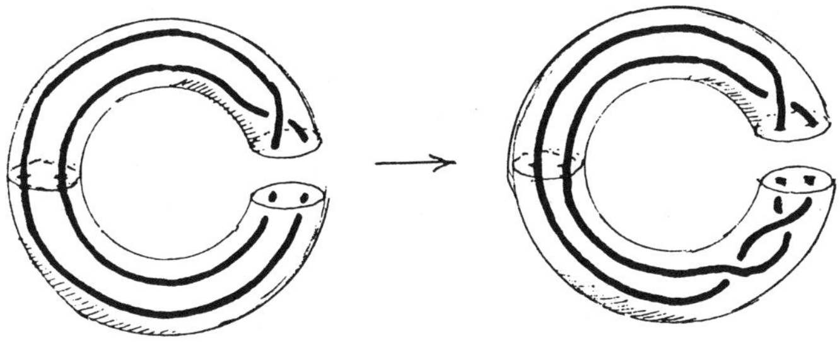

FIGURE S12.1

12.6. The Cartesian product of the Möbius band by $[0, 1]$ may be represented in the form of a cylinder from which a smaller concentric cylinder has been deleted and the points of the lateral surface of the smaller cylinder are identified as shown in Fig.S12.2. After this identification, all the points of the lateral surface of the smaller cylinder, except those of the base circle, become interior points. Hence the boundary is homeomorphic to a sphere with two attached Möbius bands, i.e., a Klein bottle.

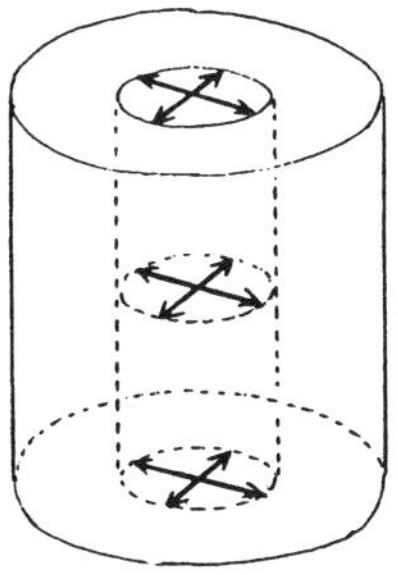

FIGURE S12.2

12.7. Let $\sigma : M^n \to M^n$ be an involution without fixed points. Consider the manifold $M^n \times [0, 1]$ and identify the points $(x, 0)$ and $(\sigma(x), 0)$ for all $x \in M^n$. All the points of the boundary $M^n \times \{0\}$ become interior points, so that the resulting manifold has the boundary $M^n \times \{1\}$.

12.8. (a) $\sigma(x_0, x_1, \dots x_{2n}, x_{2n+1}) = (-x_1, x_0, \dots, -x_{2n+1}, x_{2n})$.

(b) $\sigma(z_0, z_1, \dots z_{2n}, z_{2n+1}) = (-\overline{z}_1, \overline{z}_0, \dots, -\overline{z}_{2n+1}, \overline{z}_{2n})$.

(c) The closure of any one-parameter subgroup is the torus; the torus contains the circle; the circle contains an element a of order 2. Put $\sigma(g) = ag$. Then $\sigma(g) \neq g$ and $\sigma^2(g) = a^2 g = g$.

12.9. By using barycentric subdivision of the given triangulation K, construct a cellular decomposition K' such that $a_i(K') = a_{n-i}(K)$. Then compare the alternating sums

$$\sum_i (-1)^i a_i(K) = \sum_i (-1)^i a_i(K').$$

12.10. By attaching two copies of the manifold W^{n+1} along the identity homeomorphism of their boundary, we get its *double*, the manifold N^{n+1}. Check that $\chi(M^n) \equiv \chi(N^{n+1}) \pmod 2$. If n is odd, then $\chi(M^n) = 0$; if n is even, then $\chi(N^{n+1}) = 0$.

12.11. Check that $\chi(\mathbb{R}P^{2n}) = 1$ and $\chi(\mathbb{C}P^{2n}) = 2n + 1$.

Chapter VI

15.1. When the orientation of the curve J is reversed, the crossing with $\varepsilon = 1$ is transformed as shown in Fig.S15.1. For crossings with $\varepsilon = -1$, the transformation is similar.

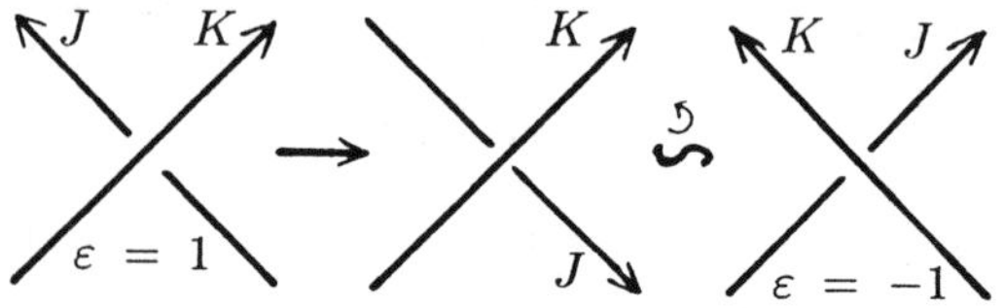

FIGURE S15.1

15.2. The link $L = \{J, K\}$ can be positioned between two parallel planes Π_1 and Π_2 (Fig.S15.2).

FIGURE S15.2

Let us compute $\mathrm{lk}(J, K)$ for the diagram of the link L in the plane Π_1, and compute $\mathrm{lk}(K, J)$ for the diagram of L in Π_2. Fig.S15.3 displays the relevant crossings in those diagrams. Clearly, the results of the two computations will be the same.

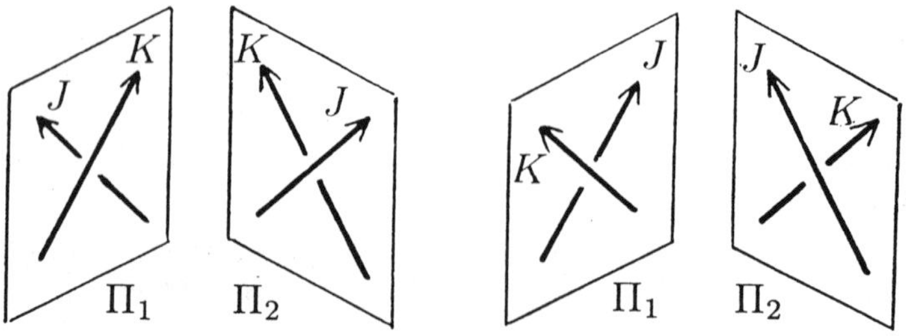

FIGURE S15.3

15.3. Let us arbitrarily orient the given curves. It is easy to check that the linking numbers will be, respectively, (a): ± 1; (b): ± 2; (c): ± 3; (d): ± 1. (The ambivalence of the sign is due to the ambivalence in the choice of the orientation.) In all four cases the linking number is nonzero.

15.4. Consider an arbitrary curve K codirected with J. The part of the picture where J and K are parallel (Fig.S15.4,a) may be transformed as shown in Fig.S15.4,b or as shown in Fig.S15.4,c. Hence $\mathrm{lk}(J,K)$ may be replaced by $\mathrm{lk}(J,K)+1$ or $\mathrm{lk}(J,K)-1$.

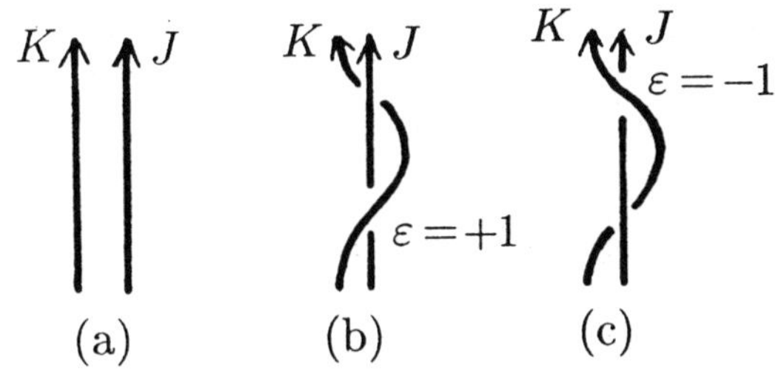

FIGURE S15.4

15.5. The required parallels are shown in Fig.S15.5.

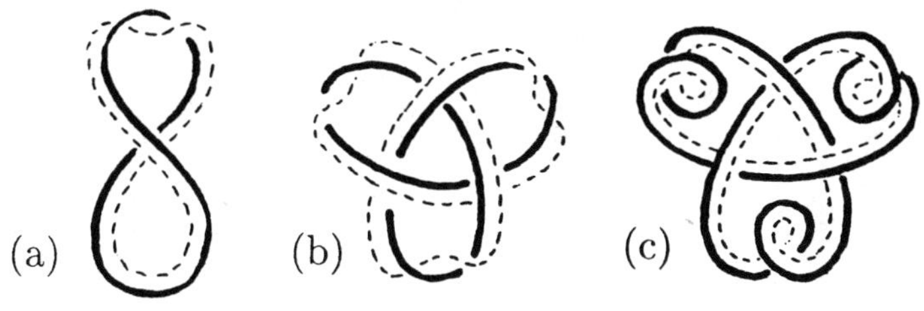

FIGURE S15.5

16.1. Let us represent the first link as shown in Fig.S16.1.

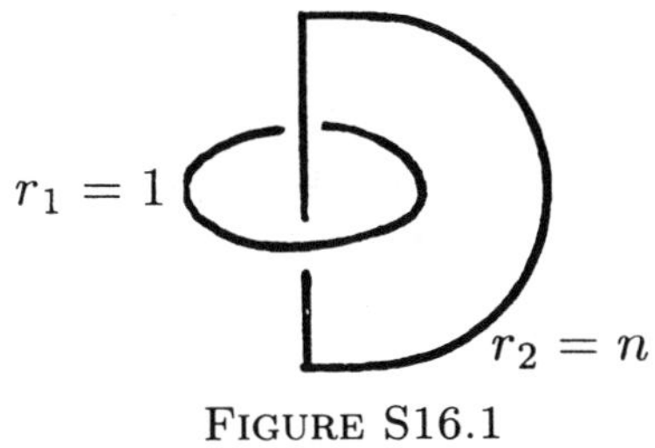

FIGURE S16.1

Perform a twist of -1 revolution. The result will be new framings

$$r_1' = 1/(-1+1) = \infty \quad \text{and} \quad r_2' = n-1.$$

But surgery along any curve with framing ∞ is redundant.

16.2. (a) For arbitrary $r_1 \in \mathbb{Q} \cup \infty$ and $r_2 \in \mathbb{Z}$, we can carry out the transformations shown in Fig.S16.2. Indeed, the linking number of the knot and the circle is equal to zero; therefore $r_2' = r_2 + \mathrm{lk}^2(K, J) = r_2$. Putting $r_1 = \infty$, we obtain the required transformation.

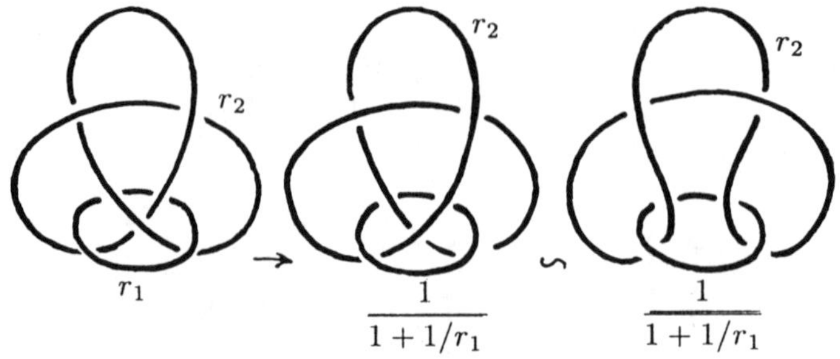

FIGURE S16.2

(b) For arbitrary $r_1 \in \mathbb{Q} \cup \infty$ and $r_2 \in \mathbb{Z}$, we can carry out the transformations shown in Fig.S16.3. Indeed, the linking number of the knot and the circle is equal to ± 2; hence $r_2' = r_2 - \mathrm{lk}^2(K, J) = r_2 - 4$. Putting $r_1 = \infty$, we get the required transformation.

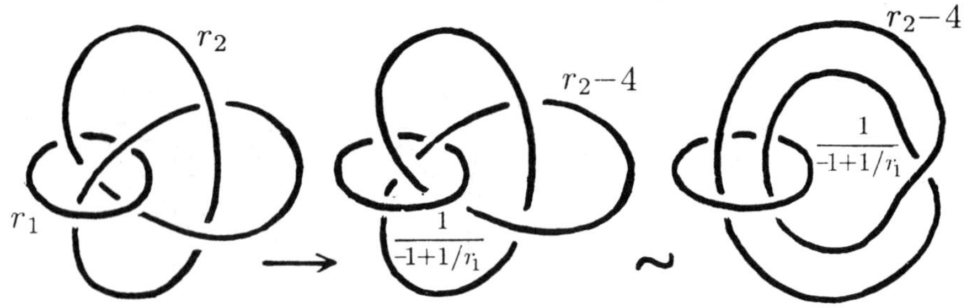

FIGURE S16.3

18.1. The quotient group $I^*/I^{*'}$ is obtained from the group I^* by adding the relation $xyx^{-1}y^{-1} = 1$, i.e., $xy = yx$. Therefore

$$\begin{aligned} I^*/I^{*'} &= \{x, y | xyx = yxy, yx^2y = x^3, xy = yx\} \\ &= \{x, y | x = y, y^2 = x\} = \{x | x^2 = x\} = \{x | x = 1\}. \end{aligned}$$

18.2. The generators a and b may be expressed in terms of x and y, and conversely x and y may be expressed in terms of a and b. Therefore it suffices to prove that the relations $xyx = yxy$ and $yx^2y = x^3$ are equivalent to the relations $a^5 = b^3 = (ba)^2$.

Suppose that $xyx = yxy$ and $yx^2y = x^3$. Then, by the definition of a and b,

$$(ba)^2 = xyx(xyx) = xyxyxy = b^3 \quad \text{and} \quad (ba)^2 = x(yx^2y)x = x^5 = a^5.$$

Conversely, suppose that $a^5 = b^3 = (ba)^2$. Then $a^4 = bab$ and $b^2 = aba$, and therefore

$$xyx = ba = a^{-1}b^2 = yxy \quad \text{and} \quad yx^2y = a^{-1}bab = a^{-1}a^4 = a^3 = x^3.$$

18.3. The curve J for the right trefoil with framing -1 can be obtained from a similar curve J for the right trefoil with framing $+1$ (see Fig.18.5) by adding two twists x^{-1}. Therefore the addition of a spanning disk to J produces the relation $yxzx^{-4} = 1$, i.e., $yx(xyx^{-1})x^{-4} = 1$. Thus, $\pi_1(M^3_-) = \{x, y | xyx = yxy, yx^2y = x^5\}$. Substituting $a = x$, $b = xy$, we get

$$\pi_1(M^3_-) = \{a, b | a^7 = b^3 = (ba)^2\}.$$

18.4. Suppose $G = \pi_1(M^3_-)$. Then

$$G/G' = \{x, y | x = y, y^2 = x^3\} = \{x | x^2 = x^3\} = \{x | x = 1\}.$$

19.1. See Fig.S19.1.

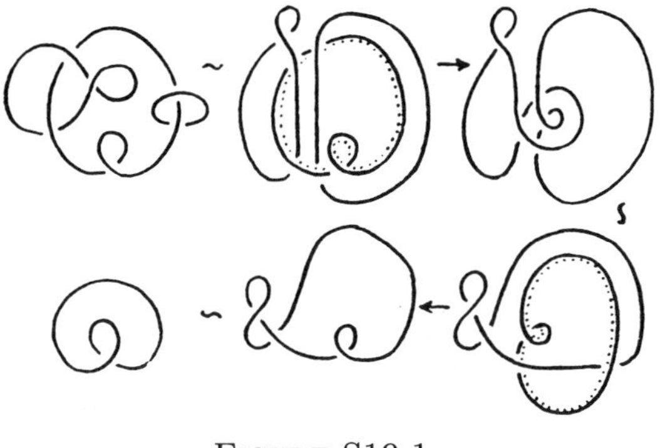

FIGURE S19.1

Chapter VII

20.1. Assume the converse. Then the covering manifold contains a closed disorienting path, i.e., a closed curve γ that cannot be covered by coordinate neighborhoods with compatible orientations. By general position, we can assume that γ does not pass through any preimage of a branch point. By compactness, we can take a disorienting covering of γ each element of which projects homeomorphically to M^2. Then $p(\gamma)$ is a disorienting curve in M^2, contrary to the condition of the problem.

21.2. (a) According to Theorem 20.2, $\chi(M^2_g) = n\chi(M^2_h)$, where n is the number of sheets of the covering p. Therefore $2 - 2g = n(2 - 2h)$.

(b) Fig.S21.1 shows the covering of the sphere with $h = 3$ handles by the sphere with $g = 7$ handles. (In this case $g - 1 = 3(h - 1)$.) The map p is obtained by identifying the points of M^2_g that correspond to each other under the rotation by $2\pi/3$. In the general case the construction is similar.

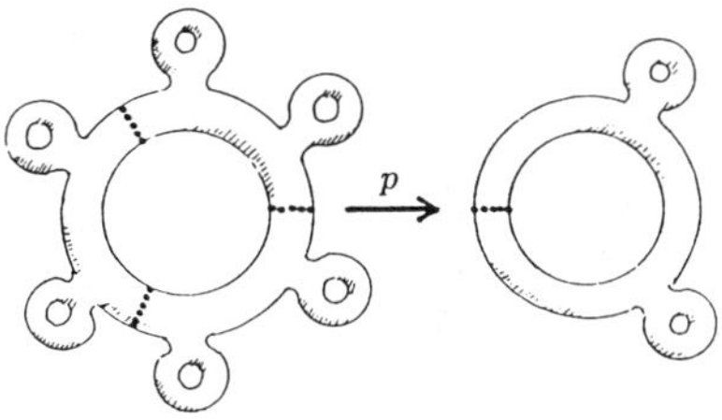

FIGURE S21.1

21.3. Let a be the number of points in the inverse image of branch points. According to formula (21.5),

$$\chi(D^2) = n(\chi(D^2) - 1) + a.$$

Since $\chi(D^2) = 1$, we obtain $a = 1$.

22.1. According to Theorem 22.3, there exists a branched covering $p_1 : M^3 \to S^3$ whose branch set is the 1-skeleton of the tetrahedron T. Represent the sphere S^3 as $\mathbb{R}^3 \cup \infty$. We can assume that T is a regular tetrahedron in $\mathbb{R}^3$. Let l be a symmetry axis of the tetrahedron passing through a vertex and the center of the opposite face (Fig.S22.1).

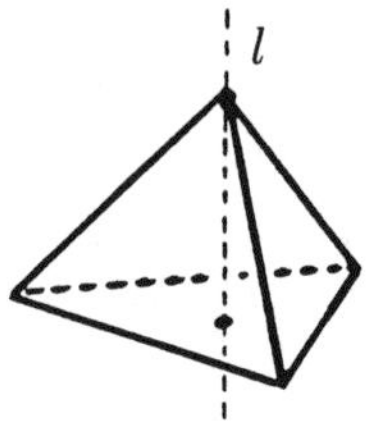

FIGURE S22.1

Identify the points of $\mathbb{R}^3$ that correspond to each other under rotations by 120° about l. The result will still be $\mathbb{R}^3$. This identification is a covering $p_2 : S^3 \to S^3$ branching along the circle $l \cup \infty$. The map $p = p_2 \circ p_1 : M^3 \to S^3$ is a branched covering. Its branch set is the circle $l \cup \infty$ and the one-dimensional complex obtained from the 1-skeleton of the tetrahedron under the described identification (see Fig.S22.2).

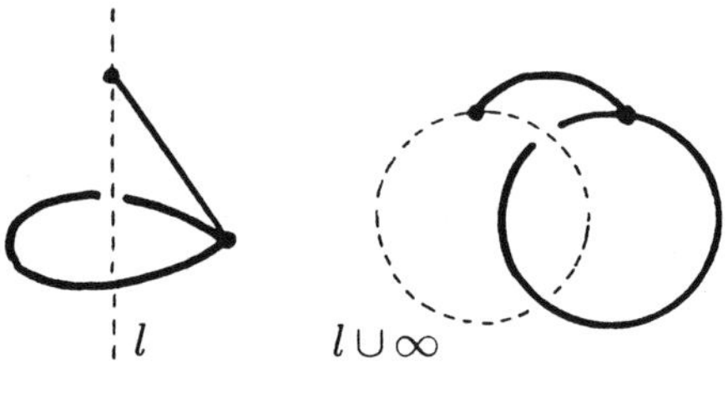

FIGURE S22.2

22.2. Let M^3 be the manifold obtained from S^3 by surgery along the framed link shown in Fig.S22.3,a. According to Proposition 22.8, there exists a two-sheeted covering $p : M^3 \to S^3$ branching along the trefoil. Destroying one circle with framing -1, we pass from the framed link in Fig.S22.3,a to the one in Fig.S22.3,b. Surgery along this last framed link produces the lens space $L(3, 1)$.

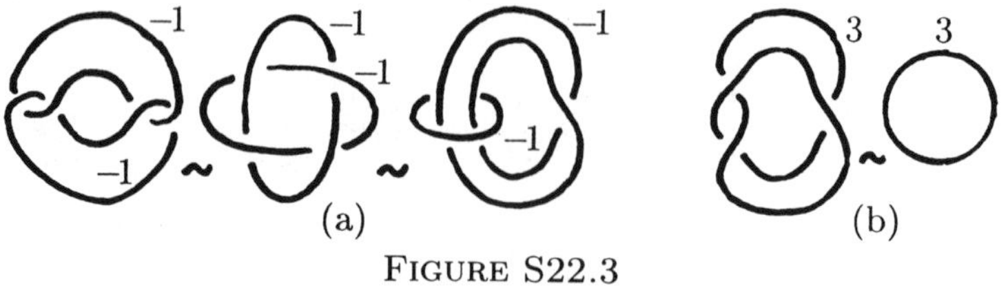

FIGURE S22.3

22.3. Attach two solid tori to each other along the homeomorphism of their boundaries that interchanges the parallels and meridians. The result will be the sphere S^3, and there exists an isotopy that exchanges T_1 and T_2 (see 8.6). Clearly, if for the solid torus T_1 the curve is of the form $p\alpha + q\beta$, then for the solid torus T_2 this curve is of the form $q\alpha + p\beta$.

23.1. The covering homeomorphisms for the twists of the base by π and 2π are shown in Fig.S23.1 and Fig.S23.2. The restrictions of these homeomorphisms to the boundary are rotations by $2\pi/3$ and $4\pi/3$ respectively.

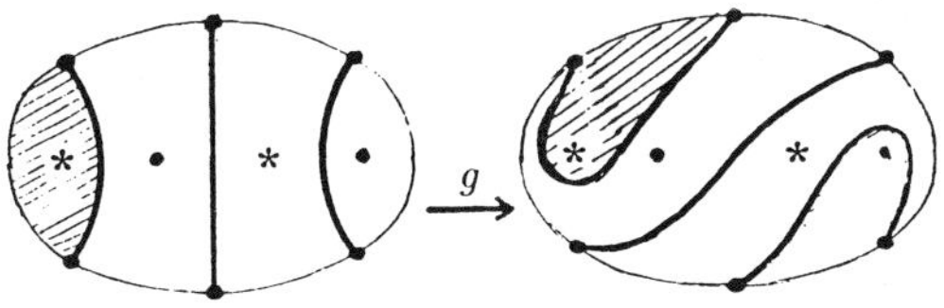

FIGURE S23.1

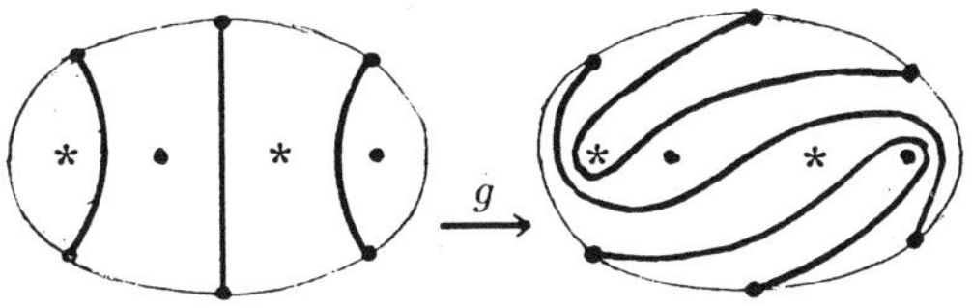

FIGURE S23.2

24.1. The crossing that appears or disappears under the first Reidemeister move is necessarily monochromatic. All the essentially different correct colorings of the crossings occuring under the second and third Reidemeister move are shown in Fig.S24.1. This figure shows that each correct coloring of the given diagram induces exactly one coloring of the resulting diagram (by the corresponding move).

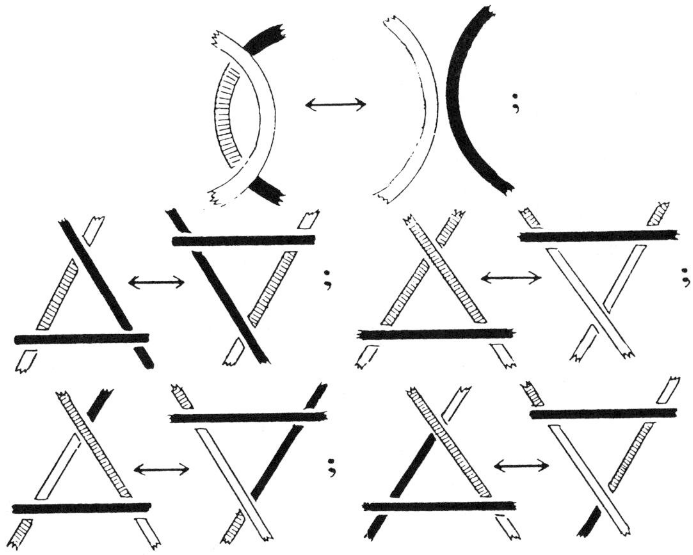

FIGURE S24.1

24.2. If $p : M^3 \to S^3$ is a branched covering and $h : S^3 \to S^3$ is a homeomorphism isotopic to the identity, then $h \circ p : M^3 \to S^3$ is a covering whose branch set diagram is obtained from the branch set diagram of p by plane isotopies and Reidemeister moves.

24.3. All the given moves reduce to the one shown in Fig.24.3 (see the main text) and to Reidemeister moves (Fig.S24.2).

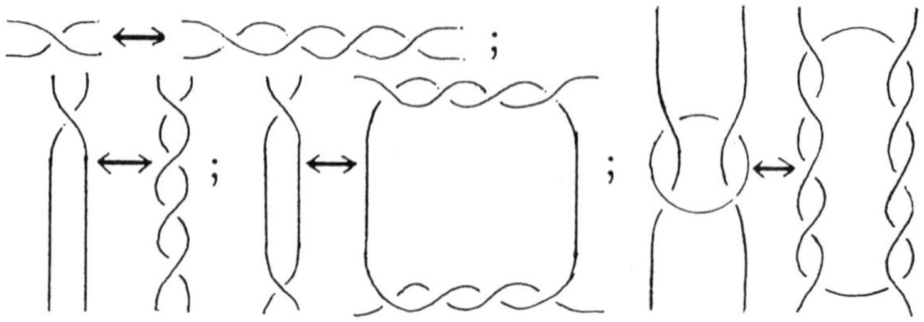

FIGURE S24.2

Chapter VIII

26.1. (a) Number the distinguished points of the disk D^2 clockwise. The arcs of any basis diagram join points of opposite parity. Indeed, suppose an arc joins points of the same parity. Then it splits the remaining points into two sets, each of which has an odd number of points. On the other hand, the points of either of these sets may be split into pairs joined by arcs. This is a contradiction.

Consider the disk D^2 with $2(n+1)$ distiguished points on the boundary. From the point marked 1, an arc of the basis diagram may join this point to any of the points $2, 4, \ldots, 2(n+1)$. The arc joining the points 1 and $2k$ cuts D^2 into two disks with $2(k-1)$ and $2(n-k+1)$ marked points. In these disks our nonintersecting arcs may be constructed in c_{k-1} and c_{n-k+1} different ways. Hence,

$$c_{n+1} = \sum_{k=1}^{n+1} c_{k-1}c_{n-k+1} = \sum_{i=0}^{n} c_i c_{n-i}.$$

(b) Consider the series

$$c(x) = \sum_{i=0}^{\infty} c_i x^i, \quad \text{where} \quad c_i = \dim S(D^2, 2i).$$

According to part (a) of this problem,

$$\begin{aligned} xc^2(x) &= \sum_{i,j=0}^{\infty} c_i c_j x^{i+j+1} = \sum_{k=0}^{\infty} x^{k+1}\Big(\sum_{i=0}^{k} c_i c_{k-i}\Big) \\ &= \sum_{k=0}^{\infty} c_{k+1}x^{k+1} = c(x) - c_0 = c(x) - 1. \end{aligned}$$

Solving the quadratic equation $c^2(x) - c(x)/x + 1/x = 0$, we obtain

$$c(x) = \frac{1 \pm \sqrt{1-4x}}{2x} = \frac{1 \pm (1 - 2x + \ldots)}{2x}.$$

The solution $c(x) = x^{-1} - 1 + \dots$ is unsatisfactory, so

$$c(x) = \frac{1 - \sqrt{1-4x}}{2x}.$$

Using Newton's binomial formula, we get

$$(1-4x)^{1/2} = 1 - \sum_{n=0}^{\infty} \frac{2^n(1\cdot 3\cdot 5\cdot \ldots \cdot(2n-1))}{1\cdot 2\cdot 3\cdot \ldots \cdot (n+1)} x^{n+1} = 1 - 2x\sum_{n=0}^{\infty} \frac{(2n)!}{n!(n+1)!} x^n.$$

Therefore,

$$c_n = \frac{1}{n+1}\binom{2n}{n}.$$

29.1. Destroying two crossing points, we can obtain the relation pictured in Fig.S29.1.

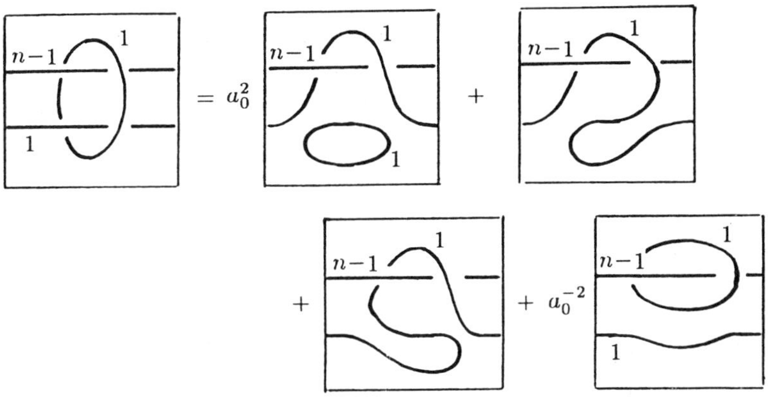

FIGURE S29.1

In the first diagram in the right-hand side, destroy the circle that has appeared. Then the first three diagrams are proportional to one and the same diagram, and so they may be replaced by one diagram with the coefficient $a_0^2(-a_0^2 - a_0^{-2}) + 1 + 1 = 1 - a_0^4$ (Fig.S29.2).

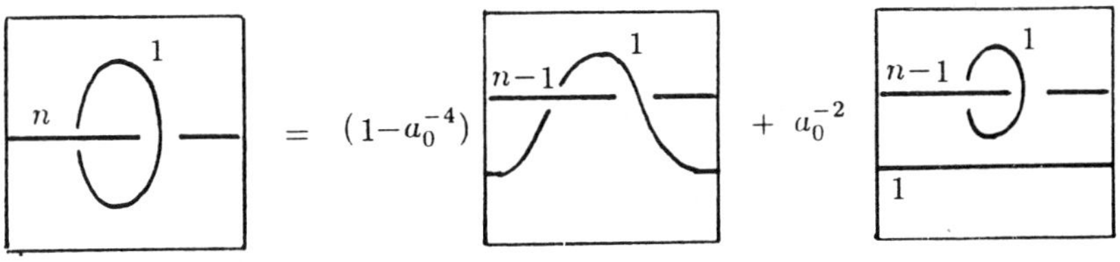

FIGURE S29.2

A similar destruction of two crossing points in the diagram with coefficient $1 - a_0^4$ yields the relation pictured in Fig.S29.3.

For the solution of the problem under consideration, i.e., in the presence of the diagram $f^{(n)}$, we can disregard the diagram with the coefficient $1 - a_0^{-4}$, since it belongs to the subalgebra A_n, for which $f^{(n)}A_n = 0$. Indeed, in the further decomposition of this diagram, the diagram 1_n (consisting of n parallel strings) will never

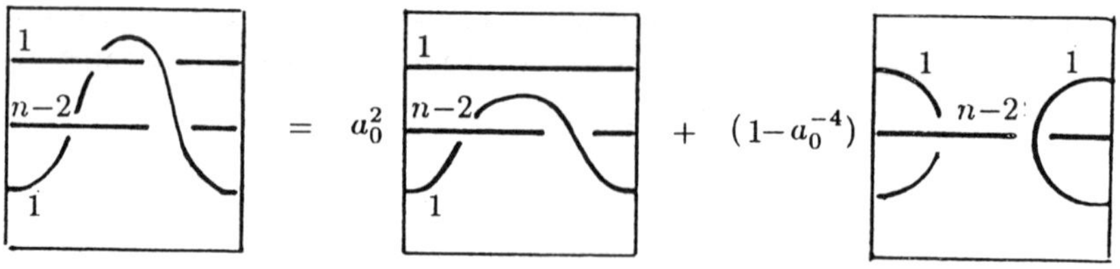

FIGURE S29.3

appear. Therefore, as the result of successive downward moves of the nonhorizontal string, we obtain the sequence of equalities shown in Fig.S29.4. Together with the relation in Fig.S29.2, they lead to the relation in Fig.S29.5.

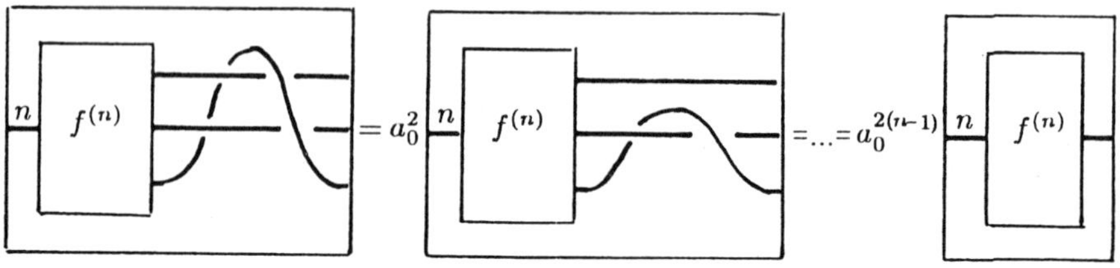

FIGURE S29.4

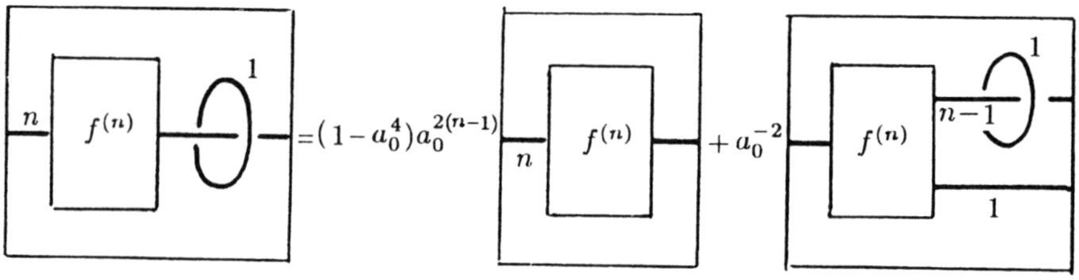

FIGURE S29.5

Although the relation we have obtained is not reccurent, it can be used to obtain the relation in Fig.S29.6, in which the sequence $\{c_n\}$ satisfies the following reccurent relation:

(*) $$c_n = (1 - a_0^4)a_0^{2(n-1)} + a_0^{-2}c_{n-1}.$$

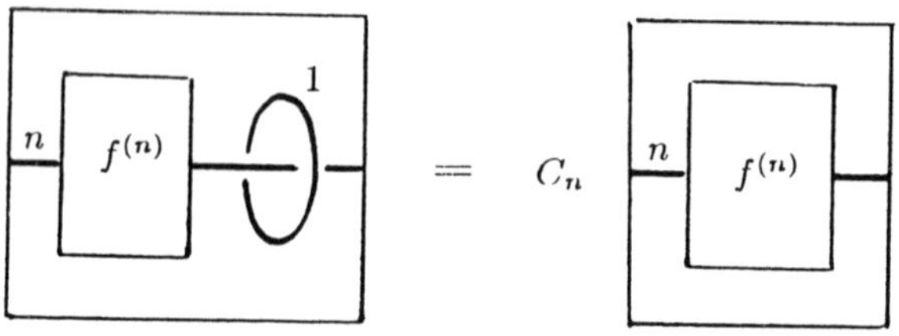

FIGURE S29.6

This is a consequence of the following obvious remark: in the situation shown in Fig.S29.7,a, elements of the algebra A_{n-1} also lead to trivial diagrams, as in the situation pictured in Fig.S29.7,b.

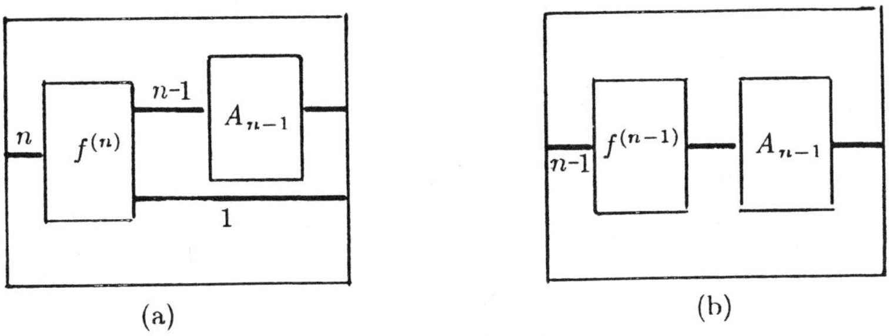

FIGURE S29.7

Since $f^{(0)}$ is the empty diagram, we have $c_0 = -(a_0^2 + a_0^{-2})$. It is easy to verify that the sequence $c_n = -(a_0^{2(n+1)} + a_0^{-2(n+1)})$ satisfies both this initial condition and the reccurent relation (*).

29.2. By definition,

$$\langle\omega\rangle_U = \langle\sum_{n=0}^{r-2} \Delta_n S_n(\alpha)\rangle_U = \sum_{n=0}^{r-2} \Delta_n \langle S_n(\alpha)\rangle_U.$$

The element $\langle S_n(\alpha)\rangle_U \in S(\mathbb{R}^2)$ is obtained from $S_n(\alpha)$ by replacing α^k by k circles; hence $\langle S_n(\alpha)\rangle_U = \Delta_n$. Thus,

$$\langle\omega\rangle_U = \sum_{n=0}^{r-2} \Delta_n^2 = (a_0^2 - a_0^{-2})^{-2} \sum_{n=0}^{r-2} (a_0^{2(n+1)} - a_0^{-2(n+1)})^2.$$

Having in mind the relation $a_0^{4r} = 1$, we obtain

$$\sum_{n=0}^{r-2} a_0^{4(n+1)} = \frac{a_0^{4r} - a_0^4}{a_0^4 - 1} = -1, \qquad \sum_{n=0}^{r-2} a_0^{-4(n+1)} = -1.$$

Therefore,

$$\sum_{n=0}^{r-2} (a_0^{2(n+1)} - a_0^{-2(n+1)})^2 = -2r.$$

29.3. The destruction of one crossing point leads to the relation shown in Fig.S29.8.

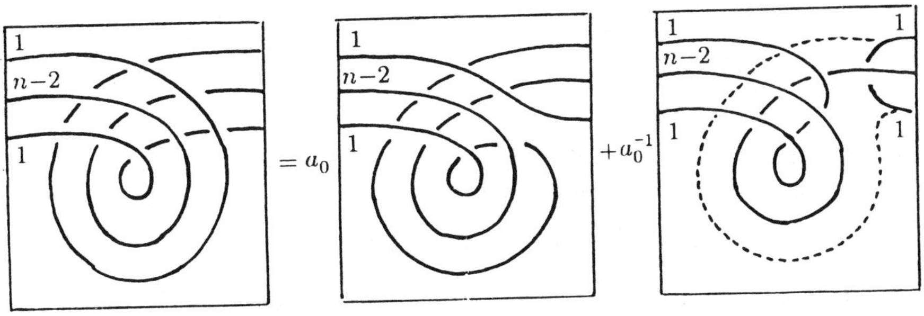

FIGURE S29.8

The diagram with the coefficient a_0^{-1} lies in the subalgebra A_n; hence in the presence of the element $f^{(n)}$ it vanishes. In the diagram with coefficient a_0, let us destroy one more crossing point. The result will be the relation shown shown in Fig.S29.9.

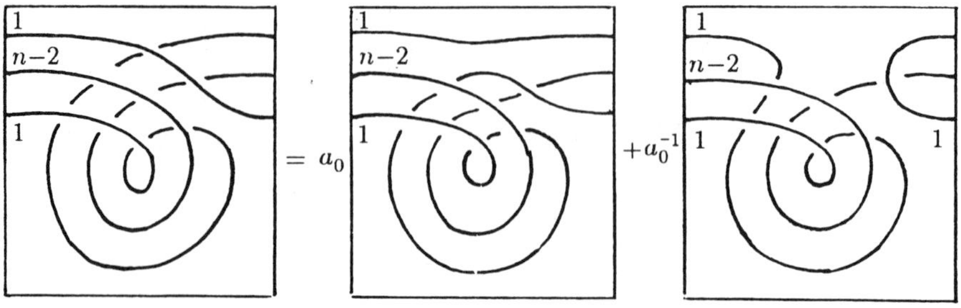

FIGURE S29.9

The diagram with the coefficient a_0^{-1} in the presence of the element $f^{(n)}$ also vanishes. By using the second and third Reidemeister moves and a plane isotopy, we can transform the diagram with coefficient a_0 so that it will appear as in Fig.S29.10,a.

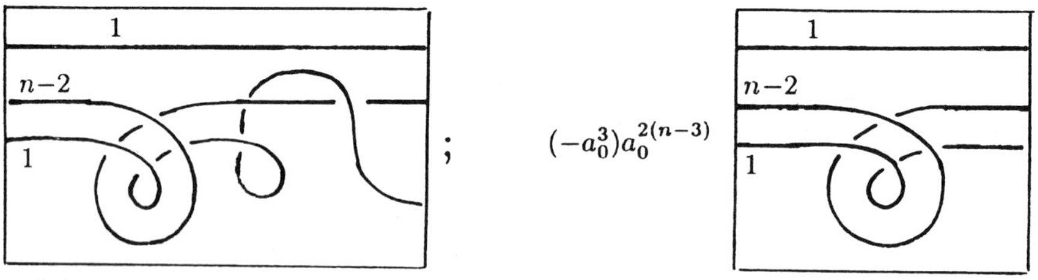

(a) (b)

FIGURE S29.10

In solving Problem 29.1, we obtained the relation represented in Fig.S29.4. Using it and destroying the loop, we can replace the diagram we obtained by the one shown in Fig.S29.10,b. As the conclusion of our graphic calculations, we see that in the presence of the element $f^{(n)}$ the relation displayed in Fig.S29.11 is valid.

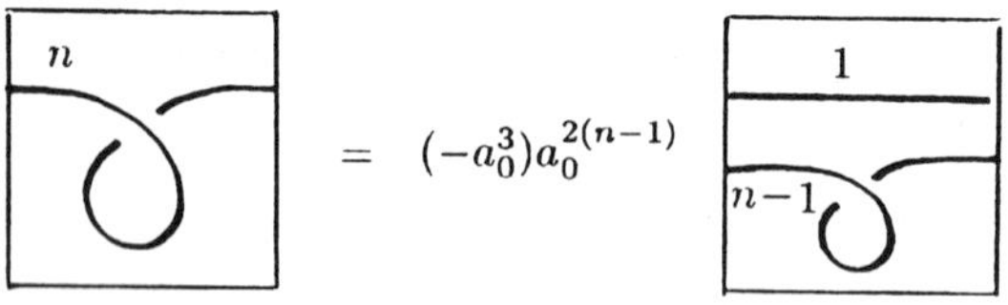

FIGURE S29.11

Applying this reccurrent relation to the diagram from the statement of the problem, we see that its image in $S(\mathbb{R}^2)$ is equal to

$$(-a_0)^{3n} a_0^{2(1+2+\cdots+(n-1))} = (-1)^n a_0^{n^2+2n}.$$

In the case under consideration $n = r - 2$, $a_0^r = -1$, and r is odd. Hence the desired value is $-(a_0^r)^{r-2} = -(-1)^{r-2} = 1$.

References

[AB] Y. Aharonov and D. Bohm, *Significance of electromagnetic potentials in quantum theory*, Phys. Rev. (2) **115** (1959), 485–491.

[AF] E. Artin and R. Fox, *Some wild cells and spheres in three-dimensional space*, Ann. of Math. (2) **49** (1948), 979–990.

[Ale1] J. W. Alexander, *A lemma on systems of knotted curves*, Proc. Nat. Acad. Sci. USA **9** (1923), 93–95.

[Ale2] ———, *Topological invariants of knots and links*, Trans. Amer. Math. Soc. **20** (1923), 275–306.

[Ale3] ———, *A matrix knot invariant*, Proc. Nat. Acad. Sci. USA **19** (1933), 272–275.

[Arn1] V. I. Arnold, *The cohomological ring of the colored braid groups*, Math. Notes Acad. Sci. USSR **5** (1969), 227–232.

[Arn2] ———, *On topological invariants of algebraic functions*, Trans. Moscow Math. Soc. **21** (1971), 30–52.

[Arn3] ———, *Topological invariants of algebraic functions.* II, Functional Anal. Appl. **4** (1970), 91–98.

[Art1] E. Artin, *Theorie der Zöpfe*, Abh. Math. Sem. Univ. Hamburg **4** (1925), 47–72.

[Art2] ———, *Theory of braids*, Ann. of Math. (2) **48** (1947), 101–126.

[Ash] C. W. Ashley, *The Ashley Book of Knots*, Doubleday, New York, 1944.

[Ati1] M. F. Atiyah, *The Geometry and Physics of Knots*, Cambridge Univ. Press, Cambridge, 1990.

[Ati2] ———, *The Jones–Witten invariants of knots*, Séminaire Bourbaki 1989/90, Exposé 715, Astérisque, no. 189–190, Soc. Math. France, Paris, 1990, pp. 7–16.

[Ati3] ———, *On framing of 3-manifolds*, Topology **28** (1990), 1–7.

[AW] Y. Akutsu and M. Wadati, *Knots, links, braids and exactly solvable models in statistical mechanics*, Comm. Math. Phys. **117** (1988), 243–259.

[Bax] R. J. Baxter, *Exactly Solved Models in Statistical Mechanics*, Academic Press, New York, 1982.

[BHMV] C. Blanchet, N. Habegger, G. Masbaum, and P. Vogel, *Topological quantum field theories derived from the Kauffman bracket*, Topology **34** (1995), 883–927.

[Bir1] J. S. Birman, *Braids, Links and Mapping Class Groups*, Ann. of Math. Studies, vol. 82, Princeton Univ. Press, Princeton, NJ, 1974.

[Bir2] ———, *New points of view in knot and link theory*, Bull. Amer. Math. Soc. (N.S.) **28** (1993), 253–287.

[BL] J. Birman and X.-S. Lin, *Knot polynomials and Vassiliev invariants*, Invent. Math. **111** (1993), 225–275.

[BM] J. Baez and J. P. Muniain, *Gauge Fields, Knots and Gravity*, World Sci. Publ., Singapore, 1994.

[BN] D. Bar-Natan, *On the Vassiliev knot invariants*, Topology **34** (1995), 423–475.

[Bri] E. Brieskorn, *Die Monodromie der isolierten Singularitäten von Hyperflächen*, Manuscr. Math. **2** (1970), 103–161.

[BZ] G. Burde and H. Zieschang, *Knots*, de Gruyter, Berlin, 1986.

[Cal] G. Călugăreanu, *L'intégrale de Gauss et l'analyse des nœuds tridimensionnels*, Rev. Roumaine Math. Pures Appl. **4** (1959), 5–20.

[Car] P. Cartier, *Construction combinatoire des invariants de Vassiliev*, C. R. Acad. Sci. Paris Sér. I Math. **316** (1993), 1205–1210.

[CDL] S. V. Chmutov, S. V. Duzhin and S. K. Lando, *Vassiliev knot invariants.* I–III, Singularities and Bifurcations (V. I. Arnold, ed.), Advances in Soviet Math, vol. 21, Amer. Math. Soc., Providence, RI, 1994, pp. 117–147.

[CF] R. H. Crowell and R. H. Fox, *Introduction to Knot Theory*, Blaisdell, Waltham, MA, 1963.

[Con] J. H. Conway, *An enumeration of knots and links and some of their algebraic properties*, Computational Problems in Abstract Algebra (Proc. Conf., Oxford, 1967; J. Leech, ed.), Pergamon Press, New York, 1970, pp. 329–358.

[Cou] R. Courant, *Differential and Integral Calculus.* Vol. II, Interscience, New York, 1953.

[CP] V. Chari and A. Pressley, *A Guide to Quantum Groups*, Cambridge Univ. Press, Cambridge, 1994.

[Cr1] L. Crane, *2-D physics and 3-D topology*, Comm. Math. Phys **135** (1991), 615–640.

[Cr2] ———, *Conformal field theory, spin geometry and quantum gravity*, Phys. Lett. B **259** (1991), 243–248.

[CY] L. Crane and D. Yetter, *A categorical construction of 4D topological quantum field theories*, Quantum Topology (L. Kauffman and R. Baadhio, eds.), Ser. on Knots and Everything, vol. 3, World Sci. Publ., Singapore, 1992, pp. 120–130.

[De1] M. Dehn, *Über die Topologie des dreidimensionalen Raumes*, Math. Ann. **69** (1910), 137–168.

[De2] ———, *Die bieden Kleeblattschlingen*, Math. Ann. **102** (1914), 402–413.

[Dow] J. S. Downing, *Decomposing compact 3-manifolds into homeomorphic handlebodies*, Proc. Amer. Math. Soc. **24** (1970), 241–244.

[Dri1] V. G. Drinfeld, *Quantum groups*, Proc. Internat. Congr. Math. (Berkeley, 1986), Vol. 1, Amer. Math. Soc., Providence, RI, 1987, pp. 798–820.

[Dri2] ———, *Hopf algebras and the quantum Yang-Baxter equation*, Soviet Math. Dokl. **32** (1985), 254-258.

[Dro] Yu. V. Drobotukhina, *An analogue of the Jones polynomial for links in $\mathbb{R}P^3$ and a generalization of the Kauffman-Murasugi theorem*, Leningrad Math. J. **2** (1991), 613–630.

[DT1] C. H. Dowker and M. B. Thistlethwaite, *On the classification of knots*, Math. Rep. Acad. Sci. Canada **4** (1982), 129–131.

[DT2] ———, *Classification of knot projections*, Topology Appl. **16** (1983), 19–31.

[Duh] P. Duhornoy, *From large cardinals to braids via distributive algebra*, J. Knot Theory Ramifications **4** (1995), 33–79.

[Far] M. Sh. Farber, *Presentations of knot modules*, Izv. Akad. Nauk Azerbaĭdzhan. SSR Ser. Fiz.-Tekhn. Mat. Nauk **2** (1981), no. 2, 105–111. (Russian)

[FH...] P. Freyd, D. Yetter, J. Hoste, W. B. R. Lickorish, K. C. Millett, and A. Ocneanu, *A new polynomial invariant of knots and links*, Bull. Amer. Math. Soc. (N.S.) **12** (1985), 239–246.

[FM] R. H. Fox and J. W. Milnor, *Singularities of 2-spheres in 4-space and equivalence of knots*, Bull. Amer. Math. Soc. **63** (1957), 406.

[Fox] R. H. Fox, *A quick trip through knot theory*, Topology of 3-Manifolds (M. K. Fort Jr., ed.), Prentice-Hall, Englewood Cliffs, NJ, 1962, pp. 120–167.

[FP] F. Frankl and L. Pontrjagin, *Ein Knotensatz mit Anwendung auf die Dimensionstheorie*, Math. Ann. **102** (1930), 785–789.

[FR] R. P. Fenn and C. P. Rourke, *On Kirby's calculus of links*, Topology **18** (1979), 1–15.

[FRT] L. D. Faddeev, N. Yu. Reshetikhin, and L. A. Takhtajan, *Quantization of Lie groups and Lie algebras*, LOMI Preprint E-14-87, Steklov Math. Inst., Leningrad, 1987; rev. aug. version, Leningrad Math. J. **1** (1990), 193–225.

[Fu] Jürgen Fuchs, *Affine Lie Algebras and Quantum Groups*, Cambridge Univ. Press, Cambridge, 1992.

[Gar] F. A. Garside, *The braid group and other groups*, Quart. J. Math. Oxford Ser. (2) **20** (1969), 235–254.

[Gau] K. F. Gauss, *Zur mathematischen Theorie der electrodynamischen Wirkungen*, manuscript, first published in his *Werke*, Vol. 5, Königl. Ges. Wiss. Göttingen, Göttingen, 1877, p. 605.

[Gaw] K. Gawędzki, *Conformal field theory*, Séminaire Bourbaki 1989/90, Exposé 704, Astérisque, no. 177/178, Soc. Math. France, Paris, 1989, pp. 1–31.

[GH] C. McA. Gordon and W. Heil, *Simply-connected branched coverings of* S^3, Proc. Amer. Math. Soc. **35** (1972), 287–288.

[GL] C. McA. Gordon and J. Luecke, *Knots are determined by their complements*, J. Amer. Math. Soc. **2** (1989), 371–415.

[Gor] C. McA. Gordon, *Some aspects of classical knot theory*, Knot Theory, Lecture Notes in Math., vol. 685, Springer-Verlag, Berlin, 1978, pp. 1–60.

[Hak1] W. Haken, *Theorie der Normalflächen*, Acta Math. **105** (1961), 245–375.

[Hak2] ———, *Über das Homöomorphieproblem der 3-Mannigfaltigkeiten.* I, Math. Z. **80** (1962), 89–120.

[Hay] Miles Van Valzah Hayes, *A Unified Field Theory*, Stinehour Press, Lunenburg, VT, 1964.

[Haz] M. Hazewinkel, *Multiparameter quantum groups and multiparameter R-matrices*, Acta Appl. Math. **41** (1995), 57–98.

[Hem] G. Hemion, *The Classification of Knots and 3-dimensional Spaces*, Oxford Univ. Press, Oxford, 1992.

[Hemp] J. Hempel, *Construction of orientable 3-manifolds*, Topology of 3-manifolds (M. Fort, ed.), Prentice-Hall, Englewood Cliffs, NJ, 1962, pp. 207–212.

[HLM] H. M. Hilden, M. T. Lozano, and J. M. Montesinos, *Universal knots*, Bull. Amer. Math. Soc. (N.S.) **8** (1983), 449–450.

[HLMW] H. M. Hilden, M. T. Lozano, J. M. Montesinos, and W. C. Witten, *On universal groups and 3-manifolds*, Invent. Math. **87** (1987), 441–456.

[HMT] H. M. Hilden, J. M. Montesinos, and T. Thickstun, *Closed oriented 3-manifolds as 3-fold branched coverings of* S^3 *of special type*, Pacific J. Math. **65** (1976), 65–76.

[Jim1] M. Jimbo, *A q-difference analogue of* $U(\mathfrak{g})$ *and the Yang-Baxter equation*, Lett. Math. Phys. **10** (1985), 63–69.

[Jim2] ———, *Quantum R-matrix for the generalized Toda system*, Comm. Math. Phys. **102** (1986), 537–547.

[Jon1] V. F. R. Jones, *A polynomial invariant for knots via von Neumann algebras*, Bull. Amer. Math. Soc. (N.S.) **12** (1985), 103–111.

[Jon2] ———, *Hecke algebra representatons of braid groups and link polynomials*, Ann. of Math. (2) **126** (1987), 335–388.

[Jon3] ———, *On knot invariants related to some statistical mechanics models*, Pacific J. Math. **137** (1989), 311–334.

[Kas] Ch. Kassel, *Quantum Groups*, Springer-Verlag, Berlin, 1995.

[Kau1] L. H. Kauffman, *On Knots*, Ann. of Math. Studies, vol. 115, Princeton Univ. Press, Princeton, NJ, 1987.

[Kau2] ———, *State models and the Jones polynomial*, Topology **26** (1987), 395–407.

[Kau3] ———, *From knots to quantum groups (and back)*, Hamiltonian Systems, Transformation Groups and Spectral Transform Methods (Proc. CRM Workshop, Montréal, 1989; J. Harnad and J. E. Marsden, eds.), Centre de Recherches Math., Univ. Montréal, Montréal, 1990, pp. 161–176.

[Kau4] ———, *Knots and Physics*, World Sci. Publ., Singapore, 1991.

[KB] J. Kania-Bartoszynska, *Examples of different manifolds with the same invariants of Witten and Reshetikhin-Turaev*, Topology **32** (1993), 47–54.

[Kea] C. Kearton, *Factorization is not unique for 3-knots*, Indiana Univ. Math. J. **28** (1979), 451–452.

[Kel] W. Thompson (Lord Kelvin), *Hydrodynamies*, Proc. Roy. Soc. Edinburgh **41** (1867), 94–105.

[Kin] S. Kinoshita, *On the Alexander polynomials of 2-spheres in a 4-sphere*, Ann. of Math. (2) **74** (1961), 518–531.

[Kir] R. Kirby, *A calculus for framed links in* S^3, Invent. Math. **45** (1978), 35–56.

[KN] S. Kobayashi and K. Nomizu, *Foundations of Differential Geometry*, Interscience, New York, 1963.

[Kne] H. Kneser, *Geschlossene Flächen in dreidimensionalen Mannigfaltigkeiten*, Jahresber. Deutsch. Math.-Verein. **38** (1929), 248–260.

[Kon1] M. Kontsevich, *Feynman diagrams and low-dimensional topology*, First European Congr. Math. (Paris, 1992), Vol. I, Progr. Math., vol. 120, Birkhäuser, Basel, 1992, pp. 97–121.

[Kon2] ———, *Vassiliev's knot invariants*, I. M. Gelfand Seminar (S. Gelfand and S. Gindikin, eds.), Advances in Soviet Math., vol. 16, part 2, Amer. Math. Soc., Providence, RI, 1993, pp. 137–150.

[KS] R. Kirby and M. G. Scharlemann, *Eight faces of the Poincaré homology 3-sphere*, Geometric Topology (Proc. Georgia Topology Conf., 1977; J. C. Cantrell, ed.), Academic Press, New York, 1979, pp. 113–146.

[KSk] P. P. Kulish and E. K. Sklyanin, *Solutions of the Yang-Baxter equation*, J. Soviet Math. **19** (1982), 1596–1620; reprinted in *Yang-Baxter Equation in Integrable Systems* (M. Jimbo, ed.), World Sci. Publ., Singapore, 1990.

[Lic1] W. B. R. Lickorish, *A representation of orientable combinatorial 3-manifolds*, Ann. of Math. (2) **76** (1962), 531–540.

[Lic2] ———, *A finite set of generators for the homotopy group of a 2-manifold*, Proc. Cambridge Philos. Soc. **60** (1964), 769–778.

[Lic3] ———, *A finite set of generators for the homotopy group of a 2-manifold (corrigendum)*, Proc. Cambridge Philos. Soc. **62** (1966), 679–681.

[Lic4] ———, *The skein method for three-manifold invariants*, J. Knot Theory Ramifications **2** (1993), 171–194.

[Lic5] ———, *Distinct 3-manifolds with all $SU(2)_q$ invariants the same*, Proc. Amer. Math. Soc. **117** (1993), 285–292.

[Lu] Ning Lu, *A simple proof of the fundamental theorem of Kirby calculus on links*, Trans. Amer. Math. Soc. **331** (1992), 143–156.

[Lus] G. Lusztig, *Introduction to Quantum Groups*, Birkhäuser, Boston, MA, 1993.

[Mak1] G. S. Makanin, *The conjugacy problem in the braid groups*, Soviet Math. Dokl. **9** (1968), 1156–1157.

[Mak2] ———, *On normalizers in the braid groups*, Math. USSR Sb. **15** (1971), 167–175.

[Mar1] A. A. Markoff [Markov], *Über die freie Äquivalenz der geschlossenen Zöpfe*, Recueil Math. Moskau [Mat. Sb.] **1 (43)** (1936), 73–78.

[Mar2] ———, *Foundations of the Algebraic Theory of Braids*, Trudy Mat. Inst. Steklov. **16** (1945). (Russian; 8-page English summary)

[Mar3] ———, *Undecidability of the homeomorphism problem*, Dokl. Akad. Nauk. SSSR **121** (1958), 218–220. (Russian)

[Mat1] S. V. Matveev, *Transformations of special spines and the Zeeman conjecture*, Math. USSR Izv. **31** (1988), 423–434.

[Mat2] ———, *Construction of a complete algebraic knot invariant*, Preprint, Chelyabinsk Univ., Chelyabinsk, 1981 = Manuscript No. 736–81, deposited at VINITI, 1981. (Russian) R.Zh.Mat. **1981**, 5A492.

[Mat3] ———, *Distributive groupoids in knot theory*, Math. USSR Sb. **47** (1984), 73–83.

[MF] S. V. Matveev and A. T. Fomenko, *Algorithmic and Computer Methods in Three-dimensional Topology*, Izdat. Moskov. Gos. Univ., Moscow, 1991. (Russian)

[Mie] E. W. Mielke, *Knot wormholes in geometrodynamics?*, General Relativity and Gravitation **8** (1977), 175–196.

[Mil1] J. W. Milnor, *On the total curvature of knots*, Ann. of Math. (2) **52** (1950), 248–257.

[Mil2] ———, *On the 3-dimensional Brieskorn manifolds $M(p,q,r)$*, Knots, Groups and 3-manifolds (L. P. Neuwirth, ed.), Ann. of Math. Studies, vol. 84, Princeton Univ. Press, Princeton, NJ, 1975, pp. 175–225.

[Moi1] E. E. Moise, *Affine structures in 3-manifolds.* V: *The triangulation theorem and Hauptvermutung*, Ann. of Math. (2) **57** (1952), 547–560.

[Moi2] ———, *Affine structures in 3-manifolds.* VII: *Invariance of the knot-types; local tame imbedding*, Ann. of Math. (2) **59** (1954), 159–170.

[Mon1] J. Montesinos, *A representation of closed, orientable 3-manifolds as 3-fold branched coverings of S^3*, Bull. Amer. Math. Soc. **80** (1974), 845–846.

[Mon2] ———, *Surgery on links and double branched covers of S^3*, in Knots, Groups and 3-manifolds (L. Neuwirth, ed.), Ann. of Math. Studies, vol. 84, Princeton Univ. Press, Princeton, NJ, 1975, pp. 227–259.

[Mon3] ———, *Three-manifolds as 3-fold branched coverings of S^3*, Quart. J. Math. Oxford Ser. (2) **27** (1976), 85–94.

[Mont] Susan Montgomery, *Hopf Algebras and Their Action on Rings*, CBMS Regional Conf. Ser. Math., no. 82, Amer. Math. Soc., Providence, RI, 1993.

[Mor1] H. R. Morton, *Problems*, Braids (J. Birman and A. Libgober, eds.), Contemporary Math., vol. 78, Amer. Math. Soc., Providence, RI, 1988, pp. 557–574.

[Mor2] ———, *Threading knot diagrams*, Math. Proc. Cambridge Philos. Soc. **99** (1986), 247–260.

[MT] W. M. Menasco and M. Thistlethwaite, *The Tait flyping conjecture*, Bull. Amer. Math. Soc. (N.S.) **25** (1991), 403–412.

[Mun] J. R. Munkres, *Elementary Differential Topology*, Ann. of Math. Studies, vol. 54, Princeton Univ. Press, Princeton, NJ, 1966.

[Mur1] K. Murasugi, *The Jones polynomial and classical conjectures in knot theory*, Topology **26** (1987), 187–194.

[Mur2] ———, *Jones polynomials and classical conjectures in knot theory*. II, Math. Proc. Cambridge Philos. Soc. **102** (1987), 317–318.

[Ni1] J. Nielsen, *Untersuchungen zur Topologie der geschlossenen zweiseitigen Flächen*, Acta Math. **50** (1927), 189–358.

[Ni2] ———, *Die Struktur periodischer Transformationen von Flächen*, Danske Vid. Selsk. Mat.-Fys. Medd. **15** (1937), no. 1, 1–77.

[Pap] C. D. Papakyriakopoulos, *On Dehn's lemma and the asphercity of knots*, Ann. of Math. (2) **66** (1957), 1–26.

[Pra] V. V. Prasolov, *Intuitive Topology*, Amer. Math. Soc., Providence, RI, 1994.

[Prz] J. H. Przytycki, *Elementary conjectures in classical knot theory*, Quantum Topology (L. Kauffman and R. Baadhio, eds.), Ser. on Knots and Everything, vol. 3, World Sci. Publ, Singapore, 1993, pp. 292–320.

[PT] J. H. Przytycki and P. Traczyk, *Invariants of Conway type*, Kobe J. Math. **4** (1987), 115–139.

[Rei] K. Reidemeister, *Knot Theory*, Chelsea, New York, 1948.

[RF] V. A. Rokhlin and D. B. Fuchs, *Beginner's Course in Topology: Geometric Chapters*, "Nauka", Moscow, 1977; English transl., Springer-Verlag, Berlin, 1984.

[Roh] V. A. Rohlin, *A three-dimensional manifold is the boundary of a four-dimensional one*, Dokl. Akad. Nauk SSSR **81** (1951), 355–357. (Russian)

[Rol] D. Rolfsen, *Knots and Links*, Publish or Perish, Inc., Berkeley, CA, 1976.

[Rou] C. Rourke, *A new proof that* Ω_3 *is zero*, J. London Math. Soc. **31** (1985), 373–376.

[RT1] N. Y. Reshetikhin and V. Turaev, *Ribbon graphs and their invariants derived from quantum groups*, Comm. Math. Phys. **127** (1990), 1–26.

[RT2] ———, *Invariants of three-manifolds via link polynomials and quantum groups*, Invent. Math. **103** (1991), 547–597.

[Rub] J. H. Rubinstein, *The solution to the recognition problem for* S^3, Lectures, Haifa, Israel, May 1992.

[Sch] H. Schubert, *Die eindeutige Zerlegbarkeit eines Knoten in Primknoten*, Sitzungsber. Heidelberg. Akad. Wiss. Math.-Nat. Kl. **1949**, no. 3, 57–104.

[Seg] G. Segal, *Two-dimensional conformal field theories and modular functors*, Ninth Internat. Congr. Math. Phys. (Swansea, 1988; B. Simon et al., eds.), Adam Hilger, Bristol, 1989, pp. 22–37.

[Sos] A. B. Sossinsky, *Feynman diagrams and Vassiliev invariants*, IHES Prepint **13/92** (1992).

[ST] H. Seifert and W. Threlfall, *Lehrbuch der Topologie*, Teubner, Leipzig, 1934.

[Sum] D. W. Sumners, *Lifting the curtain: using topology to prove the hidden action of enzymes*, Notices Amer. Math. Soc. **42** (1995), 528–537.

[SW] A. Shapiro and J. H. C. Whitehead, *A proof and extension of Dehn's lemma*, Bull. Amer. Math. Soc. **64** (1958), 174–178.

[Tai] P. G. Tait, *On knots*. I, II, III, Trans. Roy. Soc. Edinburgh **28** (1879), 145–190; **32** (1887), 327–342, 493-506; reprinted in his *Scientific Papers*, Vol. 1, Cambridge Univ. Press, Cambridge, 1898, pp. 273–347.

[Thi] M. B. Thistlethwaite, *Knot tabulations and related topics*, Aspects of Topology (I. M. James and E. H. Kronheimer, eds.), Cambridge Univ. Press, Cambridge, 1985, pp. 1–76.

[Tho] A. Thompson, *Thin position and the recognition problem for* S^3, Math. Res. Lett. **1** (1994), 613–630.

[Tur1] V. G. Turaev, *The Yang-Baxter equation and invariants of links*, Invent. Math. **92** (1988), 527–553.

[Tur2] ———, *Algebras of loops on surfaces, algebras of knots, and quantization*, Braid Group, Knot Theory and Statistical Mechanics (C. N. Yang and M. L. Ge, eds.), Adv. Ser. Math. Phys., vol. 9, World Sci. Publ., Singapore, 1989, pp. 59–96.

[Tur3] ———, *Quantum Invariants of Knots and 3-manifolds*, de Gruyter, Berlin, 1994.

[TV] V. G. Turaev and O. Viro, *State sum invariants of 3-manifolds and quantum 6j-symbols*, Topology **31** (1992), 865–902.

[Vas1] V. A. Vassiliev, *Cohomology of knot spaces*, Theory of Singularities and Its Applications (V. I. Arnold, ed.), Advances in Soviet Math., vol. 1, Amer. Math. Soc., Providence, RI, 1990, pp. 23–70.

[Vas2] ———, *Complements of Discriminants of Smooth Maps: Topology and Applications*, rev. ed., Amer. Math. Soc., Providence, RI, 1994.

[Vog] P. Vogel, *Representations of links by braids: A new algorithm*, Comment. Math. Helv. **65** (1990), 104–113.

[Wald] F. Waldhausen, *On irreducible 3-manifolds which are sufficiently large*, Ann. of Math. (2) **87** (1968), 56–88.

[Wall] A. H. Wallace, *Modifications and cobounding manifolds*, Canad. J. Math. **12** (1960), 503–528.

[Wei] N. M. Weinberg, *Sur l'équivalence libre des tresses fermées*, C. R. (Dokl.) Acad. Sci. URSS **23** (1939), 215–216.

[Whi] H. Whitney, *On regular closed curves in the plane*, Compositio Math. **4** (1937), 276–284.

[Wit1] E. Witten, *Physics and geometry*, Proc. Internat. Congr. Math. (Berkeley, 1986), Vol. 1, Amer. Math. Soc., Providence, RI, 1987, pp. 267–303.

[Wit2] ———, *Quantum field theory and the Jones polynomial*, Comm. Math. Phys. **121** (1989), 351–399.

[Wit3] ———, *Gauge theory, vertex models, and quantum groups*, Nuclear Phys. B **330** (1990), 225–246.

[Wu] F. Y. Wu, *The Potts model*, Rev. Modern Phys. **54** (1982), 235–268.

[Zee] E. C. Zeeman, *Twisting spun knots*, Trans. Amer. Math. Soc. **115** (1965), 471–495.

[Zi1] H. Zieschang, *Über die Nielsensche Kürzungsmethode in freien Produkten mit Amalgam*, Invent. Math. **10** (1970), 4–37.

[Zi2] ———, *Finite Groups of Mapping Classes of Surfaces*, Lecture Notes in Math., vol. 875, Springer-Verlag, Berlin, 1981.

[ZVC] H. Zieschang, E. Vogt, and H. D. Coldewey, *Surfaces and Planar Discontinuous Groups*, Lecture Notes in Math., vol. 835, Springer-Verlag, Berlin, 1980.

Index

The numbers given in this index are subsection numbers, not page numbers

Selected Titles in This Series

(Continued from the front of this publication)

118 **Kenkichi Iwasawa,** Algebraic functions, 1993
117 **Boris Zilber,** Uncountably categorical theories, 1993
116 **G. M. Fel′dman,** Arithmetic of probability distributions, and characterization problems on abelian groups, 1993
115 **Nikolai V. Ivanov,** Subgroups of Teichmüller modular groups, 1992
114 **Seizô Itô,** Diffusion equations, 1992
113 **Michail Zhitomirskiĭ,** Typical singularities of differential 1-forms and Pfaffian equations, 1992
112 **S. A. Lomov,** Introduction to the general theory of singular perturbations, 1992
111 **Simon Gindikin,** Tube domains and the Cauchy problem, 1992
110 **B. V. Shabat,** Introduction to complex analysis Part II. Functions of several variables, 1992
109 **Isao Miyadera,** Nonlinear semigroups, 1992
108 **Takeo Yokonuma,** Tensor spaces and exterior algebra, 1992
107 **B. M. Makarov, M. G. Goluzina, A. A. Lodkin, and A. N. Podkorytov,** Selected problems in real analysis, 1992
106 **G.-C. Wen,** Conformal mappings and boundary value problems, 1992
105 **D. R. Yafaev,** Mathematical scattering theory: General theory, 1992
104 **R. L. Dobrushin, R. Kotecký, and S. Shlosman,** Wulff construction: A global shape from local interaction, 1992
103 **A. K. Tsikh,** Multidimensional residues and their applications, 1992
102 **A. M. Il′in,** Matching of asymptotic expansions of solutions of boundary value problems, 1992
101 **Zhang Zhi-fen, Ding Tong-ren, Huang Wen-zao, and Dong Zhen-xi,** Qualitative theory of differential equations, 1992
100 **V. L. Popov,** Groups, generators, syzygies, and orbits in invariant theory, 1992
99 **Norio Shimakura,** Partial differential operators of elliptic type, 1992
98 **V. A. Vassiliev,** Complements of discriminants of smooth maps: Topology and applications, 1992 (revised edition, 1994)
97 **Itiro Tamura,** Topology of foliations: An introduction, 1992
96 **A. I. Markushevich,** Introduction to the classical theory of Abelian functions, 1992
95 **Guangchang Dong,** Nonlinear partial differential equations of second order, 1991
94 **Yu. S. Il′yashenko,** Finiteness theorems for limit cycles, 1991
93 **A. T. Fomenko and A. A. Tuzhilin,** Elements of the geometry and topology of minimal surfaces in three-dimensional space, 1991
92 **E. M. Nikishin and V. N. Sorokin,** Rational approximations and orthogonality, 1991
91 **Mamoru Mimura and Hirosi Toda,** Topology of Lie groups, I and II, 1991
90 **S. L. Sobolev,** Some applications of functional analysis in mathematical physics, third edition, 1991
89 **Valeriĭ V. Kozlov and Dmitriĭ V. Treshchëv,** Billiards: A genetic introduction to the dynamics of systems with impacts, 1991
88 **A. G. Khovanskiĭ,** Fewnomials, 1991
87 **Aleksandr Robertovich Kemer,** Ideals of identities of associative algebras, 1991
86 **V. M. Kadets and M. I. Kadets,** Rearrangements of series in Banach spaces, 1991
85 **Mikio Ise and Masaru Takeuchi,** Lie groups I, II, 1991

(See the AMS catalog for earlier titles)